JOHN VON NEUMANN:
SELECTED LETTERS

HISTORY OF MATHEMATICS VOL 27

JOHN VON NEUMANN:
SELECTED LETTERS

MIKLÓS RÉDEI
Editor

AMERICAN MATHEMATICAL SOCIETY | LONDON MATHEMATICAL SOCIETY

2000 *Mathematics Subject Classification*. Primary 00A99, 01A70.

For additional information and updates on this book, visit
www.ams.org/bookpages/hmath-27

Library of Congress Cataloging-in-Publication Data

Von Neumann, John, 1903–1957.
[Correspondence. Selections]
John von Neumann: Selected letters / Miklós Rédei, editor.
p. cm. — (History of mathematics, ISSN 0899-2428 ; v. 27)
Includes bibliographical references.
ISBN 0-8218-3776-1 (acid-free paper)
Softcover ISBN 978-1-4704-6863-7
1. Von Neumann, John, 1903–1957–Correspondence. 2. Mathematicians–United States–Correspondence. 3. Mathematics. I. Title: Selected letters. II. Rédei, Miklós. III. Title. IV. Series.

QA29.V66A4 2005
510′92–dc22 2005048258

---

11 10 9 8 7 6 5 4 3 2 27 26 25 24 23 22

Contents

Preface

The present volume is a collection of selected letters written by John von Neumann (1903, Budapest, Hungary; 1957, Washington, D.C., U.S.A.). Apart from a short volume containing von Neumann's letters (written and published in Hungarian) to Rudolf Ortvay [**36**], the present volume is the first and only substantial published collection of letters by von Neumann.

While substantial, this is a strongly *selected* volume since von Neumann had written hundreds more letters. Many of those not selected for the volume are not suitable for publication for one reason or another, many could in principle have been included however. The guiding principle of selection was that the letters published here should contribute to our understanding of John von Neumann as a scientist – broadly interpreted – and as a public figure. The volume also should be interesting for historians of science, especially of mathematics and mathematical physics. Accordingly, letters of exclusively private nature or content are not included in the volume. Also excluded from the collection are letters of reference written on behalf of colleagues – irrespective of whether the persons involved are still alive. Some of von Neumann's letters are still classified hence not accessible for scholarly research, and there may exist letters in private property, not available in archives, which could have been included in the volume had they been known to me. Keeping the volume within a reasonable size also restricted the number of letters that could be included. Thus the selection has been both disciplined and determined by contingent factors.

The letters published here vary in content, style and length. Some are on very general issues that are easy to understand but some concern very technical topics in mathematics and mathematical physics and, naturally, not all technical letters are self-contained. The "Introductory Comments" are intended to facilitate reading the collection, especially the technical letters, by giving brief and concise reviews of the related technical background and by putting some of the letters into perspective. The organization of the introductory comments is orthogonal to the organization of the rest of the volume: while von Neumann's letters are arranged in alphabetic order of the addressees (given an addressee the letters are arranged in order of their dates) the comments are thematically grouped.

Some of the letters are not dated completely. When the year of writing cannot be found on the letter, but can be guessed to be year 19XY, we use the notation [19XY ?]. If no educated guess can be made, then the year of the date is replaced by four question marks: [????].

In reproducing the letters, the original text and punctuation of von Neumann has been followed carefully – even when the original contains obvious typos or misspelled words. In some cases, for instance when the typos and misspellings might lead to misunderstandings, footnotes have been inserted to clarify the text.

All editorial footnotes inserted into the letters end with the word "The Editor". Authors of other footnotes are given explicitly.

With few exceptions the letters are published here for the first time. The originals of the letters are deposited in the following archives:

- John von Neumann Papers, Manuscript Division, Library of Congress, Washington, D.C., U.S.A.
- Oswald Veblen Papers, Manuscript Division, Library of Congress, Washington, D.C., U.S.A.
- Harvard University Archives, Pusey Library, Cambridge, Massachusetts, U.S.A.
- Archives of Scientific Philosophy, Special Collections, Hillman Library, University of Pittsburgh, Pittsburgh, Pennsylvania, U.S.A.
- "Philosophisches Archiv" of the Library of the University of Konstanz, Konstanz, Germany
- John Hay Library, Brown University, Providence, Rhode Island, U.S.A.
- Library of the Institute for Advanced Study, Princeton, U.S.A.
- Institute Archives, California Institute of Technology, Pasadena, California, U.S.A.
- S. Ulam Papers, American Philosophical Society Library, Philadelphia, Pennsylvania, U.S.A.
- "Staatsbibliothek zu Berlin", Berlin, Germany
- Department of Manuscripts and Rare Books, Library of the Hungarian Academy of Sciences, Budapest, Hungary
- Gödel Archive, Manuscripts Division, Department of Rare Books and Special Collections, Princeton University Library, Princeton, U.S.A.

The "originals" are often genuinely original, hand-written or typed letters and sometimes carbon copies of typed letters. In the present volume no distinction will be made between these cases and the term "original" will be used indiscriminately to refer to any type of document deposited in an archive. The majority of the original letters are deposited in the John von Neumann Papers of the Library of Congress. At the end of each letter the archive is given in which the document used for the volume is deposited, together with the bibliographic data of a previous publication, if any.

I wish to thank the curators, archivists and librarians that helped me in locating and collecting the letters in these archives:

Dr. Len Bruno, Curator of scientific part of Manuscript Division, and the whole library staff of the Manuscript Division of the Library of Congress helped me during my numerous visits to the Manuscript Division of the Library of Congress and called my attention to the Oswald Veblen, Harry Wexler and Robert Oppenheimer papers at the Library of Congress. Lisa Coats and Marcia Tucker located and made available for inspection and copying the von Neumann documents out of thousands of other archival items in the Institute for Advanced Study Archives of the Historical Studies-Social Science Library of the Institute for Advanced Study in Princeton. J. Cox, Curator and Valerie-Anne Lutz, Assistant Curator at the Library of the American Philosophical Society helped to search the S. Ulam Papers. Dr. Brigitte Parakenings at the "Philosophisches Archiv" of the Library of the University of Konstanz, Germany and Brigitte Arden, Associate Curator of the Special Collections of Hillman Library, University of Pittsburgh, made accessible the Archives of

Scientific Philosophy and the Archive for the History of Quantum Physics. Bonnie Ludt, Curator of the Institute Archives of the California Institute of Technology provided an item from the H.P. Robertson Papers. Dr. Mark N. Brown, Curator, John Hay Library, Brown University, made available von Neumann's letters to M. Stone.

Some letters were written in Hungarian or German, these have been translated into English for the volume; in such cases the name of the translator(s) are also given at the end of those letters. I wish to thank especially A. Lenard, retired professor of mathematics at Indiana University, for making available his translation of von Neumann's letters to Ortvay. I was helped with some of the translations by K. Wehmeier; I wish to thank him for his help. A number of other colleagues also have generously helped me in one way or another in preparing the volume. I am very thankful to all: G. Bana, F. Conway, R. Dudau, G. Hon, T. Matolcsi, D. Petz, J. Siegelman and M. Stöltzner. I also thank L. Cole, the copy editor of the manuscript, and S. Gelfand, acquisition editor of the American Mathematical Society for their editorial help and patience.

The copyright of all unpublished von Neumann papers and letters is with Marina von Neumann Whitman, the literary heir of John von Neumann. I would like to thank Marina Whitman not only for her permission to publish the letters but also for her encouragement and support to carry out the project of editing and publishing the volume. In addition to Marina Whitman's permission to publish, a number of specific permissions related to individual letters are listed separately after this Preface.

The archival research was supported by two grants from the Hungarian National Science Foundation (OTKA) under grant numbers T 025841 and T 035234. The archival research in the Library of Congress was greatly facilitated by the hospitality of the Center for Hellenic Research of Harvard University, Washington, D.C., where I was staying during some of my visits to the Library of Congress. The archival research leading to the current volume started during my stay as a Senior Resident Fellow in the Dibner Institute for History of Science and Technology at MIT in the academic year 1997-1998. Substantial parts of the writing and editing took place while I was staying in the Department of Logic and Philosophy of Science, University of California at Irvine during the winter and spring terms of the 2002-2003 academic year and during my stay in the summer of 2004 in the Philosophy, Probability and Modelling Group in the University of Konstanz; the Konstanz group is supported by the Alexander von Humboldt Foundation (through a Sofja Kovalevskaja Award). I wish to thank all the above organizations and institutions for their help and for their hospitality.

Miklós Rédei

Budapest, August 2005

Foreword by P. Lax

Everybody who ever met von Neumann was astounded by the speed, power, depth and range of his thinking. In a film about him, distributed on VHS by the MAA, the physicist Hans Bethe remarks, only half in jest, that von Neumann's brain seemed like an upward mutation in the human species.

To gain a measure of von Neumann's achievements, consider that had he lived a normal span of years, he would certainly have been a recipient of a Nobel Prize in economics. And if there were Nobel Prizes in computer science and mathematics, he would have been honored by these, too. So the writer of these letters should be thought of as a triple Nobel laureate or, possibly, a $3\frac{1}{2}$-fold winner, for his work in physics, in particular, quantum mechanics.

This collection of letters opens a window on von Neumann's way of thinking, his interests, his relation to people, and his personality. One can make a few general observations. He was exquisitely polite, as shown for instance in his letter to Dixmier. Even in his letter of resignation from the German Mathematical Society, written in 1935 to Blaschke, he maintains an ice-cold politeness. On the other hand, his letter to Johannes Stroux, President of the German Academy in East Berlin, declining an offer of membership, is tinged with genuine regret.

Von Neumann was a dedicated liberal. In his 1949 letter to Lyman Spitzer, then chairman of the Scientists' Committee on Loyalty Problems, von Neumann brands as pernicious a suggestion that clearance be required for unclassified research. He states eloquently that the training of talented people in science should be done with absolutely no consideration for anything other than scientific ability.

Von Neumann had a deep interest in world affairs, especially the coming of World War II. His judgments, expressed in letters to Veblen and Ortvay, are prescient, displaying his great powers of analysis, and profound knowledge of history.

Von Neumann wrote with a light touch; his letters are full of levity, tinged occasionally with black humor. In a letter to Schrödinger, he points out that the quantum entanglement of separated particles does not contradict the principle of locality, and illustrates his point with a charming off-color joke.

Von Neumann liked to avoid generalities and dive right into the details of whatever he was dealing with. In writing to Cirker, President of Dover publications, about an enlarged edition of his book on quantum mechanics, he calculates to three figure accuracy the portion of new material. In a letter to Veblen he estimates carefully the cost of setting up Math Reviews. Then in a letter to T.V. Moore at Standard Oil he analyses in excruciating detail the efficient management of a fleet of 18 tankers, adding at the end "I realize that the above description is rather sketchy."

Von Neumann was very receptive to ideas of others and was generous with his praise, as in his letter to Tannaka. He had unbounded admiration for Gödel. When he first learned of Gödel's incompleteness theorem, he realized that it implied that the consistency of a system of axioms cannot be proved within the system. Since Gödel had reached the same conclusion, von Neumann didn't publish his own work. Later he was instrumental in rescuing Gödel from Europe and getting a position for him at the Institute for Advanced Study.

Von Neumann was ready to help those who needed it. When his student Halperin was accused of espionage, he came to his defense. There is a touching correspondence with Ortvay about getting high school mathematics texts in Hungarian for an immigrant working man in Chattanooga, Tennessee.

The heart of this collection are the letters about mathematics, such as rings of operators, about quantum mechanics and quantum logic, superconductivity, computing and computers, neurology, etc. Both von Neumann and G. D. Birkhoff were deeply concerned about ergodicity in statistical mechanics; in some of these letters von Neumann expresses his annoyance with Birkhoff for unfairly scooping him on the ergodic theorem. Looking back, neither von Neumann's mean ergodic theorem nor Birkhoff's pointwise ergodic theorem can be applied in statistical mechanics, for it is very difficult to verify the hypothesis of metric transitivity. One of the difficulties is that many physical systems are not metrically transitive; this can be demonstrated by applying KAM theory, or even by numerical calculations that were not available in the time of Birkhoff and von Neumann. The ingenious theorem of Oxtoby and Ulam, that all measure preserving transformations are ergodic except for a set of first category, is irrelevant. Transformations that describe physical processes are highly differentiable and so form a set of first category in the maximum norm.

So why hasn't statistical mechanics collapsed? Jack Schwartz and Arthur Wightman have pointed out that in statistical mechanics ergodicity is an overkill; it would show that the time average of every continuous function equals its phase-space average. But in statistical mechanics we are not interested in every continuous function, only in those that have thermodynamic significance. This is a very special class of functions; the equality of their time and phase-space average is due to other reasons, such as their high degree of symmetry.

In a footnote to a letter to Rudolf Ortvay in Hungary, written in February 1939, von Neumann remarks that in the U.S. the uranium-barium disintegration is thought to be of great importance. Clearly at this date von Neumann must have been thinking of the theory of the nucleus, not of nuclear weapons.

The tragedy of von Neumann's early death has robbed mathematics and the sciences of a natural leader and an eloquent spokesman. These letters make it possible for the present generation to catch a glimpse of the most scintillating mind of the twentieth century. Those who wish to learn more about von Neumann and about his mathematics should look at the outstanding Memorial Issue of the Bulletin of the AMS, vol. 64, no. 8, 1958, and at Norman Macrae's authoritative biography, distributed by the AMS.

Peter D. Lax

New York, 2005

Introduction by Marina von Neumann Whitman

As John von Neumann's daughter, it is a privilege and a pleasure to write some introductory words to this collection of his letters so thoughtfully and insightfully collected and edited by Professor Redei. Those letters — perhaps the majority — that deal with mathematical subjects are beyond my reach. But that leaves many that gave me new insights into my father, glimpses into his life and personality that I was never fully aware of while he was alive.

The letters reminded me forcibly of what a remarkable polymath my father was. I knew, of course, of his contributions not only to mathematics but to physics, economics, and perhaps neurobiology as well. And I remembered that his knowledge of ancient and Byzantine history often left specialists in the field goggle-eyed. But it hadn't dawned on me until I read letters concerning organizational issues at Harvard, the University of Pennsylvania, and the Institute itself, that he had remarkable insights into such practical matters as well. I had forgotten also what a courtly, formal, unfailingly polite European gentleman he was, whether refusing an invitation from the World Affairs Council of Boston, gently informing a proud mother that her son's paper was worthless, or apologizing for his tardiness in delivering a promised paper. In light of his disdain for the mathematical apparatus of classical economics, his letter refusing to write a review of Samuelson's Foundations of Economic Analysis reflects an impressive adherence to the old adage: "if you can't say something good about someone, don't say anything at all."

Having lived through a good bit of history myself by now, I am astounded at the prescience my father displayed in his letters of the 1930s regarding the increasing certainty of a war in Europe that would ultimately engulf the entire world, the certainty that the U.S. would intervene on the side of England, and the likely progress and outcome of that war. From the vantage point of the 21st century, his 1939 analysis of the role and motives of the United State is truly remarkable: "I admit that the U.S.A. could be imperialist. I would not be surprised if in 20 years it would become so. But today it is not yet. I think that in Europe the U.S.A. is understood just as little as Europe is understood here." For John von Neumann, his knowledge of history and his coolly logical analysis of current conditions combined to produce a Cassandra-like view of the future that was both a gift and a curse.

The letters also reveal, more or less in passing, how incredibly busy and peripatetic my father was; he seemed always to be about to embark on a trip or just returned from one. This perpetual motion was attributable not only to his cosmopolitan background but to the fact that, throughout much of his career, he led a double life: as an intellectual leader in the ivory tower of pure mathematics and as a man of action, in constant demand as an advisor, consultant and decision-maker to what is sometimes called the military-industrial complex of the United States. My own belief is that these two aspects of his double life, his wide-ranging activities as well as his strictly intellectual pursuits, were motivated by two profound convictions. The first was the overriding responsibility that each of us has to make full use of whatever intellectual capabilities we were endowed with. He had the scientist's passion for learning and discovery for its own sake and the genius's ego-driven concern for the significance and durability of his own contributions. The second was the critical importance of an environment of political freedom for the pursuit of the first, and for the welfare of mankind in general.

I'm convinced, in fact, that all his involvements with the halls of power were driven by his sense of the fragility of that freedom. By the beginning of the 1930s, if not even earlier, he became convinced that the lights of civilization would be snuffed out all over Europe by the spread of totalitarianism from the right: Nazism and Fascism. So he made an unequivocal commitment to his home in the new world and to fight to preserve and reestablish freedom from that new beachhead.

In the 1940s and 1950s, he was equally convinced that the threat to civilization now came from totalitarianism on the left, that is, Soviet Communism, and his commitment was just as unequivocal to fighting it with whatever weapons lay at hand, scientific and economic as well as military. It was a matter of utter indifference to him, I believe, whether the threat came from the right or from the left. What motivated both his intense involvement in the issues of the day and his uncompromisingly hardline attitude was his belief in the overriding importance of political freedom, his strong sense of its continuing fragility, and his conviction that it was in the United States, and the passionate defense of the United States, that its best hope lay.

Now, a little bit about John von Neumann's legacy, from the vantage point of his daughter and only child. In particular, I will focus on his concerns during the last year or two of his life. Especially toward the end, but even before, he became deeply concerned about the question of his ongoing legacy, in two respects. One had to do with the durability of his work, his intellectual contributions; he was surprisingly insecure about whether his work would still be recognized "in a hundred years". Well, the hundred years he had in mind aren't up yet, but he might be reassured to know that the royalties I still receive on books he wrote in the 1930s, 40s, and 50s vastly exceed anything I receive on my own much more recent publications.

Interestingly enough, despite his prescience in matters of state, my father wasn't a very accurate prophet regarding what turns the practical applications of his pioneering work would take. For example, he clearly expected that the computer would have its impact primarily on scientific research and military work. He was particularly interested in its role in advancing the accuracy of weather forecasting, and ultimately, climate modification. I don't think progress in this area has gone nearly as far or as fast as he hoped and expected. Similarly, I think he anticipated that the theory of games would have a more immediate impact on military and business decision-making than in fact it did. He might have found it a bit ironic that when finally, in 1994, the role of game theory in economics was recognized with a Nobel Prize, the prize went not to the inventors of the basic theory (who were both long-since dead) but to the developers of an important advance — the analysis of equilibria in the theory of non-cooperative games. Unsurprisingly, one of the three winners was another Hungarian, John C. Harsanyi.

On the other hand, if anyone had ever told him that the company I used to work for, General Motors, would produce and utilize literally millions of computers each year (each of the roughly eight million vehicles the company produces each year contains several, not to mention the ones in its plants and offices), I believe he would have been startled. The notion of adults fulminating against computers as corrupters of youth in the form of video games would have amused and perhaps secretly pleased the playful, childlike aspect of his personality.

In fact, my father foresaw the inadequacy not only of his own scientific forecasts but of such forecasts in general. In a 1955 article in *Fortune* magazine, he said: "All experience shows that technical changes profoundly transform political and social relationships. Experience also shows that these transformations are not a priori predictable, and that most contemporary first guesses concerning them are wrong."

The second focus of John von Neumann's concern about his earthly legacy was, to put it simply, me. I was his only offspring and, toward the end of his life, he became acutely conscious that all his eggs were in one basket, genetically speaking (if biological inaccuracy can be forgiven for the sake of metaphor). So he put tremendous pressure on me to perform up to the peak of my abilities, and made clear his displeasure with the path I appeared to be taking. I had married young, right out of college, and he thought that this was a bad beginning, simply because he feared (and it was a reasonable fear, in the 1950s) that a woman who married young was very probably reducing her chances of making a significant intellectual or professional contribution. Statistically he was right, of course, but I like to think that in this particular case he was wrong. I'm no John von Neumann, obviously, but I have had a reasonably successful and highly rewarding career as an academic economist, a presidential adviser, and a corporate executive. And, in all these careers, I have been mindful of his insistence that it is immoral not to make maximum use of one's intellectual capacities.

Beyond me is the next generation, the grandchildren whose accomplishments he couldn't foresee because he died far too early, before they were born or even contemplated. I'm deeply sorry that my father never got to know the grandson who has translated a ten-year-old's dream of "someday finding a cure for cancer" into a career as a molecular biologist/biochemist doing research on the chemistry of intercellular message transmission as a professor at the Harvard Medical School. I'm equally sorry that he could never know the granddaughter who is a physician and a teacher of physicians at the Yale Medical School. It's too early yet to tell about the generation beyond, the great-grandchildren who are only eight and five years old, but there is no question that they have been indoctrinated from birth on the importance of the life of the mind. John von Neumann would have felt reassured and gratified, I believe, by the choices his progeny and his progeny's progeny have made, to do what he considered most important, that is, utilize our intellectual capacities up to their limits to fulfill whatever potential we have. He would have felt both relieved and vindicated that the forces of democracy and freedom, although as ever under threat, have gained ground against the forces of intolerance and totalitarianism, including recapturing the land of his birth, Hungary. He would have been surprised and perhaps amused, as well as disappointed in some areas, at the twists and turns his contributions have taken in affecting our everyday lives. He would have felt reassured that we are all still here to reflect on such matters in

2005, given the fears he expressed in a the 1955 *Fortune* article I referred to that mankind might destroy itself before 1980. And he would be gratified to know that the Mathematical Societies of the two nations on whose battle against Armageddon he focused his enormous physical and intellectual energies during the Second World War are jointly publishing a collection of letters that will provide remarkable insights into the man and his work for a generation that never knew him.

Marina von Neumann Whitman

April, 2005

Photographs

John von Neumann in the mid-1930s

Admiral Lewis Strauss (Head of the AEC) swears in John von Neumann for the Atomic Energy Commission in 1955

John von Neumann receiving the Medal of Freedom from President Eisenhower on February 15, 1956

John, his wife Klari, and their dog Inverse (around 1952)

John von Neumann at the IAS computer (around 1952)

List of specific permissions

The Editor is grateful to the following persons and institutions who have granted permission to reproduce von Neumann's letters in this volume:

- Marina von Neumann Whitman, for her granting permission to publish all the letters in the volume.
- Archives of the California Institute of Technology for their permission to publish the letter to P. Robertson (January 16, 1932).
- Archives of the Institute for Advanced Study, Princeton, for permission to publish the letters to N. Aronszajn (July 15, 1948), I. Kaplansky (March 1, 1950) and F. Aydelotte (March 23, 1940).
- Nancy Collins Asher, for her permission to publish the following letters to G. Birkhoff: (Jan. 15, 1935) (Jan. 19, [1935 ?]), (November 6, [1935 ?]), (Nov. 13., [1935 ?]), (Nov. 21, 1935), (Nov. 27, [1935 ?]).
- "Staatsbibliothek zu Berlin", Germany, for granting permission to reproduce the letters to P. Jordan (December 11, 1949), (January 12, 1950).
- John Hay Library, Brown University, Providence, Rhode Island, U.S.A., for permission to reproduce letters to M. Stone (October 8, 1930), (January 22, 1932), (March 31, 1935), (April 19, 1935), (October 4, 1938).
- Department of Manuscripts and Rare Books, Library of the Hungarian Academy of Sciences, Budapest, Hungary, for permission to reproduce the letters to R. Ortvay (April 17, 1934), (March 17, [1938 ?]), (January 26, 1939), (February 2, 1939), (February 26, 1939), (March 29, [1939 ?]), (July 18, [1939 ?]).
- S. Feferman, J. W. Dawson, Jr., W. Goldfarb, C. Parsons, W. Sieg and Oxford University Press for their permission to publish the letters to Gödel in W. Sieg's translation (November 20, 1930), (November 29, 1930), (January 12, 1931), (February 14, 1933).
- P. Mancosu, for his permission to use his translation of the letter to Carnap (June 7, 1931) and Archives of Scientific Philosophy, Special Collections, Hillman Library, University of Pittsburgh, Pittsburgh, Pennsylvania, U.S.A. for permission to publish this letter to Carnap.
- American Philosophical Society Library, Philadelphia, Pennsylvania, U.S.A. for permission to publish the letters to S. Ulam (October 3, 1936), (February 3, 1937), (October 4, [1937 ?]), (March 15, [1939 ?]), (April 2, 1942), (November 9, 1943).
- Springer Publishing Company for their permission to publish the letters to F.B. Silsbee (July 2, 1945) and to H.D. Kloosterman (March 25, 1953), (April 10, 1953)

Introductory Comments

1. Von Neumann's life and career in light of his letters

John von Neumann (known in Hungary as Neumann János and having the full name Margittai Neumann János Lajos) was born on December 28, 1903 in Budapest, Hungary to a well-to-do family. His father, Max von Neumann, a successful banker, made his fortune during the calm and economically prosperous years of the Austro-Hungarian Empire that followed the so-called "self rule" in 1867 that secured Hungary's semi-independence within the monarchy.

John von Neumann had a first-rate education. This education started by home schooling and included language instruction in the form of the presence of German speaking maids in the household in downtown Budapest, where von Neumann was raised. (German being von Neumann's first second language, his German remained superior to his English until about the mid thirties – this is clearly shown by the linguistic superiority of his later letters to the ones he wrote in the early thirties.)

When the time came in 1914, von Neumann enrolled in the famous, expensive, private Lutheran high school ("Ágostai Hitvallású Evangélikus Főgimnázium") in Budapest. His talent in mathematics was recognized there by László Rátz, von Neumann's high school mathematics teacher. Rátz asked for (and got) permission from von Neumann's father to arrange for tutoring von Neumann in mathematics by faculty members of the Technical University in Budapest, insisting at the same time that von Neumann attends regular mathematics classes, which von Neumann did. As a result of this private tutoring von Neumann had been prepared already in high school to become a professional mathematician: his first paper (coauthored with M. Fekete, von Neumann's tutor) got published in 1922 [**Neumann 1922**]; yet, after graduating from high school in 1921, the von Neumann's made the decision to enroll him in the chemical engineering program of the Eidgenössiche Technische Hochschule (ETH) in Zürich, Switzerland. Chemical engineering was a very popular field at the time and, in addition, a chemical engineer had a far better chance of landing a job than did a mathematician, a consideration that weighed heavily in the eyes of von Neumann's practical father. To prepare for the entrance examination at ETH, von Neumann went first to Friedrich Wilhelm University in Berlin to study chemistry in a non-degree course. He took the entrance exam to ETH successfully in 1923, was admitted to ETH and studied chemical engineering in Zürich until 1926. However, simultaneously with his studies in Berlin and at ETH, von Neumann also enrolled in the mathematics program in Budapest, and he finished his formal university studies in 1926 by receiving his degree in chemical engineering from ETH and his Ph.D. in mathematics (with a thesis on axiomatic set theory) in Budapest.

Von Neumann remembered fondly the time he spent in Zürich: In his letter to an old acquaintance he writes:

> I am also very glad to hear from you from Zurich and to know that you and your family are well, and to be reminded of the town in which I passed several very delightful years as a student at E.T.H.
> (von Neumann to de Horvath, September 20, 1955)

In 1926 von Neumann went to Göttingen on a Rockefeller Fellowship to work as Hilbert's assistant. Göttingen was not only one of the centers of mathematics but it also was a mecca of theoretical physics, and in Göttingen von Neumann could familiarize himself with the latest developments concerning quantum mechanics (see the comments on quantum mechanics in Section 5).

In 1926 von Neumann handed in his application for "habilitation" in mathematics to the Friedrich Wilhelm University in Berlin, Germany. His application was reviewed by E. Schmidt and I. Schur, which was followed by a positive decision by scholars such as L.G.E.M. Bieberbach, E. Hopf, M. von Laue, R. von Mises, W. Nernst, M. Planck and W. Schottky. As a result von Neumann became a non-tenured faculty ("Privatdozent") in Berlin in 1927. In 1929 he left Berlin and went to Hamburg, where he did not stay long: in 1929 he was invited by Princeton University:

> As you know, I was invited by Princeton University for one semester in 1929. In January of 1930 I was offered a permanent professorship, which I did not accept. Upon this they entrusted me to substitute for 5 years in the "Jones professorship for mathematical physics" on a one semester per year basis. In January of 1933 the then beginning "Institute" [of Advanced Study (IAS)] invited me to occupy a permanent chair. Before me, Einstein and two professors at the University (the topologist J.W. Alexander, and O. Veblen) had been appointed.
> (von Neumann to Ortvay, April 17, 1934)

Contrary to the still widespread belief that von Neumann had left Europe for fear of political prosecution, he did not: he was keen on emphasizing that he had taken up residency in the U.S.A. before the political situation in Europe became unbearable and that he therefore never considered himself a refugee scientist (von Neumann's letter to M.R. Davie, May 3, 1946). During the congressional hearing related to von Neumann's appointment as Atomic Energy Commissioner he lists the following reasons why he left Europe:

> ... conditions in Hungary were rather limited ... the thing I was doing had a better field in America, ... I was much more in sympathy with the institutions in America, ... I expected World War II and I was apprehensive that Hungary would be on the Nazi side, and I did not want to be caught dead on that side. [**62**, p. 38]

Before leaving for Princeton, von Neumann married Marietta Kövesi in 1930. The couple settled in Princeton, their only child, daughter Marina, was born there in 1935. In 1937 von Neumann's marriage ended in divorce. In reply to a letter from Ulam, his close friend, von Neumann writes:

> ... many thanks for your letter of Sept. 30., and particularly for what it contained about my "domestic" complications. I am sorry that things went this way – but at least I am not particularly responsible for it. I hope that your optimism is well founded – but

> since happiness is an eminently empyrical proposition, the only thing I can do, is to wait and see ...
> (von Neumann to Ulam, October 4 [1937 ?])

He did not wait long: in 1938 he married Klara Dan, a Hungarian woman, right after she had divorced from her husband. Von Neumann reports on this turn in his life to Veblen:

> In my matrimonial affairs there is at last some palpable progress: The 6 month's "waiting period" ended during the last week of September, the formal-legal proceedings were started on September 27, and they are going now at such a pace, that I expect that the divorce will be completed around October 20. Getting married involves some more red tape, but I expect, that we will be married before Nov. 1, and able to sail around Nov. 10-15. I suppose, that I need not tell you, how glad I am that things are now in this shape.
> (von Neumann to Veblen, October 3, 1938)

Being a non-refugee scientist, von Neumann regularly visited Europe and Hungary in particular during the thirties, while he was already a member of the IAS. He reported on his trips to Europe in letters to U.S. colleagues. One of his major correspondents during the early thirties was Oswald Veblen, von Neumann's colleague at the IAS. Von Neumann's letter to Veblen (April 3, 1933) shows that von Neumann was keenly aware of the deteriorating political conditions in Germany and he had soon drawn some consequences: protesting the German Mathematical Society's political prosecution of one of its members von Neumann resigned from his membership from the Society (letter to Blaschke, January 28, 1935). His last letter in this collection to Veblen from pre-war Europe (October 3, 1938) clearly shows that von Neumann followed the advice he had received from the U.S. consulate: get out of Europe while you can.

His letters to Veblen and to Ortvay also prove that von Neumann anticipated the second world war and he tried early to get involved: in 1937, right after he became a U.S. citizen, he applied to become a reserve lieutenant. He took and passed the necessary examinations but his application was finally rejected on the grounds that he was over age (the age limit was 35). His contribution to war efforts of the U.S. took place on the level of scientific advising: His first consulting position was with the Ballistic Research Laboratory in Aberdeen (Maryland) (from 1937 on). During the war von Neumann became increasingly involved in military-related research and consulting. He did work on high explosives; in his letter to the Editor of the *Evening Star* (May 29, 1947) he recalls in a popular, non-technical form some results of his activity concerning blasts. From 1943 on von Neumann also participated in the Manhattan Project as a consultant. One of his major contributions to the development of the atomic bomb was to help the "implosion lens method" work. This method was used to assemble the fissile material above critical mass by squeezing it with the help of carefully arranged ordinary explosives; von Neumann, relying on his expertise on shock waves and traditional high explosives, saw that the method, if sufficiently refined, could work, and he convinced the Manhattan Team to try this design.

His "exceptionally outstanding service" to the United States and its Navy during the war was acknowledged by awarding him the *Medal for Merit* (October,

1946) and the *Distinguished Service Award* (July, 1946) – the letter to Duncan (December 18, 1947) contains the complete text of citations. Von Neumann was proud of receiving the Medal for Merit:

> I need not tell you that I am very highly gratified by this recognition, although I cannot help feeling, over and above the normal knowledge of personal inadequacy that all I did during the war were per se very interesting and stimulating intellectual pursuits. Actually, the war introduced me to great parts of mathematical physics and applied mathematics which I had neglected before, and I feel that I received intellectually a good deal more than I gave.
> (von Neumann to Strauss, December 9, 1946)

Other awards von Neumann received include: the Medal of Freedom (1956, presidential award), the Enrico Fermi Award (1956) and, for his achievements in pure mathematics: the Bôcher prize (1938). Of this latter prize von Neumann writes:

> This prize is given by the American Mathematical Society once every 5 years for the best paper published in America during the last 5 years in Analysis or its border fields. It is called "the Bôcher Prize", the name comes from the American mathematician Maxime Bôcher, who was, about one generation ago, full professor at Harvard and the leading light there before the times of Birkhoff - Osgood. This was the fifth time the prize was awarded, the previous recipients have been: 1923: G.D. Birkhoff – 1924: E.T. Bell and S. Lefschetz (jointly) – 1928: J.W. Alexander – 1933: M. Morse and N. Wiener (jointly) - 1938: myself. The prize is $100 ...
> (von Neumann to Ortvay January 26, 1939)

Von Neumann was also honoured by being the American Mathematical Society's Gibbs Lecturer (1937), he was American Mathematical Society Colloquium Lecturer (1937), Vanuxem Lecturer (Princeton University, 1953) and he also served as the president of the American Mathematical Society (1951-1952). He received honorary doctorates from Princeton (1947), University of Pennsylvania (1950), Harvard University (1950), Univesity of Istanbul (1952), Case Institute of Technology (1952), Institute of Politechnics (Munich, Germany, 1953) and Columbia University (1954).

After the war von Neumann received offers from a number of universities: Harvard, Columbia, UCLA and MIT. MIT's offer included the promise to fund an electronic computer project at MIT, making available MIT's extensive engineering know-how. Computer design was in the focus of von Neumann's academic interest after the war and he was urging the IAS' leadership to make IAS the home of an electronic computer development project. His efforts and persistency paid off: IAS, with government support and with participation of Radio Corporation of America (RCA) decided to carry out a computer project with von Neumann as director (1945-1955). Having secured the IAS' support for the computer project von Neumann declined MIT's offer. Reading von Neumann's letter to G.B. Harrison, Dean of MIT (November 20, 1945), in which von Neumann declines the offer, one cannot help having the feeling that this decision was a very difficult one for him and that he was regretful when he wrote:

> Owing to my long association with the Institute for Advanced Study, and owing to the fact that I had always urged the importance of such a project, I felt that the Institute for Advanced Study had the first call on my services if it could offer me a promising opportunity to carry out such a program. ... I must therefore decline your very flattering offer, although I do this with great regret.
> (von Neumann to G.B. Harrison, November 20, 1945)

The regret, which is also expressed in his letter to Wiener (November 20, 1945), can be interpreted as anticipation of what later became increasingly obvious, namely that the IAS was not the ideal place to carry out the engineering-heavy computer project. More generally, there is circumstantial evidence that the IAS was not the most suitable institution for von Neumann after the war: its ivory-tower-like intellectual climate was not entirely ready to accommodate von Neumann's increasingly application oriented interests; in addition, von Neumann's very extensive and diverse consulting activity not only in government but also in the private sector does not seem to have been welcomed by the IAS either. This is shown by von Neumann's carefully worded and thoughtful letter to R. Oppenheimer (February 19, 1948), in which he replies to Oppenheimer's attempt to regulate IAS' members' activities, including consulting and vacations. Von Neumann opposes formal regulations and emphasizes that

> ... a certain contact with the strivings and problems of the world that surrounds us is desirable and even necessary in certain important parts of the sciences...
> (von Neumann to Oppenheimer, February 19, 1948)

It is clear from this letter that von Neumann and Oppenheimer (then the director of the IAS) had different views as to how a research institution should be run. Their differences went beyond administrative issues: von Neumann disagreed with many of Oppenheimer's views on politics and atomic weapons development; yet, in the famous 1954 security hearings that resulted in revoking Oppenheimer's security clearance, von Neumann rose to Oppenheimer's defense, declaring that he never considered Oppenheimer as a security risk.

After the war von Neumann's military and governmental consulting activity expanded both in volume and in significance tremendously: he was serving on a number of very influential committees that shaped post war U.S. military policy. Based on documents in the von Neumann Papers of the Library of Congress, Aspray [**3**] compiled a list of von Neumann's consulting contracts: the list contains 20 major advisory positions held by von Neumann around 1954 [**3**, pp. 246-247]. In his letter to H.H. Rankin (August, 1954) von Neumann lists what he considered as his most significant governmental advisory board appointments at the time: consultant to the U.S. Atomic Energy Commission; consultant to the Armed Forces Special Weapons Project; member of the Scientific Advisory Board of the U.S. Air Force; member of the Scientific Advisory Committee of a classified Air Force project connected with the Office of the Special Assistant (Research and Development) of the Secretary of the Air Force; member of the Technical Advisory Board on Atomic Energy connected with the Office of the Assistant Secretary of Defense (Research and Development) and member of the Scientific Advisory Committee of the U.S. Army Ordnance Corps, Ballistic Research Laboratories, Aberdeen Proving Ground.

The letter to T.V. Moore (April 9, 1953), at Standard Oil Development Co., in which von Neumann describes a problem in operational research and gives suggestions for a numerical investigation of the problem, which is in itself an interesting problem, is an example of his industrial consulting. (The von Neumann Archive in the Library of Congress contains a number of lengthy reports on results of analysis of technical problems related to oil exploration.)

While von Neumann never had a full-time teaching position in his life, he had very definite views on how the instruction in science and engineering should be organized at a research university. As one can infer from the letters to V. Bush (November 15, 1949, December 6, 1949 and January 25, 1950) his advice was also sought in connection with educational matters: Von Neumann apparently served on a panel ("Harvard Panel") that was supposed to give all sorts of advice concerning instruction and organization at Harvard University. The composition and authority of this panel are unclear but these letters and the letter to Gottschalk (November 30, 1955) indicate how von Neumann was thinking about certain aspects of higher education. One point he makes is that

> ... the Harvard School of Engineering should not and cannot be a professional school in the ordinary sense of the word. It should, therefore, be run on a higher scholastic level than what is practical for the main part of the educational effort in the traditional professional schools. If this principle is accepted, it should follow that no concessions are to be made in teaching the fundamental subjects: Mathematics, Physics, Chemistry. This would suggest to rely for the teaching of these subjects as much as possible on the respective departments themselves, that is, to forego the convenient expedients of having either of these subjects presented in simplified and mollified "versions for engineers".
> (von Neumann to Bush, December 6, 1949)

Von Neumann also gave advice on the question of how to integrate the emerging field of computer science in the academic-departmental structure of universities: He thought that the best place for computer science is in a mathematics department (letter to Gottschalk and Rademacher, November 30, 1955), emphasizing at the same time that the administrative unification of these fields cannot be fruitful if forced. He also anticipated that substantial education will be needed to train computer literate people.

That von Neumann considered advising a very substantial part of his professional activity is shown by his letter to P.A. Dodd (February 24, 1956), in which he negotiated the details of his appointment as professor at large in the University of California at Los Angeles for the period after his planned stepping down as Atomic Energy Commissioner:

> In the past I have done a good deal of industrial and government consulting – of course, always in the area of my scientific and intellectual interests. I would expect to continue to do so in the future. Am I right in assuming that such consulting meets with the approval of the University of California and is in conformity with its policies?
> (von Neumann to Dodd, February 24, 1956)

Von Neumann's attitude towards industrial and government consulting is shown by several other letters in the collection. The letter to S. MacLane (May 17, 1948) is perhaps the most revealing from the point of view of how von Neumann was thinking about consulting, especially military-government advising: von Neumann considered it a moral obligation to do military-government consulting in spite of knowing that large organizations (such as the Army) are only moderately effective in using the advice scientists can deliver.

The impact and importance of von Neumann as a military adviser is summarized by L. Strauss, who in an interview conducted for the "John von Neumann Documentary Video" (prepared and distributed by the Mathematical Association of America), describes the following scene:

> I remember one particular incident at the Walter Reed Hospital, the army hospital here in Washington, when the Defense Department felt that it had to consult him. He was their chief scientific adviser on air force matters. And I recall the extraordinary picture of sitting beside the bed of this man, in his forties, who had been an immigrant, and there surrounding him were: the Secretary of Defense, the Deputy Secretary of Defense, the Secretaries of Air, Army, Navy and the chiefs of staff.

In 1954 president Eisenhower appointed von Neumann as an Atomic Energy Commissioner. Von Neumann then moved to Washington, D.C. to take the position – officially, he was on leave from IAS from the Spring of 1955. His term as an Atomic Energy Commissioner was to end by June 1959; however, von Neumann planned to step down earlier and he did not wish to return to IAS: as the letter to J.R. Killian, the president of MIT (February 24, 1956) shows, von Neumann was again offered a position at MIT and he also was negotiating a position at UCLA (letter to P.A. Dodd, February 24, 1956). Von Neumann finally decided in March 1956 to accept the offer to become professor at large at UCLA; he was never able to take the position however: in August, 1955 he was diagnosed with cancer and soon he was confined to a wheelchair. His last academic activity was the preparation for the 1956 Silliman Lectures at Yale. He accepted the invitation in 1955 before he fell ill and continued working on the lectures even after the diagnosis. But the lectures were never delivered and the material for the lectures could not be completed (the incomplete manuscript was posthumously published [**64**]). In April 1956 von Neumann had to check in the hospital for the last time; he died in the Walter Reed Army hospital in Washington, D.C. on February 8 of 1957. He is buried in the Princeton cemetery.

It seems fair to say that if influence of a scientist is interpreted broadly enough to include impact on fields beyond science proper, then John von Neumann was probably the most influential mathematician who ever lived: not only did he contribute to almost all branches of modern mathematics and create new fields but he also changed history after the second world war by his work in computer design and by being a technical advisor to the post-war military-political establishment of the U.S.A. By making von Neumann's letters available, the present volume hopes to contribute to a better understanding of the extraordinary man that John von Neumann was.

2. Logic and foundations of mathematics

Von Neumann started his career as a mathematician with work on axiomatic set theory and he also worked on Hilbert's program aimed at proving consistency of mathematics by finitistic means [**Neumann1927e**].

The turning point in the history of Hilbert's program was the famous Second Conference for Epistemology of the Exact Sciences, which took place between the 5th and 7th of September 1930 in Königsberg. The meeting was organized jointly by the "Gesellschaft für Empirishe Philosophie" (a group of philosophers of science in Berlin) and by the Vienna Circle, a similar group of philosophers based in Vienna. Participants of the Königsberg meeting included R. Carnap, A. Heyting, J. von Neumann and K. Gödel. Carnap gave a talk on the foundations of logicism, Heyting on the intuitionistic foundations of mathematics and von Neumann on the formalist foundations of mathematics. (For a detailed account of the events related to this conference see [**16**, pp. 135, 137] and [**17**, pp. 327-330].) It was in the discussion session on September 7 of this conference that Gödel announced the first version of what became known as Gödel's first incompleteness theorem: every sufficiently rich and consistent axiom system contains meaningful statements that are undecidable within that system (they cannot be proved in the axiom system, nor can their negation).

Von Neumann grasped immediately the significance of Gödel's result for the axiomatic foundation of mathematics, pressed Gödel for further details and, as his letter to Gödel (November 20, 1930) shows, shortly after the Königsberg conference, he apparently obtained what is known as Gödel's second incompleteness theorem. The second incompleteness theorem says that the consistency of a sufficiently rich axiomatic theory (such as Peano arithmetics) cannot be proved within the system itself: the statement expressing the consistency of the system is an undecidable proposition in the system. As can be inferred from von Neumann's (November 29, 1930) letter to Gödel, Gödel, in his reply to von Neumann's (November 20, 1930) letter (a reply which seems to have been lost) told von Neumann that he had already established the second incompleteness theorem. Von Neumann therefore decided immediately:

> As you have established the theorem on the unprovability of consistency as a natural continuation and deepening of your earlier results, I clearly won't publish on this subject.
> (von Neumann to Gödel, November 29, 1930).

From the second incompleteness theorem von Neumann had drawn a very strong conclusion for the Hilbert program: "... there is no rigorous justification for classical mathematics" (von Neumann to Gödel, November 29, 1930); more generally:

> Gödel's results mean that there is no "complete" axiomatic system, not even in mathematics, and I believe that there is actually no other consistent interpretation of this complex of questions.
> (von Neumann to Ortvay, July 18, 1939)

On the interpretation of Gödel's second incompleteness theorem von Neumann strongly disagreed with Gödel: in his letter to Carnap (June 7, 1931) von Neumann writes:

> Thus I am today of the opinion that
> (1) Gödel has shown the unrealizability of Hilbert's program.

> (2) There is no more reason to reject intuitionism (if one disregards the aesthetic issue, which in practice will also for me be the decisive factor).
>
> Therefore I consider the state of the foundational discussion in Königsberg to be outdated, for Gödel's fundamental discoveries have brought the question to a completely different level. (I know that Gödel is much more careful in the evaluation of his results, but in my opinion on *this* point he does not see the connections correctly).
>
> (von Neumann to Carnap, June 7, 1931)

Whereas Gödel thought otherwise:

> I wish to note expressly that Theorem XI [the second incompleteness theorem] does not contradict Hilbert's formalistic viewpoint. For this viewpoint presupposes only the existence of a consistency proof in which nothing but finitary means of proof is used, and it is conceivable that there exists finitary proofs that *cannot* be expressed in the formalism of P [Russel's Principia plus the Peano axioms].
>
> [**15**] ([**16**, p. 195])

Strongly formulated disagreements with Gödel can also be found in von Neumann's letter to Gödel (January 12, 1931). The disagreement between Gödel and von Neumann on the implications of the second incompleteness theorem is due to the fact that von Neumann interpreted the notion of "finite methods" in connection with Hilbert's program in a much more restricted sense than did Gödel: According to von Neumann (letter to Carnap, June 7, 1931; letter to Gödel, November 29, 1931), all intuitionistic-finitistic inferences can be expressed in a formal system for which Gödel's second incompleteness theorem holds, whereas Gödel thought that this is questionable.

In spite of fundamental professional disagreements between von Neumann and Gödel about the ramifications of the second incompleteness theorem von Neumann had the highest respect for Gödel and his work: he compared Gödel to Aristotle ([**16**, p. 8] and references there), and in his letter to A. Flexner (September 27, 1939), in which von Neumann tries to help Gödel to get a visa to the U.S.A., von Neumann writes:

> The claim may be made with perfect justification that Gödel is unreplaceable for our educational program. Indeed Gödel is absolutely irreplaceable; he is the only mathematician alive about whom I would dare to make this statement. [...] I am convinced that salvaging him from the wreck of Europe is one of the great single contributions anyone could make to science at this moment.
>
> (von Neumann to Flexner, September 27, 1939)

Von Neumann's appreciation of Gödel's work is also apparent from his letter to O. Veblen (November 30, year unknown).

3. Operator algebras

The theory of operator (in particular, von Neumann) algebras is perhaps von Neumann's deepest contribution to mathematics, of which he himself was rather proud: One can sense a satisfaction in his letter to Dixmier (June 18, 1954) in which von Neumann reacts to Dixmier's suggestion that "rings of operators" should be called von Neumann algebras to acknowledge von Neumann's contribution to this field.

A von Neumann algebra $\mathcal{M}$ is a *-closed subalgebra of bounded operators on a Hilbert space $\mathcal{H}$ that are also closed in the strong operator topology (the latter meaning that if for some bounded operator Q on $\mathcal{H}$ one has $Q_n\xi \to Q\xi$ $(n \to \infty)$ for all ξ in $\mathcal{H}$ and $Q_n \in \mathcal{M}$ for all n, then $Q \in \mathcal{M}$). Already in [**Neumann1929b**] von Neumann established a major theorem concerning von Neumann algebras: the "double commutant theorem", which says that a set $\mathcal{M}$ of bounded operators on a Hilbert space $\mathcal{H}$ is closed in the strong operator topology if and only if $\mathcal{M}$ coincides with its second commutant $(\mathcal{M}')'$, where the commutant $\mathcal{M}'$ of a set of bounded operators $\mathcal{M}$ is the set of operators that commute with *every* element in $\mathcal{M}$. An immediate consequence of the double commutant theorem is that the set of projections $\mathcal{P}(\mathcal{M})$ in a von Neumann algebra $\mathcal{M}$ form a complete lattice and that the projections determine the von Neumann algebra completely by $\mathcal{P}(\mathcal{M})'' = \mathcal{M}$.

Von Neumann (partly in collaboration with F.J. Murray) published four major papers on the theory of von Neumann algebras [**Neumann1936a**], [**Neumann1937a**], [**Neumann1940**] and [**Neumann1943**]. The first paper's main result was a classification theory of von Neumann algebras that are irreducible in the sense of not containing non-trivial operators commuting with every other operator in the algebra (these von Neumann algebras are called "factors") [**Neumann1936a**]. The factorial property of $\mathcal{M}$ can be expressed by $\mathcal{M} \cap \mathcal{M}' = \{\lambda I\}$, where I is the identity operator on $\mathcal{H}$ and λ is a complex number. The set $\mathcal{B}(\mathcal{H})$ of *all* bounded operators is clearly a factor and, as his letter to M. Stone (March 31, 1935) shows, von Neumann originally conjectured that only this "type" of factor exists. Von Neumann had suggested J. Murray to prove this conjecture (see [**35**]), and, if this conjecture could have been proved true, "... that would be a worthy result and the end of the project." [**25**, p. 63]. It turned out, however, that this conjecture is false: there are five classes of factors, the different types are denoted by von Neumann as I_n, I_∞, II_1, II_∞ and III_∞. The classification of factors was given in terms of a (relative) dimension function d defined on the lattice of projections $\mathcal{P}(\mathcal{M})$ of a von Neumann algebra $\mathcal{M}$. The map d from $\mathcal{P}(\mathcal{M})$ into the set $\mathbb{R}^+ \cup \{\infty\}$ is a dimension function if

(i) $d(A) > 0$ if and only if $A \neq 0$,
(ii) $d(A) = d(B)$ if there exists an isometry $U \in \mathcal{M}$ between ranges of the projections A and B,
(iii) $d(A) + d(B) = d(A \vee B) + d(A \wedge B)$.

The (relative) dimension function d on an arbitrary factor is a generalization of the ordinary linear dimension of the closed linear subspace a projection projects to, and the ordinary dimension takes on the positive integer values $0, 1, 2, \ldots, n$ and $0, 1, 2, \ldots$, respectively, in the two well-known cases of the set of all bounded operators on a finite, n-dimensional (respectively infinite) dimensional Hilbert space. In the cases II_1, II_∞ and III_∞ the ranges of the dimension function are, respectively, the following: the unit interval $[0, 1]$, the set of non-negative real numbers $\mathbb{R}^+$ and the two element set $\{0, \infty\}$. (See [**57**] for a systematic treatment, or [**42**], [**39**] for a brief review of the Murray-von Neumann classification theory). The result of the classification theory can thus be summarized in the form of the following table:

range of d	type of factor $\mathcal{N}$	the lattice $\mathcal{P}(\mathcal{M})$
$\{0, 1, 2, \ldots, n\}$	I_n	modular
$\{0, 1, 2, \ldots, \infty\}$	I_∞	orthomodular, non-modular
$[0, 1]$	II_1	modular
$\mathbb{R}^+$	II_∞	orthomodular, non-modular
$\{0, \infty\}$	III_∞	orthomodular, non-modular

The existence of five different classes of factors was unexpected and also surprising, not least in part because the (relative) dimension function could take on not only integer values but also real numbers. In their first joint paper Murray and von Neumann [**Neumann1936a**] could construct factors in every class except the type III_∞. The problem of existence of type III_∞ factors remained a difficult open problem, and it was only years after the classification theory had been worked out that von Neumann could prove that type III_∞ factors do in fact exist [**Neumann1940**]. Von Neumann reports on obtaining this result in the letter to I. Halperin (June 1, [1939 ?]), this letter is witness to von Neumann's excitement about this result, a rare excitement that von Neumann himself calls "euphoria". This letter expresses von Neumann's hope that he can make some additional progress concerning the classification of the type II_1 factor into further subclasses.

The type II_1 factor (the one on which a dimension function with the interval $[0, 1]$ as its range exists) fascinated von Neumann from the very moment of its discovery in [**Neumann1936a**]. For a number of subtle conceptual and mathematical reasons von Neumann thought that this von Neumann algebra provides a mathematical framework for quantum theory which is more suitable than Hilbert space theory (see the remarks on quantum logic in Section 6). Finding further subclasses of the type II_1 was thus also a major problem von Neumann worked on intensely in the period from 1936-1940. The letter to I. Halperin (February 22, 1940) reports that Murray and von Neumann succeeded in showing that not all type II_1 factors can be approximated by an increasing sequence of finite dimensional matrix subalgebras of the algebra. A partial result leading to this result is described in the letter to Halperin (December 17, [1939 ?]). The letters to Halperin (June 1, [1939 ?]) and (February 22, 1940) describe the general technique of constructing von Neumann factors; this technique is known as "group measure space construction". The main idea, in modern notation, is the following.

Let $(X, \mathcal{B}, \mu)$ be a measure space and G a countable group of measure-preserving transformations. The group measure space construction yields a von Neumann algebra acting on the Hilbert space $L^2(\mu) \otimes l^2(G)$, which is regarded as a set of functions defined on G and with values in $L^2(\mu)$. For every $f \in L^\infty(\mu)$ define

$$((M_f\xi)(g))(x) = f(g^{-1}x)(\xi(g)(x)) \qquad (\xi \in L^2(\mu) \otimes l^2(G)) \tag{1}$$

and for every $g \in G$ set

$$V_g(\xi)(g')(x) = \xi(g^{-1}g')(x) \qquad (\xi \in L^2(\mu) \otimes l^2(G)) \tag{2}$$

Let $\mathcal{M}(\mu, G)$ be the von Neumann algebra generated by the operators

$$\{M_f : f \in L^\infty(\mu)\} \cup \{V_g : g \in G\}$$

Then the choice of the unit circle with the Lebesgue measure and (the powers of) an irrational rotation yields a factor of type II_1. The real line with the Lebesgue measure and the rational translations yield a factor type II_∞. Von Neumann modified this measure theoretic procedure to obtain a type III_∞ factor from the Lebesgue measure of the real line and the group of all rational linear transformations (see [**25**] for a more detailed review of the construction).

In the letters to Dixmier (November 14, 1953) and I. Kaplansky (March 1, 1950) von Neumann distinguishes different levels on which one can carry out an investigation of operator algebras: on the first, most concrete level, one considers the operators as maps acting on a particular Hilbert space. (In today's terminology, we would say that the algebras are analyzed in a concrete representation.) On the second level one regards the operators as carriers of an abstract algebraic structure, independently of any particular representation. The next level focuses on the algebraic features of the selfadjoint part of an abstract operator algebra. This level defines the topic of what is called today Jordan algebra theory, while on the fourth, highest level of abstraction, the lattice theoretic structure defined by the set of projections in the algebra is studied. Von Neumann mentions that, after publishing a paper [**Neumann1936b**] on the level of Jordan algebra theory, and after having achieved some further results which he never published, he neglected working on the level of Jordan algebra theory since he considered analyzing the lattice theoretic structure of the idempotents of an algebra "... as the really satisfactory one, and having worked it out, lost a good deal of my interest in the others." (von Neumann to Dixmier, November 14, 1953).

4. Unbounded operators

Motivated to a large extent by mathematical problems in quantum mechanics, von Neumann made pioneering work in the theory of unbounded operators [**Neumann1929a**] and [**Neumann1932d**]; specifically, he clarified the subtle differences between symmetric and selfadjoint (unbounded) operators. The unboundedness causes technical difficulties that are mainly related to the fact that unbounded operators are not everywhere defined; hence one has to be extremely careful about the domains of those operators in definitions and calculations involving them.

The lengthy and technical letter to Kemble (December 6, [????]) shows some of the subtleties of the subject and is a historical document in the long development of the problem of proving selfadjointness of Hamilton operators that occur in quantum mechanics. Unfortunately, the terminology used by von Neumann in connection with unbounded operators differs – both in his papers and in his letter to Kemble – from the one that has become accepted today. To avoid confusion and to help the reader, the basic definitions and terminology in connection with the theory of unbounded operators is recalled below with special emphasis put on the relation of today's terminology to the one used by von Neumann.

Let $\mathcal{D}(T)$ denote the domain of definition of a linear operator T defined on the Hilbert space $\mathcal{H}$. If $\mathcal{D}(T)$ is dense in $\mathcal{H}$, then there exists (uniquely) an operator T^* with definition of domain $\mathcal{D}(T^*)$ such that

$$\langle \xi, T\eta \rangle = \langle T^*\xi, \eta \rangle \qquad \xi \in \mathcal{D}(T^*), \quad \eta \in \mathcal{D}(T). \tag{3}$$

The operator T^* is called the *adjoint* of T. Let $\Gamma(T)$ denote the *graph* of an operator T, the graph is defined by

$$\Gamma(T) = \{(\xi, \eta) \ : \ \xi \in \mathcal{D}(T), \eta = T\xi\}.$$

An operator T is called *closed* if its graph $\Gamma(T)$ is closed in the following sense: If $\xi_n \in \mathcal{D}(T)$ is a sequence of elements such that $\xi_n \to \xi$ and $T\xi_n \to \eta$, then $\xi \in \mathcal{D}(T)$ and $T\xi = \eta$. It is a classical result of operator theory (Hellinger-Toeplitz theorem) that an everywhere defined closed operator is bounded.

Given two operators T_1 and T_2 one writes $T_1 \subseteq T_2$ if T_2 extends T_1. This means that the domain $\mathcal{D}(T_1)$ of the definition of T_1 is contained in the domain of definition $\mathcal{D}(T_2)$ of T_2 and $T_1(\xi) = T_2(\xi)$ for all $\xi \in \mathcal{D}(T_1)$. The relation $T_1 \subseteq T_2$ can be expressed in terms of the graphs of the two operators as $\Gamma(T_1) \subseteq \Gamma(T_2)$.

The definition of adjoint operator entails that

$$T_1 \subseteq T_2 \qquad \text{implies} \qquad T_2^* \subseteq T_1^*. \tag{4}$$

The operator T is called *symmetric* if

$$\langle \xi, T\eta \rangle = \langle T\xi, \eta \rangle \qquad \xi, \eta \in \mathcal{D}(T). \tag{5}$$

Thus T is symmetric if and only if $T \subseteq T^*$. Operator T is called *maximal symmetric* if it is symmetric and there exists no *symmetric* operator T' that extends T. The operator T is called *selfadjoint* if it coincides with its adjoint: $T = T^*$.

A selfadjoint operator is closed; hence an unbounded selfadjoint operator cannot be everywhere defined (by the Hellinger-Toeplitz theorem). If T is selfadjoint, then it is maximal symmetric because by (4) one has for a symmetric T' that extends T,

$$T \subseteq T' \subseteq T'^* \subseteq T = T^*$$

which implies $T = T'$. A maximal symmetric operator need not be selfadjoint however. The relation of maximal symmetric and selfadjoint operators was clarified by von Neumann and can be described in terms of the *defect indices*: If T is a symmetric operator, then n^+ and n^- defined below are called the defect indices of T:

$$n^+ = \dim \left[Range(T + iI) \right]^{\perp}, \tag{6}$$

$$n^- = \dim \left[Range(T - iI) \right]^{\perp}. \tag{7}$$

Von Neumann proved that a symmetric operator T is maximal symmetric if and only if *one* of its defect indices is zero and T is selfadjoint if and only if *both* of its defect indices are zero (see [**Neumann1929a**, Satz 33]). Von Neumann's terminology differs from the above however: he called symmetric operators "Hermitean" (he also used the terminology "selfadjoint") and he called selfadjoint operators "hypermaximal". (Von Neumann mentions E. Schmidt as the creator of the terminology "hypermaximal" ([**Neumann1929a**, footnote 23].) Thus the relation of the standard terminology and the one used by von Neumann can be summarized as follows:

Standard terminology	*von Neumann's terminology*
T symmetric	T *Hermitean* (equivalently: *selfadjoint*)
T maximal symmetric	T *maximal Hermitean*
T selfadjoint	T *hypermaximal*

One of the distinguishing features of selfadjoint operators is that they possess a "spectral resolution" ("Zerlegung der Einheit" in von Neumann's original terminology). This is spelled out in what is called the "spectral theorem for selfadjoint operators", which in modern terminology and notation reads as follows: For every selfadjoint operator there exists a unique spectral measure P: a map $P\colon \mathcal{B}(\mathbb{R}) \to \mathcal{P}(\mathcal{H})$ from the Borel σ algebra $\mathcal{B}(\mathbb{R})$ of real numbers into the set $\mathcal{P}(\mathcal{H})$ of projections on the Hilbert space $\mathcal{H}$ such that

$$(8) \qquad \langle \xi, T\eta \rangle = \int_{\mathbb{R}} id_{\mathbb{R}} d\mu^P_{\xi,\eta} \qquad \eta \in \mathcal{D}(T), \quad \xi \in \mathcal{H}$$

where the complex measure $\mu^P_{\xi,\eta}$ on the Borel σ algebra $\mathcal{B}(\mathbb{R})$ of real numbers is defined by

$$(9) \qquad \mathcal{B}(\mathbb{R}) \ni d \mapsto \mu^P_{\xi,\eta}(d) = \langle \xi, P(d)\eta \rangle.$$

Conversely, given a spectral measure P, every operator T defined by eq. (8) defines a (generally unbounded) selfadjoint operator. The terminology "spectral measure" is justified by the fact that the support of P coincides with the spectrum of T and (8) is the generalization of the well-known spectral representation $T = \sum_i^N \lambda_i E_i$ of a selfadjoint operator (matrix) in a finite dimensional Hilbert space in terms of eigenprojections E_i and eigenvalues λ_i (the "diagonalization" of a selfadjoint matrix). The spectral theorem was proved by von Neumann [**Neumann1929a**, Satz 36]. Unitary operators also have a spectral measure; the spectral measure is defined on the unit circle rather than on the real line — in accordance with the fact that the spectrum of unitary operators lies on the unit circle. Von Neumann reports his result on spectral theory of unitary operators in his letter to Veblen (October 3, 1938).

Since the operators representing observables in quantum mechanics are typically not everywhere defined unbounded operators, it was a major mathematical problem to clarify whether (on what assumptions) they are selfadjoint. One such operator is the one investigated in detail by von Neumann in his letter to Kemble (December 6, [????]):

> If I understand you correctly, your question is this:
>
> Given a differential operator
>
> $$(10) \qquad H = -(\frac{\partial^2}{\partial x_1^2} + \ldots + \frac{\partial^2}{\partial x_n^2}) + V(x_1, \ldots, x_n)$$
>
> when is it possible to apply the general theorems of operator-theory to it, i.e., when is it "hypermaximal"?
>
> (von Neumann to Kemble, December 6, [????])

The above differential operator gives the wave equation of an m particle system moving in a potential. Remarkably, von Neumann, the undisputed expert on operator theory, was unable to prove the selfadjointness of that Hamilton operator, as he "shamefully confesses" (letter to Kemble (December 6, [1935 ?]). A general proof of selfadjointness of typical Hamilton operators was only given in 1951 by Kato [**26**], and as P. Lax reports:

> I recall in the summer of 1951 the excitement and elation of von Neumann when he learned that Kato has proved the self-adjointness of the Schrödinger operator associated with the helium atom.
> [**30**, p. 414]

5. Quantum mechanics

5.1. Mathematical foundations of Hilbert space quantum mechanics. Von Neumann started working on quantum mechanics in 1926 when he was an assistant of D. Hilbert in Göttingen. In the academic year 1926-1927 Hilbert gave lectures on the mathematical foundations of quantum mechanics. Von Neumann and Nordheim were attending Hilbert's lectures, and Hilbert, Nordheim and von Neumann published a joint paper based on those lectures [**Neumann1927a**]. As the authors of the paper [**Neumann1927a**] themselves emphasize, the paper's treatment of the mathematical foundations of quantum mechanics was not entirely satisfactory mathematically: it is assumed in the paper that every linear operator on a Hilbert space is an integral operator but this fiction can only be maintained formally if one uses the problematic Dirac delta function, and this mathematically ill-defined entity also features in formulating the eigenvalue problem of selfadjoint operators. Von Neumann did not tolerate such mathematical nonsense and he could soon eliminate it from the theory. He did this by ground-breaking work in functional analysis: he created the notion of abstract Hilbert space as we know it today, formulated the eigenvalue problem for selfadjoint operators generally in terms of spectral measures without relying on the notion of Dirac delta function and established the theory of unbounded operators; in particular, he proved the spectral theorem for (unbounded) selfadjoint operators (see 4 of the Introductory Comments). While in Göttingen von Neumann published three papers on the mathematical foundations of quantum mechanics [**Neumann1927a**], [**Neumann1927b**], [**Neumann1927c**]. These three papers served as the basis of his book [**Neumann1932d**].

Von Neumann viewed his book much more, however, than just a clarification of the mathematical foundations of quantum mechanics: characterizing the nature and the novelty of the book in his letter to H. Cirker (October 3, 1949) he writes:

> The subject-matter is partly physical-mathematical, partly, however, a very involved conceptual critique of the logical foundations of various disciplines (theory of probability, thermodynamics, classical mechanics, classical statistical mechanics, quantum mechanics). This philosophical-epistemological discussion has to be continuously tied in and quite critically synchronised with the parallel mathematical-physical discussion. It is, by the way, one of the essential justifications of the book, which gives it a content not covered in other treatises, written by physicists or by mathematicians, on quantum mechanics.
> (von Neumann to Cirker, October 3, 1949)

Von Neumann's description of his book [**Neumann1932d**] as a philosophical-epistemological analysis is in conformity with his interpretation of mathematical physics as a discipline which is very close to philosophy: in his letter to R.O. Fornaguera (December 10, 1947), the Spanish translator of von Neumann's book *Mathematical Foundation of Quantum Mechanics* [**Neumann1932d**], von Neumann writes:

> Your questions on the nature of mathematical physics and theoretical physics are interesting but a little difficult to answer with precision in my own mind. I have always drawn a somewhat vague line of demarcation between the two subjects, but it was really more a difference in distribution of emphasis. I think that in theoretical physics the main emphasis is on the connection with experimental physics and those methodological processes which lead to new theories and new formulations, whereas mathematical physics deals with the actual solution and mathematical execution of a theory which is assumed to be correct per se, or assumed to be correct for the sake of the discussion.
>
> In other words, I would say that theoretical physics deals rather with the formation and mathematical physics rather with the exploitation of physical theories. However, when a new theory has to be evaluated and compared with experience, both aspects mix.
>
> (von Neumann to Fornaguera, December 10, 1947)

The position von Neumann takes in the above quotation concerning mathematical physics is a very moderate one: he does not see a neat separation of mathematics and theoretical physics and he takes the reflective nature of mathematical physics as its main characteristics – not mathematical exactness. "Reflective nature" means here that the immediate subject of mathematical physics is considered by him to be the physical *theory* rather than the physical world, the latter implicitly taken by von Neumann as the subject of theoretical physics. Investigating and "exploiting physical theories" is very much what philosophy of science (physics) typically does however, and this reflective nature of mathematical physics lends this discipline a philosophical character indeed (see the papers [**46**],[**54**] and [**47**] for a more detailed discussion of this point).

5.2. Uniqueness of the Schrödinger representation of the canonical commutation relation. One of von Neumann's many achievements in the field of foundations of quantum mechanics was the proof of uniqueness of the Schrödinger representation of the canonical commutation relation [**Neumann1931**].

The usual Canonical Commutation Relation (CCR) (also called *Heisenberg commutation relation*)

$$[Q, P] = iI \tag{11}$$

where Q and P are selfadjoint operators on a Hilbert space and equation (11) is supposed to hold on a dense subset of $\mathcal{H}$, can be viewed as the infinitesimal form of a commutation relation and, accordingly, it can be reformulated in terms of the one-parameter groups $\{U(a)\}$ and $\{V(b)\}$ of unitary operators determined by Q, P

as infinitesimal generators:

$$U(a) = e^{iaQ} \qquad V(b) = e^{ibP} \quad a, b \in \mathbb{R}.$$

In terms of U and V the commutation relation (11) becomes

$$U(a)V(b) = e^{iab}V(b)U(a) \quad a, b \in \mathbb{R}. \tag{12}$$

To study the commutation relation in its form (12) (the so-called Weyl-form) von Neumann introduced the two-parameter family of operators

$$S(a,b) \equiv \exp\Big(-\frac{1}{2}iab\Big)U(a)V(b) \tag{13}$$

in terms of which (12) becomes

$$S(a,b)S(c,d) = \exp\Big(\frac{1}{2}i(ad-bc)\Big)S(a+c,b+d) \tag{14}$$

and which also satisfy

$$S(-a,-b) = S(a,b)^*. \tag{15}$$

A map $(a,b) \mapsto S(a,b)$ from the two dimensional linear space $\mathbb{R}^2$ into the set of bounded operators $\mathcal{B}(\mathcal{H})$ that satisfies (14)-(15) is called the representation of the canonical commutation relation.

A specific representation of CCR is the "Schrödinger representation" defined on the Hilbert space of the square integrable functions $L^2(\mathbb{R})$ by

$$(Qf)(x) = xf(x) \qquad (Pf)(x) = -if'(x) \quad f \in L^2(\mathbb{R}). \tag{16}$$

Von Neumann proved in [**Neumann1931**] what has become known as the Stone-von Neumann theorem on the uniqueness of the representation of the canonical commutation relation: If the representation S of CCR is continuous in the weak (or, equivalently, strong) operator topology and is irreducible, then it is unique in the sense of being unitary equivalent to the Schrödinger representations given by (16).

The uniqueness result was stated by M. Stone first [**53**] with some hints as to the proof but it was von Neumann who gave the full proof in [**Neumann1931**]. Von Neumann's letter to Veblen (September 23, 1930) reports on this result, and his letter to Stone (October 8, 1930) lets us peak into his semi-formal thinking that explains the intuition behind the proof of the uniqueness theorem:

> Any proof of this theorem had to construct with the aid of P, Q or
>
> $$U(\alpha) = e^{i\alpha P} \qquad V(\beta) = e^{i\beta Q}$$
>
> some operator, which has easily identifiable properties, determining him in a unique way – and which operator on the other hand can be used to determine some vectors in Hilbert space.
> (von Neumann to Stone, October 8, 1930)

The general form of the operator determined by U and V is

$$A = \int\int a(\alpha,\beta)U(\alpha)V(\beta)\mathrm{d}\alpha\mathrm{d}\beta \tag{17}$$

with an integrable function $\mathbb{R}^2 \ni (\alpha,\beta) \mapsto a(\alpha,\beta) \in \mathbb{R}$. Using (13), the operator A defined by (17) can be rewritten in terms of $S(a,b)$, and von Neumann constructs an A which is given by

$$A = \int\int \exp\Big(-\frac{1}{4}(|a|^4 + |b|^2)\Big) S(a,b) da db. \tag{18}$$

The crucial observation is that the operator $P = \frac{1}{2\pi}A$ is a projection and, if the representation S is irreducible, then P is one-dimensional, spanned by a unit vector $\xi \in \mathcal{H}$; hence, if S and S' are two irreducible representations of CCR and ξ' is the analogously defined vector determined by S', then the map $T\colon \mathcal{H} \to \mathcal{H}$ defined by

$$TS(a,b)\xi = S'(a,b)\xi'$$

extends linearly to a unitary of $\mathcal{H}$ that intertwines between the two representations S and S'. This means that an irreducible representation of CCR is unique up to unitary isomorphism. (For an extensive review of von Neumann's uniqueness result and related subsequent developments see Summer's paper [**56**], for a comprehensive review of the theory of representation of CCR from an operator algebraic point of view see [**38**].)

5.3. Correlations between spatially separated quantum systems. In 1935 Einstein, B. Podolsky and N. Rosen published the famous "EPR paper" [**11**]. The paper's aim was to prove that one should consider the quantum mechanical description of physical reality incomplete – provided that one accepts the *principle of locality*: that the physical state of a subsystem $\mathcal{S}_1$ of a joint system $(\mathcal{S}_1 + \mathcal{S}_2)$ cannot be changed instantaneously by performing a measurement on subsystem $\mathcal{S}_2$ spatially separated from subsystem $\mathcal{S}_1$. The discussions between Einstein, Rosen and Podolsky that led to the EPR paper were taking place in Einstein's office at the Institute for Advanced Study in the spring of 1935 [**24**]. Von Neumann was Einstein's colleague at the Institute and, given that the EPR paper concerned the completeness of quantum mechanics, in which von Neumann was very much interested (he discussed completeness of quantum mechanics at length in his book [**Neumann1932d**]), one would expect that Einstein and von Neumann exchanged ideas on this issue. There is, however, no record of any exchange between von Neumann and Einstein on this subject, nor did von Neumann ever express any views on the EPR paper's argument – the only exception I am aware of being von Neumann's letter to Schrödinger published in the present volume (April 11, 1936). We know that Einstein and Schrödinger corresponded about the EPR paper in the summer of 1935: Einstein was complaining to Schrödinger about the final wording of the EPR paper, thinking that the main point was buried in it by erudition [**24**]. In his letter to Schrödinger (June, 1935, quoted in [**24**]) Einstein gave a much simpler argument purportedly showing the incompleteness of quantum mechanics – again utilizing the principle of locality. Motivated in part by the EPR paper and by his correspondence with Einstein, Schrödinger published two papers on the problem of probabilistic correlations between spatially separated quantum systems [**51**] and [**52**]. Von Neumann's letter to Schrödinger (April 11, 1936) is a direct reply to Schrödinger's second paper [**52**].

Schrödinger argues in both [**51**] and [**52**] that the presence of correlations between spatially separated quantum subsystems of a joint quantum system threatens the principle of locality and thereby is in direct contradiction with the theory of relativity. Specifically, Schrödinger considers a composite quantum system described by the tensor product Hilbert space $\mathcal{H}_1 \otimes \mathcal{H}_2$, where $\mathcal{H}_1$ and $\mathcal{H}_2$ are assumed to be identical copies of an L^2 function space describing system $\mathcal{S}_1$ and $\mathcal{S}_2$, respectively.

Schrödiner shows in [**51**] that any state vector $\Psi(x, y) \in \mathcal{H}_1 \otimes \mathcal{H}_2$ of the composite system can be written as

$$\Psi(x, y) = \sum_k a_k g_k(x) \otimes f_k(y) \qquad g_k \in \mathcal{H}_1 \quad f_k \in \mathcal{H}_2, \quad a_k \in \mathbb{C} \tag{19}$$

with $\{g_k\}$ and $\{f_k\}$ being complete sets of orthogonal (unit) vectors in the respective spaces (not all a_k are necessarily nonzero). The decomposition (19) is called the biorthogonal decomposition and it is unique (up to re-labelling of the elements g_k and f_k respectively).

Vectors f_k and g_k can be viewed as eigenvectors (with eigenvalues λ_k^F, λ_k^G) of some observables F and G of system $\mathcal{S}_1$ and $\mathcal{S}_2$, respectively. If we carry out a measurement of G on $\mathcal{S}_2$ and find eigenvalue λ_n^G, then "... we have to assign to the *first* system the wave function $g_k(x)$." [**52**, p. 450]. From the perspective of system $\mathcal{S}_1$, the state of the joint system $(\mathcal{S}_1 + \mathcal{S}_2)$ given by $\Psi(x, y)$ differs from state g_k because the state of $\mathcal{S}_1$ given by g_k is a pure state on $\mathcal{S}_1$, whereas the state given by vector $\Psi(x, y)$ is a mixed state given by the density matrix

$$\rho_1 = \sum_k |a_k|^2 P_{g_k}; \tag{20}$$

here P_{g_k} denotes the one-dimensional projection in $\mathcal{H}_1$ that projects to the one-dimensional subspace spanned by element g_k. The range $rng(\rho_1)$ of the density matrix ρ_1 is spanned by those g_k for which $a_k \neq 0$.

Schrödinger finds such instantaneous change in system $\mathcal{S}_1$'s state as a result of measurement on the spatially distant system $\mathcal{S}_2$ already troublesome enough; yet, in [**52**] he goes even further by showing that if h_i is another set of orthogonal unit vectors in $\mathcal{H}_2$ corresponding to eigenvectors of an observable H with eigenvalues λ_i^H, then the state $\Phi(x, y)$ can be rewritten as

$$\Phi(x, y) = \sum_i w_i \Big[\sum_k \alpha_{ik} g_k(x) \Big] \otimes h_i(y) \tag{21}$$

with constants w_i and α_{ik} depending on the set $\{h_i\}$ and such that the functions $g_i' = \sum_k \alpha_{ik} g_k(x)$ are normalized (but not orthogonal) and belong to $rng(\rho_1)$. Schrödinger points out that by a suitable choice of h_i *every* unit vector in $rng(\rho_1)$ can be obtained as a g_i'; on the other hand, if one carries out a measurement of H on $\mathcal{S}_2$ and finds eigenvalue λ_i^H then one has to assign state g_i' to system $\mathcal{S}_1$. This means that by choosing an appropriate observable H to measure on system $\mathcal{S}_2$ one can transform the state of system $\mathcal{S}_1$ into *any* state that lies in the range of ρ_1 (this transformation occurs with probability $|w_i|^2$). Therefore, if Ψ is such that none of the a_k is equal to zero, and, consequently, $rng(\rho_1) = \mathcal{H}_1$, then

> ... *in general* a sophisticated experimenter can, by a suitable device which does *not* involve measuring non-commuting variables, produce a non-vanishing probability of driving the system [$\mathcal{S}_1$] into any state he chooses ... [**52**, p. 446]

Schrödinger summarizes the conceptual ramification of the situation in the following words:

> 4. Indubitably the situation described here is, in present quantum mechanics, a necessary and indispensable feature. The question arises, whether it is so in Nature too. I am not satisfied about there being sufficient experimental evidence for that. Years ago I

> pointed out[1] that when two systems separate far enough to make it possible to experiment on one of them without interfering with the other, they are bound to pass, during the process of separation, through stages which were beyond the range of quantum mechanics as it stood then. For it seems hard to imagine a complete separation, whilst the systems are still so close to each other, that, from the classical point of view, their interaction could still be described as an unretarded *actio in distans*. And ordinary quantum mechanics, on account of its thoroughly unrelativistic character, really only deals with the *actio in distans* case. The whole system (comprising in our case both systems) has to be small enough to be able to neglect the time that light takes to travel across the system, compared with such periods of the system as are essentially involved in the changes that take place.
>
> Though in the mean time some progress seemed to have been made in the way of coping with this condition (quantum electrodynamics), there now appears to be a strong probability (as P.A.M. Dirac[2] has recently pointed ont on a special occasion) that this progress is futile. [**52**, p. 451]

In his letter to Schrödinger (April 11, 1936) von Neumann reacts to the above passage in Schrödinger's paper. Von Neumann's position is that presence of correlations between spatially separated quantum systems is not, in and by itself, reason for concern:

> I cannot accept your § 4. completely. I think that the difficulties you hint at are "pseudo-problems". The "action at distance" in the case under consideration says only that even if there is no dynamical interaction between two systems (e.g. because they are far removed from each other), the systems can display statistical correlations. This is not at all specific for quantum mechanics, it happens classically as well.
> (von Neumann to Schrödinger, April 11, 1936)

In the example von Neumann gives in the letter to illustrate his point the probability distributions are (tacitly) interpreted subjectively: as degrees of ignorance and it is this feature of the example by which a potential clash between the change of the distribution for $\mathcal{S}_2$ and the principle of "no action at a distance" is avoided: for if the probabilities are just measures of our ignorance then, no real physical change takes place when the probability distribution changes. (This feature of the example is displayed nicely by the joke von Neumann recalls: the real physical change occurs only in Paris; the colonel's getting cuckolded in Madagascar is just a "semantic change".)

5.4. Relativistic quantum mechanics. Von Neumann also reacts to Schrödinger's above remark about difficulties concerning relativistic quantum electrodynamics:

[1] Schrödinger's footnote: Annalen der Physik (4), 83 (1927), 961. Collected Papers (Blackie and Son, 1928), p. 141.

[2] Schrödinger's footnote: P.A.M. Dirac, Nature. 137 (1936), 298.

> And of course quantum electrodynamics proves that quantum mechanics and the special theory of relativity are compatible "philosophically" – quantum electrodynamic fails only because of the concrete form of Maxwell's equations in the vicinity of a charge. (von Neumann to Schrödinger, April 11, 1936)

Von Neumann saw two problems with a quantum theory of fields: the problem of singularities caused by point-likeness of electric charges and the problem of infinite degrees of freedom caused by the fact that fields can only be characterized by an infinite number or parameters. He addressed this latter issue in an unpublished paper written in 1937 and entitled "Quantum Mechanics of Infinite Systems" [**45**]:

> I wish to discuss some rather incomplete ideas concerning difficulties that arise in some parts of quantum mechanics. In general there have been no serious difficulties when we are dealing with a finite number of particles, but very essential difficulties arise as soon as we treat a system having an infinite number of degrees of freedom; for example, the theory of holes, which, because of the pair generation, requires an indefinite number of particles; also the *Dirac* non-relativistic theory of light and the *Pauli-Heisenberg* relativistic quantum electro-dynamics, these being equivalent to systems consisting of an infinite number of particles.
>
> In dealing with a continuum we find two types of infinity. One arises from the fact that we have an infinite space, but this does not lead to serious difficulties and can be avoided by considering a finite box, or, better, by assuming periodicity in space. The second type of infinity is much more serious. It comes from the fact that in a continuum a field quantity has an infinite number of proper values. The assumption of periodicity does not remove this difficulty at all. Neither does the assumption that space is discrete solve the essential difficulty since if we pass to the limit "lattice → continuum" we get just the continuum result which diverges.
>
> It is the fact that we have an infinite number of degrees of freedom that causes the difficulties, and we shall therefore discuss how we can change the formal part of the theory in some way so that we can treat a system having an infinite number of degrees of freedom in a less divergent way. [**Neumann2001a**, p. 249.]

In this paper von Neumann proposes an operator algebraic framework for relativistic quantum field theory. It is yet to be analyzed what the relation of von Neumann's proposal is to the (local) algebraic relativistic quantum field theory proposed by Haag, Kastler and Araki in the fifties (see [**18**] and the references therein).

To avoid the other type of divergence, which arises from point-likeness of charges, von Neumann tried to quantize space. If space (spacetime) is discrete, then this prohibits one to get arbitrarily close to a point in space; thus divergence of electromagnetic fields becomes impossible. This idea is described in his letter to Dirac (January 27, 1934):

> It should be perhaps desirable, to have operators X, Y, Z which have discrete (point) spectra, in order to avoid the difficulties connected with the point electron (in electrodynamics).

This letter describes in detail the quantization scheme and the spectrum of X, Y and Z. Von Neumann summarizes the results of his spacetime quantization attempt in his letter to Ortvay (March 17, 1938):

> (1) The x, y, z coordinates and the time t are *non-commuting* operators.
> (2) The order of magnitude of commutators is $\frac{h}{mc}$. (That is to say, this is the uncertainty associated with a simultaneous measurement of coordinates.)
> (3) The whole structure has the Lorentz-symmetry.
> (4) Each of the x, y, z coordinates has a discrete spectrum: $\pm 1/2, \pm 3/2, \ldots$
> (5) The spectrum of the time t is continuous, from $-\infty$ to $+\infty$.
> (6) When 4. and 5. are combined with 3. this comes out:
> Given four real numbers $\alpha, \beta, \gamma, \delta$, the spectrum of the operator $\alpha x + \beta y + \gamma z + \delta t$ is as follows:
> a) If $\alpha^2 + \beta^2 + \gamma^2 - \delta^2 > 0$ then it is discrete: $\pm\epsilon/2, \pm 3\epsilon/2, \ldots$, where $\epsilon\sqrt{\alpha^2 + \beta^2 + \gamma^2 - \delta^2}$.
> b) If $\alpha^2 + \beta^2 + \gamma^2 - \delta^2 < 0$ (indeed even when $=0$) then it is continuous from $-\infty$ to $+\infty$.
>
> So this is a discrete "crystalline" space with "continuous" time, which has not only spherical symmetry (even though it is a "crystal"!), but is even invariant with respect to changes of the reference system given by Lorentz transformations, and so shows the proper Lorentz-Fitzgerald contraction phenomena. (This is made possible, of course, by the non-commuting nature of the coordinates.)
> (von Neumann to Ortvay, March 17, 1938)

Dirac replied in a letter to von Neumann (February 28, 1934)[3] to von Neumann's quantization scheme of spacetime. While finding the idea interesting, Dirac considers it very problematic that the scheme described by von Neumann does not seem to be invariant with respect to the displacement of the origin of the X, Y, Z, T space, which, if true, would render the idea physically unacceptable, says Dirac. Dirac proposes in his letter a possible way to get around this difficulty, asking von Neumann whether he considered that option. Apparently there was no further correspondence between Dirac and von Neumann about this issue, which might be due to the fact that von Neumann did not consider this solution of the divergence problem very promising:

> I did not examine this model in deeper detail because I considered it very artificial and arbitrary – and so I still think today.
> (von Neumann to Ortvay, March 17, 1938)

6. Quantum logic

Quantum logic was born in the 1936 seminal paper coauthored by Garret Birkhoff and John von Neumann [**Neumann1936c**]. The publication of this paper had been preceded by an intense and extensive correspondence between Birkhoff and von Neumann. The current collection contains a large selection of the letters von Neumann wrote to Birkhoff during the preparatory phase of the paper. These letters make it possible to reconstruct some steps in the thought process that led to

[3]Unpublished, Von Neumann Papers, Library of Congress.

the notion of quantum logic that got published in [**Neumann1936c**], and this is especially interesting both conceptually and historically because the Birkhoff-von Neumann concept of quantum logic was different from what became the standard notion. The issue of quantum logic appears in a number of other letters as well (to Husimi (September 29, 1937); to Jordan (December 11, 1949); all the letters to Silsbee but especially the one dated July 2, 1945; to Ulam, (October 4, [????]) and to Veblen (July 6, 1935)). In view of the historical and conceptual significance of von Neumann's very technical letters to Birkhoff, more detailed comments are called for in connection with quantum logic. This section is based on [**48**].

It is well known that both the syntactic as well as the semantic aspects of classical propositional logic can be described completely in terms of Boolean algebras: The Tarski-Lindenbaum algebra $\mathcal{A}$ of classical propositions is a Boolean algebra and a deductive system formulated in a classical propositional logic can be identified with a filter in $\mathcal{A}$. The notions of syntactic consistency and completeness correspond to the filter being proper and being a prime (equivalently: maximal) filter, respectively. The logical notion of interpretation turns out to be a Boolean algebra homomorphism from $\mathcal{A}$ into the two element Boolean algebra, and all the semantic notions are defined in terms of these homomorphisms.

The problem of quantum logic can be formulated as the question of whether a Boolean algebra can be replaced in the above role by another, weaker algebraic structure arising from the mathematical formalism of quantum mechanics yielding thereby a "quantum logic".

The Birkhoff-von Neumann paper [**Neumann1936c**] can be viewed as one of the first papers in which the suggestion to logicize a non-Boolean lattice appears. The main idea of this logicization is summarized in von Neumann's letter to Veblen (July 6, 1935); this letter is like an extended abstract of the Birkhoff-von Neumann paper [**Neumann1936c**].

There are two natural candidates for a non-distributive lattice to be logicized: (abstract) orthomodular lattices $(\mathcal{L}, \vee, \wedge, \perp)$ and the concrete orthomodular lattice $\mathcal{P}(\mathcal{H})$ of projections on a Hilbert space $\mathcal{H}$, the Hilbert lattice. Orthomodularity of $\mathcal{L}$ means that the following condition holds:

(22)

orthomodularity. If $A \leq B$ and $A^{\perp} \leq C$, then $A \vee (B \wedge C) = (A \vee B) \wedge (A \vee C)$.

Orthomodularity is a weakening of the modularity law:

(23) *modularity: If* $A \leq B$, then $A \vee (B \wedge C) = (A \vee B) \wedge (A \vee C)$.

which itself is a weakening of the distributivity law:

(24) *distributivity:* $A \vee (B \wedge C) = (A \vee B) \wedge (A \vee C)$ for all A,B,C.

(Boolean algebras are orthocomplemented distributive lattices.)

At the time of birth of quantum logic the notion of an abstract orthomodular lattice did not yet exist; however, the canonical example of non-distributive, orthomodular lattices, the Hilbert lattice $\mathcal{P}(\mathcal{H})$, was known already, and, since this structure emerges naturally from the Hilbert space formalism of quantum mechanics, $\mathcal{P}(\mathcal{H})$ was the most natural candidate in 1935 for the propositional system of quantum logic. Von Neumann did consider $\mathcal{P}(\mathcal{H})$ as a possible propositional system of quantum logic; yet, this lattice was not his choice: The first indication that $\mathcal{P}(\mathcal{H})$ may not be a suitable candidate for a quantum propositional system is in von Neumann's letter to Birkhoff dated January 19, 1935. Von Neumann writes:

> Using the operator-description,
> $a \vee b$, $a \wedge b$ can be formed, if the
> physically significant operators form a *ring*. (← I believe this).
> This, I think should be assumed anyhow,
> even if one does not require that
> *all* operators are phys.[ically] significant. (← but I am rather doubting lately this.)
>
> But we need probably not insist on this point too much.
> (von Neumann to Birkhoff, January 19, [1935 ?])

While in January 1935 von Neumann did not intend to insist on restricting the set of physical quantities to a proper subset of all possible operators, by November 1935 he changed his mind:

> "I am somewhat scared to consider all physical quantities = bounded self-adjoint operators as a lattice"
> (von Neumann to Birkhoff, November 6, [1935 ?])

The reason why he changed his mind was the realization that the Hilbert lattice $\mathcal{P}(\mathcal{H})$ is not modular if the Hilbert space is infinite dimensional (note that "B-lattice" means modular lattice in the next quotation):

> "In any linear space H the linear subspaces $K, L, M, \ldots$ form a B-lattice $\mathcal{L}$ with the
>
> "meet" $K \cap L$: intersection of K and L in the sense of set theory
>
> "join" $K \cup L$: linear sum of K and L, i.e. set of all $f+g$, $f \in K$, $g \in L$
>
> (Proof obvious.) But in a *metric-linear space* $\overline{H}$ the lattice $\overline{\mathcal{L}}$ of all *closed-linear subspaces* $\overline{K}, \overline{L}, \overline{M}, \ldots$, for which the "join" is defined as
>
> "join" $\overline{K} \vee \overline{L}$: closure of the linear sum of K and L, i.e. the set of all condensation points of the $f+g$, $f \in K$, $g \in L$
>
> while the "meet" is as above, is *not necessarily a B-lattice! This is in particular the case in Hilbert space.* ($\overline{K} \cup \overline{L}$ and $\overline{K} \vee \overline{L}$ are identical if $\overline{K}, \overline{L}$ are both closed and *orthogonal* to each other, but not for any two closed $\overline{K}, \overline{L}$!)
>
> In fact, it is possible to find three closed-linear subspaces $\overline{K}$, $\overline{L}$, $\overline{M}$ of Hilbert space $\overline{H}$, for which
>
> (25) $$\overline{K} \subsetneqq \overline{M}, \quad (\overline{K} \vee \overline{L}) \cap \overline{M} \supsetneqq \overline{K} \vee (\overline{L} \cap \overline{M})$$
>
> (von Neumann to Birkhoff, November 6, [1935 ?]), emphasis in original.

This letter contains a detailed proof that subspaces $\overline{K}$, $\overline{L}$, $\overline{M}$ exist that satisfy (25) (and thereby violate modularity (23)). (The Birkhoff-von Neumann paper [**Neumann1936c**] just states this fact without detailed argument.)

One can see in von Neumann's letter to Birkhoff (November 6, [1935 ?]) that von Neumann's proof makes use of the theory of unbounded selfadjoint operators, utilizing the fact that one can find two unbounded selfadjoint operators X and Y

on an infinite dimensional Hilbert space such that the intersection of their domains is empty. Von Neumann emphasizes this feature of his proof:

> Examples could be constructed which make no use of operator theory, but I think that this example shows more clearly "what it's all about": It is the existence of "pathological" operators – like X, Y above – in Hilbert space, which destroys the B-lattice character.
> (von Neumann to Birkhoff, November 6, [1935 ?])

Von Neumann regarded this pathological behavior of the set of all unbounded operators on a Hilbert space a very serious problem because it prohibits adding and composing these operators in general, which entails that these operators do not form an algebra. In von Neumann's eyes this was a great obstacle to do computations with those operators, and since the selfadjoint operators are representatives of quantum physical quantities, it appeared unnatural to him that they behave so irregularly that forming an algebra from them was not possible. He pointed out this pathology several times in his published papers (see e.g. paragraph 6 of Introduction in [**Neumann1936a**]), and the pathological character of the set of all selfadjoint operators was one of the main reasons why he hoped as late as in his famous talk on "Unsolved Problems in Mathematics" in 1954 (see [**Neumann2001b**] and [**43**]) that a restricted set of operators, and therefore a specific von Neumann algebra, the type II_1 factor would be a more suitable mathematical framework for quantum mechanics than Hilbert space theory.

In this situation von Neumann saw two options:

> (I) *Either* we define the "join" by $\cup$ (as a honest linear sum), then the lattice is B, but we must admit all (not-necessarily-closed-) linear subspaces,
> (II) *or* we define the "join" by $\vee$ (closure of the linear sum), then the B-character is lost.
> (von Neumann to Birkhoff, November 6, [1935 ?])

Since $\mathcal{P}(\mathcal{H})$ is not modular and given that von Neumann wished to preserve modularity as a property of the quantum propositional system, one would think that von Neumann suggests to choose option **(I)**. But this is *not* the case. Von Neumann thinks through the consequences of choosing option **(I)** first:

> Let us first consider the alternative **(I)**. The orthogonal complement K' still has the property $K' \cup L' = (K \cap L)'$, but $K' \cap L' = (K \cup L)'$ and $K'' = K$ are lost. We have $K \cap K' = 0$, while $K \cup K'$ is everywhere dense, but not necessarily I. There is a funny relationship between K and its "closure" K''. (For instance: All probabilities in the state K are equal to those in the state K'', but "meets" ($K \cap L$ and $K'' \cap L''$, even for closed L's) may differ.)
>
> The situation is strongly reminiscent of the "excluded middle" troubles, although I did not yet compare all details with those of the class-calculus in "intuitionistic" logics.
>
> After all it is so in normal logics, too, that these troubles arise as soon as you pass to infinite systems, although I must admitt, that the difficulties there are more "optional" then[4] here.

[4]Spelling error, should be "than".

> It has to be said, finally, that even in this case **(I)** complements exist, i.e., that for every K there exists K^*'s for which $K \cup K^* = I$, $K \cap K^* = 0$, but one needs the Hamel-basis-construction to get them.
> (von Neumann to Birkhoff, November 6, [1935 ?])

So, while von Neumann evaluates alternative **(I)** as representing an option which cannot be excluded on the grounds of being either algebraically or logically extremely weird (although it is clear from the above that he did not like the asymmetric failure of De Morgan's law), he prefers option **(II)** in spite of its being seemingly counterintuitive. Here is why:

> Alternative **(II)** seems to exclude Hilbert space, if one sticks to B-lattices.[5] Still one may observe this:
>
> Consider a ring $\mathcal{R}$ of operators in Hilbert space. The idempotents of $\mathcal{R}$ form a lattice $\mathcal{L}_{\mathcal{R}}$. One sees easily, that $\mathcal{L}_{\mathcal{R}}$ is irreducible (= no direct sum), if and only if the center of $\mathcal{R}$ consists of the αI (α=complex number) only, i.e. if $\mathcal{R}$ is a ring of the sort which Murray and I considered. (We called them "factors".) $\mathcal{L}_{\mathcal{R}}$ contains $0, I$ and a complement which dualises $\cup$ and $\cap$. (Now $\cup$ corresponds to what I called $\vee$, case **(II)**.) One may ask: When is $\mathcal{L}_{\mathcal{R}}$ a B-lattice? The answer is (this is not difficult to prove): If and only if the ring $\mathcal{R}$ is finite in the classification Murray and I gave. I.e.: $\mathcal{R}$ must be isomorphic:
>
> (1) either to the full matrix-ring of a finite-dimensional Euclidean space (say n-dimensional, $n = 1, 2, \ldots$),
> (2) to one of those of our rings, in which each idempotent has a "dimensionality", the range of which consists of all real numbers $\geq 0, \leq 1$, and which is uniquely determined by its formal properties.
>
> We called (1) "Case I_n" and (2) "Case II_1".
>
> Thus for operator-lattices the B-lattice axiom
>
> $$a \leq b \quad \rightarrow \quad (a \cup b) \cap c = a \cup (b \cap c)$$
>
> leads directly to the cases $\mathrm{I}_1, \mathrm{I}_2, \ldots$ *and* II_1!
> (von Neumann to Birkhoff, November 6, [1935 ?])

Von Neumann refers here to the Murray-von Neumann classification theory of factors, which was worked out by him in collaboration with F.J Murray precisely at the time (1934-1935) when his correspondence with G. Birkhoff on quantum logic was taking place [**Neumann1936a**] (see the Section 3 in the Introductory Comments). Thus the significance of the existence of type II_1 factors is that their projection lattices are modular. Accordingly, the set of all (not necessarily bounded) selfadjoint operators that they determine are free of the pathologies which von Neumann considered undesirable.

Von Neumann's conclusion:

> This makes me strongly inclined, therefore, to take the ring of all bounded operators of Hilbert space ("Case I_∞" in our notation) less seriously, and Case II_1 more seriously, when thinking of an

[5] Recall that B-lattice means modular lattice. The Editor.

> ultimate basis of quantum mechanics."
> (von Neumann to Birkhoff, November 6, [1935 ?])

As it can be inferred from von Neumann's letter to Birkhoff of (November 13, [1935 ?]), Birkhoff suggested another idea to save the modularity of the lattice formed by some subspaces of a Hilbert space: by restricting the linear subspaces to the finite dimensional ones. Von Neumann did not consider this idea in detail, but thought that it was not an attractive one:

> Many thanks for your letter. Your idea of requiring $a \leq c \rightarrow a \cup (b \cap c) = (a \cup b) \cap c$ in Hilbert space for the finite a, b, c only is very interesting, but will it permit to differentiate between Hilbert-space and other Banach-spaces?
> (von Neumann to Birkhoff, November 13, [1935 ?])

Rather than answering this rhetorical question, von Neumann makes his famous confession reaffirming that the operator algebraic results related to classification theory of von Neumann algebras reduce the privileged status of Hilbert space quantum mechanics:

> I would like to make a confession which may seem immoral: I do not believe absolutely in Hilbert space any more. After all Hilbert-space (as far as quantum-mechanical things are concerned) was obtained by generalizing Euclidean space, footing on the principle of "conserving the validity of all formal rules". This is very clear, if you consider the axiomatic-geometric definition of Hilbert-space, where one simply takes Weyl's axioms for a unitary-Euclidean-space, drops the condition on the existence of a finite linear basis, and replaces it by a minimum of topological assumptions (completeness + separability). Thus Hilbert-space is the straightforward generalization of Euclidean space, if one considers the *vectors* as the essential notions.
>
> Now we[6] begin to believe, that it is not the *vectors* which matter but the *lattice of all linear (closed) subspaces*. Because:
>
> (1) The vectors ought to represent the physical *states*, but they do it redundantly, up to a complex factor, only.
> (2) And besides the *states* are merely a derived notion, the primitive (phenomenologically given) notion being the *qualities*, which correspond to the *linear closed subspaces*.
>
> But if we wish to generalize the lattice of all linear closed subspaces from a Euclidean space to infinitely many dimensions, then one does not obtain Hilbert space, but that configuration, which Murray and I called "case II_1." (The lattice of all linear closed subspaces of Hilbert-space is our "case I_∞".) And this is chiefly due to the presence of the rule
>
> $$a \leq c \rightarrow a \cup (b \cap c) = (a \cup b) \cap c$$
>
> This "formal rule" would be lost, by passing to Hilbert space!
> (von Neumann to Birkhoff, November 13, [1935 ?])

---

[6] With F.J. Murray, von Neumann's coauthor. The Editor.

From what has been quoted so far from von Neumann's letters it would seem then that the modular lattice of the type II_1 factor von Neumann algebra emerges in the von Neumann letters as the strongest candidate for logicization, and so one would expect this lattice to be declared in the Birkhoff-von Neumann paper to be the propositional system of quantum logic. But this is not quite the case; in fact, the published paper does *not* at all refer to the results of the Murray-von Neumann classification theory of von Neumann algebras to support the modularity postulate. Why?

Von Neumann's letters to Birkhoff also contain clues for the the answer to this question, and the answer is that von Neumann's mind moved extremely quickly from the level of abstractness of von Neumann algebras to the level of abstractness represented by continuous geometries – and this move was taking place precisely during the preparation of the quantum logic paper: in his letter to Birkhoff (November 6, [1935 ?]) von Neumann writes:

> Mathematically – and physically, too – it seems to be desirable, to try to make a general theory of dimension in *complemented, irreducible B-lattices*, without requiring "finite chain conditions". I am convinced, that by adding a moderate amount of continuity-conditions, the existence of a numerical dimensionality could be proved, which
>
> (1) is uniquely determined by its formal properties,
> (2) and after a suitable normalisation has either the range $d = 1, 2, \ldots, n$ ($n = 1, 2, \ldots,$ finite!) or $d \geq 0, \leq 1$.
>
> I have already obtained some results in this direction, which connect the notion of dimension in a very funny way with the perspectivities and projectivities in projective geometry.
>
> It will perhaps amuse you if I give some details of this. Here they are:
>
> (von Neumann to Birkhoff, November 6, [1935 ?])

And there follows a long exposition of the theory of continuous geometries in the letter. Von Neumann's letter to Veblen (November 27, [1935 ?]) describes the construction of an example of a projective geometry.

In his letter to Birkhoff written a week later (November 13, [1935 ?]), von Neumann gives an even more detailed description of his results on continuous geometry, which confirm the two conjectures (1) and (2) above completely: on every projective geometry there exists a dimension function d having the properties

$$0 \leq \ d(A) \ \leq 1, \tag{26}$$

$$d(A) + d(B) \ = \ d(A \vee B) + d(A \wedge B), \tag{27}$$

and having discrete or continuous range.[7] These results on continuous geometry do not appear in the Birkhoff-von Neumann paper on quantum logic [**Neumann1936c**], von Neumann published them separately [**Neumann1936d**] (cf. footnote 33 in the Birkhoff-von Neumann paper [**Neumann1936c**]).

Thus by the time it came to the final version of the quantum logic paper, von Neumann knew that the projection lattice of a type II_1 von Neumann algebra is just a special case of more general continuous geometries that admit well-behaving

[7] The discrepancy between Eq. (26) and the ranges mentioned under (2) above are due to different normalizations. The Editor.

probability measures, and this explains that the major postulate in the Birkhoff-von Neumann paper is formulated in the section entitled "Relation to abstract projective geometries" and reads:

> Hence we conclude that the *propositional calculus of quantum mechanics has the same structure as an abstract projective geometry.* [**Neumann1936c**] (Emphasis in the original)

Neither the Birkhoff-von Neumann paper nor the preceding correspondence separates explicitly syntax and semantics in connection with quantum logic; however, both the published paper as well as the correspondence contain statements and arguments which indicate that the "tacit" syntax of the Birkhoff-von Neumann quantum logic is not a standard one: Birkhoff and von Neumann do *not* allow to formulate the sentences $p\&q$ and $p \sqcup q$ if the sentences p and q refer to values of not comeasurable observables. To understand the next quotations note that von Neumann uses the following terminology:

von Neumann:	standard logic:
observational sentence or *statement*	sentence variable
physical quality	proposition

Concerning observational sentences, typical form of which is "The measurement M on the system Γ gives a result ξ with certainty in the set (of numbers) Σ" (von Neumann to Birkhoff, November 27, [1935 ?]), von Neumann writes:

> Last spring you observed: Why not introduce a logical operation $a\#b$, for any two (not necessarily simultaneously decidable) properties a and b, meaning this: "If you first measure a, you find that it is present, if you next measure b, you find that it is present too."
>
> This $a\#b$ cannot be described by any operator, and in particular not by a projection (=linear subspace). The only answer I could then find was this: There is no state in which the property $a\#b$ is *certainly* present, nor is any in which it is *certainly* absent (assuming that a, b are sufficiently non-simultaneously decidable = that their projection operators E, F have no common proper-functions $\neq 0$ at all).
>
> Of course, for this reason $a\#b$ is no physical quantity relatively to the machinery of quantum-mechanics. But how can one motivate this, how can one find a criterium of what is a phys.[ical]-qual.[ity] and what not, if not by the "causality" criterion: A statement describes a physical quality if and only if the states in which it can be decided with *certainty* form a complete set.
>
> I wanted to avoid discussing this rather touchy and complicated question, and withdraw to the safe – although perhaps narrow – position of dealing with "causal statements" only. Do you propose to discuss the question fully? It might become too philosophical, but I would not say that I object absolutely to it. But it is dangerous ground – except if you have a new idea, which settles the question more satisfactorily.
> (von Neumann to Birkhoff, January 15, 1935)

The issue is touchy indeed and goes to the heart of quantum logic. Von Neumann's suggestion as to how to solve this problem is contained in his letter to Birkhoff dated January 19, [1935 ?]:

> As to $a \cup a$ and $a \cap b$ for non-simultaneously observable a, b, I think that we should proceed like this: The primitive notion is $a \leq b$ ("a implies b"). More precisely: Let $a, b, c, \ldots$ be "experimental propositions", as you defined them in your letter of Jan. 17. Define $a \leq b$ ("a implies b") as you described it in your letter. Then
> (1) $a \leq a$
> (2) $a \leq b$, $b \leq c$ imply $a \leq c$
> are obvious. Now *define* $a \sim b$ by
> (3) $a \sim b$ means $a \leq b$, $b \leq a$.
> By (1),(2) this (3) is reflexive, symmetric and transitive, thus an "equivalence". Call the "experimental propositions" identified with the help of this "equivalence" "physical qualities", as you suggested it. Now $a \sim b$ means $a = b$ (for "physical qualities"), and writing $a < b$ for $a \leq b$ but $a \not\sim b$, then (1), (2), (3) gives
> (4) $a \not< a$
> (5) $a < b$, $b < c$ imply $a < c$.
> Now the existence of a c for which
> (6) $c \leq x$ is equivalent to $a \leq x$, $b \leq x$,
> and of a d for which
> (7) $x \leq d$ is equivalent to $x \leq a$, $x \leq b$,
> must be *postulated.* c, d are unique by (4), (5), ($\sim$ means now equality!), and we can define
> (6') $a \cup b = c$ from (6)
> (7') $a \cap b = d$ from (7')
> The existence of $a \cup b = c$ and $a \cap b = d$ are postulates of the same sort as the (Dedekind) modular axiom: They originate from our knowledge about actual quantum theory. I think, that this should be our actual axiomatic treatment: That is, that $a \leq b$ should be the "undefined relation" and not $a \cup b$, $a \cap b$. Do you agree with this?
>
> The comm.[utative] and assoc.[iative] laws for $a \cup b$, $a \cap b$ can now be proved, the existence of an inverse must be postulated, but is physically obvious.
>
> I would emphasize, as you mention, that for simultaneously decidable a, b – but only for these – $a \cup b$, $a \cap b$ can be formed directly. (Physically!)
> (von Neumann to Birkhoff, January 19, [1935 ?])

This is indeed the procedure the published paper follows and it amounts to accepting a "partial logic" view, which was taken up later in the papers of Czelakowski [**9**] and others (see [**8**] and the references there).

Von Neumann was never satisfied with how he had worked out quantum logic: The series of letters he wrote to F.B. Silsbee, the president of the Washington Philosophical Society, prove this in a remarkable way. The correspondence between von Neumann and Silsbee starts with Silsbee's letter (October 31, 1944, unpublished, Library of Congress) inviting von Neumann to deliver the Fourteenth Joseph Henry

Lecture scheduled for March 17, 1945. In his reply to Silsbee (November 3, 1944) von Neumann accepted the invitation with pleasure, promising at the same time to write up the lecture in a paper. In his second letter to Silsbee (February 14, 1945) von Neumann specifies the problem of relation of logic and probability in quantum mechanics as the topic of the lecture. Von Neumann did deliver his talk as planned; however, as his letter to Silsbee (June 11, 1945) shows, he did not meet the first deadline of submitting the manuscript of the planned paper entitled "Logic of quantum mechanics". He promises at the same time to deliver the paper by July 7, the latest. But he did not meet this second deadline either; in fact, the promised paper was never written, and von Neumann's subsequent letters to Silsbee (July 2, 1945, October 22, 1945, April 20, 1946 and December 23, 1946) show von Neumann agonizing over this project. Most revealing is his letter of July 2, 1945. This letter is partly a monologue of a man exhausted and torn-apart by war-related work, a very personal and moving document of the fragile, human side of a genius.[8] Von Neumann indicates however that even if he had not been disrupted by war-work, he might not have been able to write the promised paper on quantum logic:

> I have tried to live up to my promise and to force myself to write this article, and spent much more time on it than on many comparable ones which I wrote with no difficulty at all – and it just didn't work. Perhaps if I were not continually interrupted by journeys and other obligations arising from still surviving war work, I might have been able to do it – although I am not even sure of this.
>
> (von Neumann to Silsbee, July 2, 1945)

Why did it not work? The following attempt to answer this question is based on [**42**], [**43**] and [**44**]:

As we have seen above, the finite dimension function on a projective geometry with the range $[0, 1]$ on the continuous projective geometry, was for von Neumann crucially important: he interpreted it as a probability measure on the modular lattice of the quantum propositional system, thereby creating a complete analogy with classical logic and probability theory, where a Boolean algebra is both a propositional system and a random event structure on which probability measures are defined. While there is no detailed discussion of this aspect of the dimension function in the von Neumann-Birkhoff correspondence, the Birkhoff-von Neumann paper points out that properties (26)-(27) of the dimension function describe the formal properties of probability. Since property (27) is a necessary condition for a measure to be interpreted as probability understood as relative frequency in the sense of von Mises [**33**], [**34**] the modularity property of the quantum propositional system ensured a necessary condition for quantum probability to be interpreted as relative frequency.

Thus it would seem that remaining within the mathematical framework of type II_1 von Neumann algebras one can restore the harmonious classical picture: random events can be identified with the propositions stating that the event happens, and

[8] A couple of circumstances indicate that von Neumann himself may have found this letter too personal to actually mail it to Silsbee: (i) this is the only letter in the series which is handwritten, and there is no typed version or second copy of it in the archive; (ii) there is no reply to it from Silsbee; (iii) had von Neumann sent it to Silsbee, the other letters dated after July 2, 1945, would not make a lot of sense. The Editor.

probabilities can be viewed as relative frequences of the occurrences of the events. But this restored harmony is deceiving for the following reason: von Neumann and Murray showed that a dimension function d on the projection lattice $\mathcal{P}(\mathcal{N})$ of a type II_1 algebra $\mathcal{N}$ can be extended to a trace τ on $\mathcal{N}$. The defining property of a trace τ is

$$\tau(XY) = \tau(YX) \qquad \text{for all } X, Y \in \mathcal{N}. \tag{28}$$

That is to say, the trace is exactly the functional which is insensitive (in the sense of (28)) for the non-commutativity of the algebra. In other words, there are no "properly non-commutative" probability spaces – as long as one insists on the frequency interpretation of probability; hence, if one wants to maintain the idea of non-commutative probability spaces, the frequency view has to go.

It did: von Neumann abandoned the frequency interpretation after 1936. In an unfinished manuscript written about 1937 and entitled "Quantum logic (strict- and probability logics)" he writes:

> This view, the so-called 'frequency theory of probability' has been very brilliantly upheld and expounded by R. von Mises. This view, however, is not acceptable to us, at least not in the present 'logical' context.
> [**Neumann1937b**]

Instead, von Neumann embraces in this unfinished note a "logical theory of probability", which he associates with J. N. Keynes. But he never considered this "logical theory" (interpretation) of probability as fully understood: he mentions the need for an axiomatic investigation of this issue in his address to the International Congress of Mathematicians (Amsterdam, 1954) [**Neumann2001b**] as one of the open problems in mathematics (see [**43**] for a detailed discussion of this point).

7. Ergodic theorem

Von Neumann's first significant mathematical result he obtained in the U.S.A. was his so-called "mean ergodic theorem" [**Neumann1932b**], of which he remained very proud during his life. The mean ergodic theorem was a decisive breakthrough (although by far not the final solution) in the long history of the so-called "ergodic problem", which originates in Boltzmann's work on kinetic theory of gases during the second half of the 19th Century.

The mathematical framework of the ergodic problem is a dynamical system

$$(X, S, \mu, \{\alpha_t\}, t \in \mathbb{R})$$

where (X, S, μ) is a measure space with bounded measure μ and where α_t is a one-parameter family of measure preserving transformations on X. The physical interpretation of X is that it represents the constant energy hypersurface in the phase space of a physical system of many particles and α represents the time evolution of the many body system. The ergodic problem is whether (on what conditions) the "time average" (= left-hand side of (29)) of a measurable function f defined on X exists and is equal to the "phase average" (= right-hand side of (29)) of f, i.e. whether the following equality holds:

$$\lim_{T\to\infty} \frac{1}{2T} \int_{-T}^{+T} f(\alpha_t x)\mathrm{d}t = \frac{1}{\mu(X)} \int_X f \mathrm{d}\mu. \tag{29}$$

For the equality (29) to hold, the time average has to exist in the first place, and, since the time average involves a limit, one has to specify the topology in which the limit should exist. Von Neumann's work on the problem concerned precisely this problem: he proved the existence of the time average with respect to a certain topology. His specification of the topology and the proof of the theorem was strongly motivated by B.O. Koopman's idea of transforming the ergodic problem into one that is formulated in terms of operator theory:

> The possibility of applying Koopman's work to the proof of theorems like the ergodic theorem was suggested to me in a conversation with that author in the spring of 1930. In a conversation with A. Weil in the summer of 1931, a similar application was suggested, and I take this opportunity of thanking both mathematicians for the incentive which they furnished me for undertaking the investigations of this paper.
> [**Neumann1932b**] in [**Neumann1962II**, No. 12, p. 262]

Koopman's observation was that one can "lift" the action α to the Hilbert space $L^2(X, \mu)$ by defining maps V_t on $L^2(X, \mu)$ by setting $(V_t f)(x) = f(\alpha_t x)$. Koopman realized that V_t is a one parameter family of unitary operators on $L^2(X, \mu)$ and the ergodic behavior of the dynamical system can be characterized in terms of the properties of the spectrum of the generator of the unitary group V_t. Von Neumann utilized this operator theoretic re-formulation of dynamical systems and he proved that, given any unitary group V_t on the Hilbert space $L^2(X, \mu)$, the limit

$$\lim_{T\to\infty} \frac{1}{2T} \int_{-T}^{+T} V_t f \mathrm{d}t = g \tag{30}$$

exists in Hilbert space norm for any f in $L^2(X, \mu)$ and that $V_t g = g$.

Soon after von Neumann succeeded in proving his mean ergodic theorem, George David Birkhoff strengthened von Neumann's result by showing that the limit (30) exists pointwise almost everywhere. As von Neumann's letters to Robertson (January 16, 1932) and Stone (January 22, 1932) show he was irritated by how G.D. Birkhoff behaved in connection with his discovery of the pointwise ergodic theorem: Birkhoff rushed with the publication of his result, which thus got published before von Neumann's - in spite of the fact that von Neuman had obtained his mean ergodic theorem before Birkhoff succeeded in proving his. In addition, von Neumanm thought, apparently together with a number of other mathematicians that G.D. Birkhoff had not given sufficient credit to him:

> His quotation of my result is, according to the judgement of Eisenhart, Alexander, Lefschetz, Koopman, Stone, absolutely insufficient. (I mention only these 5 names, because they all spontaneously told me, how dissatisfied they are, without any attempt on my side, to talk about this matter.) The reason they give is, that it does not show to any person, uninformed about the real history of these things, who of Birkhoff and myself got the other started; that which one of us attacked the unsolved q.E.H, and which one found an independent new proof, after he knew that it was solved, and what the necessary and sufficient conditions for its truth are.
> (von Neumann to Robertson, January 16, 1932)

As far as I am aware the letters to Robertson and Stone are the first hard evidence of the friction between von Neumann and G.D. Birkhoff, a friction that has been rumored in the community of historians of science ever since the thirties. In his letter to Robertson (January 16, 1932) von Neumann tells the story of the priority dispute between him and Birkhoff in detail; the letter to Stone (January 22, 1932) is a shorter summary of the story. The priority dispute between von Neumann and George Birkhoff is discussed at length in [**61**].

In a second paper [**Neumann1932c**], published shortly after the one containing the mean ergodic theorem, von Neumann argues that his mean ergodic theorem is sufficient for the purposes of statistical mechanics and he also tries to set the historical record straight by writing:

> Mr. G.D. Birkhoff, to whom we communicated these results orally in October, 1931, has subsequently succeeded in establishing the above surmise [existence of time average] by means of an extremely astute method of his own in the domain of point set theory; he has proved, i.e. the existence and equality of the numerical limits ... except on a set of measure zero ...
> [**Neumann1932c**] in [**Neumann1962II**, No. 13, p. 275]

Note that, to be potentially useful for foundations of classical statistical mechanics, the time average of a function should not only exist but must equal the phase average, i.e. (29) must hold; however, von Neumann did *not* prove that equality (29) holds in the typical cases of dynamical systems (nor did Birkhoff).

If (29) holds for an L^2 function, then f is α invariant and constant (μ almost everywhere). The dynamical system is defined to be *ergodic* if the constant functions are the only α invariant L^2 functions. Ergodicity is equivalent to metric transitivity (also called: metric indecomposability): the dynamical system $(X, S, \mu, \{\alpha_t)\}_t, t \in \mathbb{R})$ is called metrically transitive if any measurable α invariant set has measure 0 or $\mu(X)$, i.e. if $A \in S$ and $\alpha[A] \subseteq A$ implies $\mu(A) = 0$ or $\mu(A) = \mu(X)$. Whether metric transitivity holds for the dynamical systems occurring in statistical mechanics, is an extremely difficult problem, which remains largely open even today (see the survey [**55**] for a review of some results in the field). One should also note that the question of whether ergodicity is relevant at all for foundations of classical mechanics is itself an issue about which there is no consensus in the physics community [**10**]. Von Neumann thought that ergodicity was important: summarizing the physical significance of his ergodic theorem in his letter to Ortvay (February 2, 1939) he writes:

> Today's formulation of the ergodic theorem is that, apart from a set of measure 0, the "Zeitmittelwert" [time average] does exist as a limit and = the "Raummittelwert" [phase average]. All "disorder" assumptions are implied by the single hypothesis that we are not in that measure-0 set. No separate "H-theorem" is needed.
> (von Neumann to Ortvay, February 2, 1939)

Ergodicity (metric transitivity) is sometimes referred to as "quasi-ergodicity". Von Neumann himself uses the term "quasi-ergodic" in the title of his paper ("Proof of the quasi-ergodic hypothesis", [**Neumann1932b**]), and he says explicitly in the first sentence of the paper that he considers "quasi-ergodicity" and "ergodicity" as synonyms; furthermore, it is clear from his paper that by "ergodicity" he means metric indecomposability. Yet, the title of von Neumann's paper and his terminology in his letters to Robertson (January 16, 1932) and Stone (January 22, 1932) are potentially misleading for two reasons: first, because, as stressed above, von Neumann did not prove the quasi-ergodicity (= metric transitivity) of dynamical systems; second, because "quasi-ergodicity" is sometimes used to refer to the following *topological transitivity property* of dynamical systems: the trajectory of μ almost every point in X is topologically dense in X. It should be stressed that quasi-ergodicity in this latter sense is different from ergodicity understood as metric transitivity: ergodicity concerns just the measure theoretic aspects of the dynamic (X need not even have any topology for ergodicity to make sense). Furthermore, even if X does have a topology, quasi-ergodicity in the topological sense *does not* imply ergodicity (metric transitivity) in general, not even if X is a compact metric space (see [**19**, pp. 26-27] and [**32**, pp. 104-105]). The converse is true, however: If X is a (compact) topological space and μ is a Borel measure on X such that $\mu(A) > 0$ for every open set A, then ergodicity implies that the orbit $\{\alpha_t x\}$ of μ almost every x is everywhere dense in X [**58**, p. 132].

While the problem of whether time averages and space averages of phase functions are equal can be traced back at least to Boltzmann's work, it is a controversial issue what precisely Boltzmann's original notion of ergodicity was. According to the uncritical textbook story, Boltzmann's original idea was what became known as "ergodic hypothesis": that the dynamics of the physical systems is such that the trajectory of the phase point under the dynamical evolution α fills up the whole phase space $X \subseteq \mathbb{R}^n$, at least in the case of the kinetic theory of gases, and when it became clear that for topological reasons the trajectory of a physical dynamical system cannot fill up the whole phase space, the "ergodic hypothesis" was weakened into the "quasi-ergodic" hypothesis, understood in the sense of topological transitivity of almost every points. The textbook story is questioned in [**6**] and [**40**]. For a summary of the developments of ergodicity in the period when von Neumann was making his contribution see Mackey's review [**31**].

8. Computer science

Von Neumann is best known in the general public for his work in computer design. His activity in this field is very well documented in the work of W. Aspray [**3**]. Aspray's book contains quotations from many of von Neumann's letters concerning computer science; thus the current volume contains just a small selection of von Neumann's letters related to computers; the letters illustrate in what way he was far ahead of most of his contemporaries in his envisioning what sort of entity an electronic computer is.

Von Neumann's interest in computers and computing technics arose out of war-related work: In 1943 von Neumann was in England on a war assignment and reported from there in a letter to Veblen (May 21, 1943, not included in the volume) that he had recently developed "an obscene interest in computation" (quoted in [**3**, p. xv]. His interest was motivated by the need to solve numerically partial differential equations that were crucial in understanding explosions. Of special importance were the computations needed for the development of the atomic bomb in Los Alamos. During the search for computing machines that could be used in addition to the computing laboratory at Los Alamos to do numerical computations for the Manhattan Project, von Neumann became aware of the ENIAC (Electronic

Numerical Integrator and Computer) computer project at the Moore School at the University of Pennsylvania. ENIAC was the first truly electronic — as opposed to electromechanical — computer. ENIAC was capable of multiplying two ten-digit number in 1/300 seconds ([**3**, p. 35]).

Von Neumann became a consultant to the ENIAC project in 1944. The ENIAC project was at an advanced level at the time when von Neumann became consultant to the ENIAC team of P. Eckert, the chief engineer, and J. Mauchly, the mathematical adviser: the architecture of ENIAC had already been determined together with the main characteristics of the machine. Von Neumann came right in time however to play a decisive role in determining the logical design of *all* computers: The Eckert-Mauchly group started discussing the principles of a computer that was to supersede ENIAC: the EDVAC machine (Electronic Discrete Variable Arithmetic Computer). Von Neumann's major contribution to the EDVAC project was writing the famous "First Draft of a Report on the EDVAC" [**63**]. While this work became the source of personal friction between von Neumann and S. Goldstine (the Army's liaison officer to the ENIAC project) on one hand and Eckert and Mauchly on the other (see [**3**] for the details of the story), the report is considered the most important document on computers ever written: it details the general logical design of stored-program computers, and it does this in an abstract language, independent of any specific hardware that might implement the units. Von Neumann's abstract discussion of computer architecture shows what is commonplace now: von Neumann was not simply an outstanding and creative computer scientist but a visionary thinker in connection with computers.

Convinced of the great importance of computers for science, von Neumann made plans in 1945 to establish a computer development project at the IAS. In the letter to Overbeck (November 28, 1945) von Neumann gives a general, concise description of the computer project (including the main characteristics of the planned device) he directed in the Institute for Advanced Study in the years 1945-1953.

Of special interest is von Neumann's letter to Strauss (October 20, 1945), in which von Neumann lays out his arguments as to why, and what kind of a computer project should be carried out at the Institute for Advanced Study. Von Neumann thought that one should build a machine

> ... not for use on specific applied mathematical or physical or engineering problems, but with the purpose of experimentation with the machine itself in order to develop new approximation and computing methods, and generally to acquire the mathematical and logical forms of thinking which are necessary for the really efficient operation of such a device, with the methods it will have brought into existence.
> (von Neumann to Strauss, October 20, 1945)

In his letter to V. Bush (November 15, 1949) von Neumann points out some of the pitfalls of computer design, and analyzes the overall situation of computer development in the U.S.A. at the time. One of his overall critical remarks is that

> It seems to me that very few people have a clear idea of how a computing machine functions as a whole, and how, in addition, the entire computing machine is merely one component of a greater whole, namely, of the unity formed by the computing machine, the mathematical problems that go with it, and the type of planning

> which is called for by both.
> (von Neumann to Bush, November 15, 1949)

Von Neumann points out three consequences of failing to comprehend the computer "as a whole":

(1) All sorts of "deformed" machines are being developed.
(2) Lack of understanding of what constitutes an "all purpose" machine.
(3) Fear that the computers will do themselves out of business rapidly.

As to (1), von Neumann mentions that many computers that were developed simultaneously with the IAS machine were deformed in the sense of not having the proper balance between different functions and characteristics:

> It would be easy to name examples of machines which have arithmetical speeds out of proportion with their precisions, or with their memory access times, logical capacities out of proportion with their memory capacities, in which certain special operations are over-instrumented to the detriment of other equally important ones, etc. etc.
> (von Neumann to Bush, November 15, 1949)

In connection with (2) von Neumann writes:

> ... claims, that a "one-purpose" machine, or an "all purpose" machine with certain very special and very extreme characteristics is needed for a particular group of applications, are made much too easily – on the basis of altogether inadequate analyses of the essentials of the situation.
> (von Neumann to Bush, November 15, 1949)

As his letter to McKinsey (February 18, 1949) shows, von Neumann saw clearly that all computers under development were universal, "all purpose" machines in the sense of Turing and that therefore

> The crucial question is that of the time required for the solution. In order to judge the suitability of any given problem for either kind of machine, and in order to decide whether either kind has serious advantages over the other in connection with a given problem, time-estimates must be made.
> (von Neumann to McKinsey, February 18, 1949)

In his letter to McKinsey (February 18, 1949) von Neumann discusses the time needed to handle numerically combinatorial-logical problems as opposed to "problems of a mathematical-analytical" character, concluding that

> ... ordinary mathematical-analytical problems are never of a high logical type: They are usually of the next type after that one of the arithmetical fundamental variable. Logical-combinatorial problems ... on the other hand, are almost always of a higher type.
> (von Neumann to McKinsey, February 18, 1949)

Von Neumann finds totally unfounded the fear that computers, because they are too fast, will put themselves out of business. Von Neumann's letter to Hurd (October 13, 1950) contains the abstract of a talk in which he explicitly addresses this issue. Von Neumann thinks that the size of numerical problems gets adjusted to the speed of the available machines:

> ... the size adjusted itself essentially automatically so that the problem-solution time became longer, but not prohibitively longer, than the planning and coding time. For faster machines, the same automatic mechanism will exert pressure towards problems of larger size, and the equilibrium between planning and coding time on one hand, and problem-solution time on the other, will again restore itself on a reasonable level once it will have been really understood how to use these faster machines.
> (von Neumann to Hurd, October 13, 1950)

Also, he points out in his letter to Hurd (October 13, 1950) that some problems already require extremely fast machines. The problems he mentions are in harmony with his letters to Strauss (October 24, 1945) and to R.S. Burington (January 19, 1951) in which fields and problem areas are discussed where von Neumann thought no advance was possible without numerical computation with high speed electronic computing machines. (He mentions aerodynamics, hydrodynamics, elasticity and plasticity, optics, electrodynamics, dynamical meteorology, sorting problems, high explosives, missile design, atomic weapons.)

An important method of numerical computing is the Monte Carlo method, which requires generating a large number of random numbers. In several letters von Neumann describes the method of producing random numbers (Householder (February 3, 1948), Hurd (December 3, 1948)). For a brief history of the Monte Carlo method, see [**3**].

9. Game theory

"Of the many areas of mathematics shaped by his genius, none shows more clearly the influence of John von Neumann than the Theory of Games." [**27**, p. 100]. Von Neumann published two decisive works on game theory: one is his relatively short paper from 1928 [**Neumann1928a**] containing the famous Minimax Theorem, the other is the monumental treatise *Theory of Games and Economic Behavior* [**Neumann1944b**] written in cooperation with economist O. Morgenstern. Von Neumann regarded the Minimax Theorem the heart of the whole theory: "As far as I can see, there could be no theory of games on these bases without that theorem." [**Neumann1953**].

Non-technically formulated, the Minimax Theorem of 1928 states that in any zero-sum, two person game there exists a uniquely determined gain for each player that can be achieved by their choosing an appropriate *mixed* strategy. To explain and state the Minimax theorem semi-formally, let us introduce the following notation: Let S and T denote the *pure* strategies of two players A and B in a zero-sum two person game, and let $L(a,b)$ denote the loss of player B if B chooses strategy $b \in T$ provided A chooses strategy $a \in S$. (A pure strategy is a prescription that specifies a player's first move and the move the player is supposed to make as a response of any given move of the opponent.) Players A and B can always achieve the gains G_A and G_B given by

$$G_A = Max_{a \in S}\Big(Min_{b \in T} L(a,b)\Big), \tag{31}$$

$$G_B = Min_{b \in T}\Big(Max_{a \in S} L(a,b)\Big). \tag{32}$$

A fair game would be such that there exist strategies a^*, b^* such that

$$G_A = G_B = L(a^*, b^*)$$

(optimal strategies). In general, no such optimal strategies a^* and b^* exist. The Minimax Theorem states that there exist *mixed* strategies, however, for which the *average* gains of A and B are equal: If $M(S)$ and $M(T)$ denote the set of probability measures over the set of respective pure strategies and $K(\alpha, \beta)$ is the expected value of $L(a, b)$ with respect to the probability measures (mixed strategies) $\alpha \in M(S)$ and $\beta \in M(T)$, then

$$(33) \qquad Max_{\alpha \in M(S)}\ Min_{\beta \in M(T)} K(\alpha, \beta) = Min_{\beta \in M(T)}\ Max_{\alpha \in M(S)} K(\alpha, \beta).$$

The Minimax Theorem was proved by von Neumann for an arbitrary but *finite* set of strategies in 1928. In his letter to Kaplansky (February 1, 1945) von Neumann describes briefly how he would treat the case of infinite strategies.

Von Neumann got into a priority dispute with M. Fréchet about who the initiator of game theory really was: Fréchet published a short note [**12**] in the journal *Econometrica* in 1953 together with a commentary [**13**] that accompanied the translation from French into English of three notes of E. Borel on game theory. Fréchet claims that

> ... in reading these notes of Borel's I discovered that in this domain, *as in so many others*, Borel had been the *initiator*.
> [**12**, p. 95]. (Emphasis in original.)

As his strongly worded letter to Morgenstern (January 2, 1953) shows, von Neumann was irritated by this evaluation of Borel by Fréchet and he published a "Communication on the Borel notes" in the same issue of *Econometrica* [**Neumann1953**]. In this communication von Neumann writes:

> (2) ... By surmising, as he [Borel] did, the incorrectness of that [Minimax] theorem, Borel actually surmised the impossibility of the theory as we know it.
> (3) In view of this, Borel did hardly "initiate" the theory. I developed my ideas on the subject before I read his papers, whose negative conclusions on the decisive point (the "minimax theorem"), which alone makes the concepts in question unambiguously useful) would have been primarily discouraging."
> [**Neumann1953**, p. 124]

Von Neumann's letter to Deming (January 23, 1945) contains some brief evaluative comments by von Neumann on their work with Morgenstern together with some plans they had for future research.

At the insistence of Morgenstern (see [**27**, p. 108]) the second edition (published in 1947) of the book on game theory [**Neumann1944b**] contained a detailed proof of existence of the so-called "utility functions" u: Given a set X of options with a binary relation $x \preceq y$ interpreted as a person's preference of option y over option x, it is proved in [**Neumann1944b**] that if $\preceq$ is a complete ordering, then (under suitable and natural additional conditions on $\preceq$) there exists (essentially uniquely) a real valued utility function u on X such that if $x \preceq y$, then $u(x) \leq u(y)$. In his letter to Wold (October 28, 1946) von Neumann discusses whether one needs to introduce non-numerical utility functions, especially if one drops the requirement that the preference relation $\preceq$ is a complete ordering. Von Neumann is skeptical of the idea:

> I think that it will be difficult to get away from some rigidly determined (i.e., "up to a linear transformation" uniquely determined) connection between utilities and numbers in any *complete* system of concepts: Probability is hard to eliminate from the world with which we are dealing, and probability seems to be an essentially numerical concept.
> (von Neumann to Wold, October 28, 1946)

In his letter to Haberler (October 31, 1947) von Neumann declines reviewing "Dr. Samuelson's book on the mathematical foundations of economics". Although no exact reference is given in this letter to Samuelson's work, the book in question is probably [**50**]. Von Neumann explains his unwillingness to do the review by lack of time needed to study the work carefully; however, it is likely that he did not like Samuelson's approach to mathematical economy: As Leonard points out:

> It thus becomes evident that the theory of games was intended to constitute a radical rapture with the Hicks-Samuelson variant of neoclassical economics. Von Neumann completely rejected the latter's underlying physical metaphor of classical mechanics and the associated mathematics based on the differential calculus. [**28**, p. 755]

In his two letters to L.B. Tuckerman (August 25, 1950) and (September 25, 1950) von Neumann estimates the number of all valid games in chess (10^{80}), and, using chess, gives an example of a two-player game, for which one can prove that one of the players can certainly win without knowing how to do this.

Letter to N. Aronszajn

July 15, 1948

Professor N. Aronszajn
Harvard University
Cambridge, Massachusetts

Dear Professor Aronszajn,

Many thanks for your letter of July 8th.

I regret that I must have misunderstood your remarks about Dixmier. I did write a note to him, thanking him for the papers he sent me, but I had not originally understood that he was sending his manuscript in order that I could submit it for publication in America. His letter of April 24th does not make this clear. The paper, by the way, seems excellent to me, and I shall be very glad to help to get it published here. I am writing with this same mail to Dixmier to this effect. I am sorry I caused all this delay, and I am very much obliged to you that you kept the matter in mind and thereby prevented me from not taking care of this matter, which I would have exceedingly regretted.

I think I now remember better what I did or did not prove in the 1930's regarding the theorem that every operator must have a non-trivial invariant subspace.[9] The state in which I had the problem when I stopped working on it was this:

(a) I did prove that every totally continuous operator in Hilbert space has a non-trivial invariant subspace;

(b) In view of this, I surmised, but was unable to prove, that the same is true for every operator.

The things that you write me about representing bounded operators as parts of other bounded operators defined in suitable over-space are very interesting. I have one difficulty in connection with your definition: The operator $L = PH$ seems to have values in the under-space, but it is defined in the over-space. Should it not be an operator which is also defined in the under-space? How do you adjust this

[9]In their famous paper, N. Aronszajn and K.T. Smith: "Invariant subspaces of completely continuous operators", Annals of Mathematics, **60** (1954) 345-350, the authors write: "Some years ago, J. von Neumann informed the first author of this paper that in the early thirties he proved the existence of proper invariant subspaces for completely continuous operators in a Hilbert space; the proof was never published. In 1950 the first author found a proof of the theorem in this case which used orthogonal projections and hence could not be extended directly to Banach spaces. It was verbally verified with J. von Neumann that it was essentially the same proof as he had previously found" (p. 345). The present letter is the one in which von Neumann informs Aronszajn about his (unpublished) result. The Editor.

situation? If you could give me more precise details of this matter, I would probably be able to orient myself with respect to it and the literature that I know.

Much to my regret, I am leaving for the West in a few days and I expect to return to Princeton only toward the end of September. Where do you expect to be about that time? At any rate, I hope that we shall remain in touch with each other during my absence. The things that you tell me interest me very much. My mail will reach me with only a short delay via the Institute for Advanced Study. I am,

Sincerely yours,

John von Neumann

Original in Institute for Advanced Study Archives, Historical Studies-Social Science Library, language of original: English

Letters to F. Aydelotte

March 23, 1940

Dr. Frank Aydelotte
Institute for Advanced Study
Princeton, New Jersey

Dear Dr. Aydelotte:

The Dutch physicist, P. Debye, who has been Director of the Physics Institute of the Kaiser Wilhelm Gesellschaft in Berlin (supported by the Rockefeller Foundation), has been sent abroad by the German authorities in order to free his Institute for secret war work. When one of us met him at dinner the other evening, he made no secret of the fact that this work is essentially a study of the fission of uranium. This is an explosive nuclear process which is theoretically capable of generating 10,000 to 2,000,000 times more energy than the same weight of any known fuel or explosive. There are considerable deposits of uranium available near Joachimsthal, Bohemia, as well as in Canada. It is clear that the Nazi authorities hope to produce either a terrible explosive or a very compact and efficient source of power. We gather from Debye's remarks that they have brought together in this Institute the best German nuclear and theoretical physicists, including Heisenberg, for this research – this in spite of the fact that nuclear and theoretical physics in general and Heisenberg in particular were under a cloud, nuclear physics being considered to be "Jewish physics" and Heisenberg a "White Jew".

There is a difference of opinion among theoretical physicists about the probability of reaching practical results at an early date. This, however, is a well-known stage in the pre-history of every great invention. The tremendous importance of the utilization of atomic energy, even if only partly successful, suggests that the matter should not be left in the hands of the European gangsters, especially at the present juncture of world history.

Work of the sort which the German physicists are supposed to be doing has been going on for some time at Columbia University under Professor Fermi and Dr. Szilard, but at a slow rate because the expense of the experiments exceeds a normal departmental budget. Some effort, not entirely successful, has been made to enlist the help of the United States Government, but this process is slow and cumbersome and has met serious obstacles. It seems to us, therefore, that the problem is one which might well be brought to the attention of the Rockefeller Foundation, which would be in a position to act in a simple and direct manner. We are not going to suggest a very definite way of attacking the problem, but suppose that if the officers of the Foundation were interested they would consult with physicists who are familiar with the practical questions involved.

We have quoted Professor Debye rather freely in writing to you, but obviously we should have to be very cautious in using his name any further. In any case his only role was unintentionally to stimulate us to bring up a question which we have had on our minds for several months, without knowing what, if anything, to do about it.

Very truly yours,

John von Neumann, Oswald Veblen

Original in Frank Aydelotte's administrative correspondence, Institute for Advanced Study Archives, Historical Studies-Social Science Library, language of original: English

---

July 23, 1941

Dr. Frank Aydelotte
Stockbridge
Massachusetts

Dear Doctor Aydelotte:

On your last visit in Princeton you asked me to repeat in writing some of the things we discussed concerning the mathematical niveau and structure of various universities.

I think that it is beyond doubt that the Princeton group – by this I mean the Institute, plus Princeton University, plus temporary (alas) members of the Institute as Gödel and Siegel – is the strongest and most widely interested group of mathematicians existing anywhere at this moment. The other principal center of mathematical science in this country is Cambridge, Massachusetts, – again taking Harvard and Massachusetts Institute of Technology together. That group contains four or five unquestionably first rate men who cover a wide variety of subjects. Next important centers which should be mentioned are, I think, Chicago, Johns Hopkins, Michigan, Columbia. But the Institute and Harvard are probably the leading ones. If one should judge by sustained first-rate quality over a longer period (the last thirty years), then the first place probably should be shared by Princeton University and Harvard.

In the international field the greatest mathematical centers in our generation were Cambridge (England), Göttingen, and Moscow. There was also a center of very high mathematical quality in existence in Poland, which of course has been completely broken up and destroyed by the war. The last ten years have also shown a very promising development in Japan, but it might be too early to judge that as yet. French mathematics has lost much of the important role it had in previous generations, but the last ten years saw a considerable renascence,[10] although more under outside influence than in continuation of the French tradition. You

[10]Renaissance. The Editor.

understand, of course, how subjective all these views are; also how arbitrary and unreliable all comparisons of quality and (*horribile dictu !*) quantity in science are. But if you had to compare the present establishment in Princeton with the best international centers – that is, Cambridge (England), Göttingen and Paris, in their respective primes – I would say this: I think that Princeton is comparable to any of them in pure mathematics, but that they were probably better integrated in the direction of applications. By applications I mean both mathematical physics and applied mathematics proper.

I hope you won't mind if I conclude with one more disclaimer of reliability of such comparisons. Perhaps you will nevertheless find them not completely useless.

Sincerely yours,

John von Neumann

Original in Von Neumann Papers, Library of Congress, language of original: English

Letter to E.F. Beckenbach

March 24, 1949

Professor E.F. Beckenbach
Chairman, Committee on Arrangements
National Bureau of Standards
Institute for Numerical Analysis
University of California
Los Angeles, California

Dear Professor Beckenbach,

I am in receipt of your letter of March 18, in which you asked me for suggestions concerning a Symposium on the Construction and Applications of Conformal Maps, to be held at the Institute for Numerical Analysis in Los Angeles, probably on June 24-25, 1949.

Not having worked in this field, the remarks I can make on the subject are limited. I think that the subject is interesting, and I believe that it would be very useful to have a synoptic view of the constructive methods which exist in this field, of their effectivity and of their suitability for mathematical computing. I assume that the point of departure of such investigations has to be an investigation of all known proofs of existence for conformal mapping. There are several essentially different ones, and all of them are, or after a moderate amount of manipulation become, "constructive" in the Brouwerian sense. They are, therefore, in principle, suited for computational exploitation. It is far from clear, however, how much work each one of them implies absolutely, as well as relatively, (that is, compared to the others). It is also far from clear what complexity-characteristics of the area to be mapped contribute much or contribute little to the work involved in finding an (approximate) conformal mapping on the circle. I think, therefore, that one way in which the conference could be very useful is this: It could formulate as explicitly as possible the principles which govern this subject, find out which groups are actively interested in this work, and lay out research programs. A particular line which probably should get emphasis is the conformal mapping of polygons.

I realize that these remarks, all of which refer to one item of your multigraphed agenda, are not as specific as one might want. If I knew more about your detailed plans and aims, I could try to do better in this respect – although as I mentioned above, I am very much aware of the fact that I am not a specialist in this field.

Sincerely yours,

John von Neumann

Original in Von Neumann Papers, Library of Congress, language of original: English

Letter to H. Bethe

May 16, 1941

Professor Hans A. Bethe
Physics Department
Columbia University
New York City

Dear Bethe:

This last Tuesday I saw Wilson and Kistiakowski and mentioned to them our conversation. Both were very glad to learn about it. In the course of our discussions it appeared that your report with Teller is, among other things, a very important source of information for the behavior of gases at high temperatures. Since the temperatures occurring in shock waves go up to $25,000^\circ$, it was felt that an extension of your tabulations would be very desirable. I was charged to ask you whether you, possibly again in collaboration with Teller, would consider undertaking the extension of your work in this sense.

More precisely: To obtain the heat capacity of air up to about $25,000^\circ$ including the effect of dissociation. It would be desirable if such a calculation gave C_V both with and without dissociation. In the former case C_V would also be needed as function of pressure up to say 10,000 atmospheres. Perfect gas may be assumed, except for effect of dissociation. Density's range need not be greater than ten times normal. (One atmosphere, 300° absolute = normal.)

It would be a very great service indeed if you could undertake this work. If you should need any technical help I suppose it could be obtained from the National Defense Research Council.

I am extremely glad that I had the opportunity of discussing matters with you personally, and I hope that we shall soon see each other again, especially since we are now so near in space.

With best greetings, I am

Cordially yours,

John von Neumann

Original in Von Neumann Papers, Library of Congress, language of original: English

Letters to G. Birkhoff

The Institute for Advanced Study
School of Mathematics
Fine Hall
Princeton, New Jersey

Jan. 15, 1935

Dear Garrett,

Many thanks for your letter, its suggestions and the outline of the first part[11]. I would like to see the first (physical) part itself as soon as possible, because I think that an *à fond* discussion of the matter is very difficult without it.

My first, of course superficial reaction is, that you are right in asserting that "physical qualities" are more fundamental than "statements" or predictions, but ... But it was just an objection from you, which induced me to stress the latter ones. Last spring you observed: Why not introduce a logical operation $a\#b$, for any two (not necessarily simultaneously decidable) properties a and b, meaning this: "If you first measure a, you find that it is present, if you next measure b, you find that it is present too."

This $a\#b$ cannot be described by any operator, and in particular not by a projection (=linear subspace). The only answer I could then find was this: There is no state in which the property $a\#b$ is *certainly* present, nor is any in which it is *certainly* absent (assuming that a, b are sufficiently non–simultaneously decidable = that their projection operators E, F have no common proper-functions $\neq 0$ at all).

Of course, for this reason $a\#b$ is no physical quantity relatively to the machinery of quantum-mechanics. But how can one motivate this, how can one find a criterium of what is a phys.[ical]-qual.[ity] and what not, if not by the "causality" criterion: A statement describes a physical quality if and only if the states in which it can be decided with *certainty* form a complete set.

I wanted to avoid discussing this rather touchy and complicated question, and withdraw to the safe – although perhaps narrow – position of dealing with "causal statements" only. Do you propose to discuss the question fully? It might become too philosophical, but I would not say that I object absolutely to it. But it is dangerous ground – except if you have a new idea, which settles the question more satisfactorily.

Or did I misunderstand you?

[11] Von Neumann refers here to the draft of the manuscript of the paper G. Birkhoff, J. von Neumann: "The Logic of Quantum Mechanics", Collected Works, Vol. IV. No. 7; see Section 6 of the Introductory Comments. The Editor.

Please let me have more details of this matter!

I think that the "outline" of the first part, which you enclosed, is very good, and I am *very* eagerly looking forward to see your manuscript, resp. this part of it.

Mariette's[12] skiing was a "roaring success", she is more enthusiastic about skiing, than she ever was before. I cannot say so much about my trip to Nassau – Nassau was beautyful, but the ship was packed to 200% capacity. So it seems, that your policy – of skiing – was the right one.

Hoping to hear from you (even before we meet again in Cambridge).

Yours as ever

John

Original in Papers of Garrett Birkhoff, Harvard University Archives, Accession 13493, language of original: English

---

The Institute for Advanced Study
School of Mathematics
Fine Hall
Princeton, New Jersey

Jan. 19, [1935 ?]

Dear Garrett,

we were in Philadelphia on Jan. 16/17, so I received your two letters of Jan. 16. and 17. simultaneously. I see now, that I did not understand fully what you suggested in your letter of Jan. 10., and there is indeed no essential disagreement between us. Do I understand you correctly, that your idea of "hypothetical" vs. "physically significant" "physical qualities" is this: If a "state" is represented by a direction in n-dimensional orthogonal space (as a simplified substitute for Hilbert space), and so so by a point of $n-1$-dimensional projective space P_{n-1}, then a "physical quality" is any subset S of P_{n-1}. It is "physically significant" if S is a linear subspace of P_{n-1}, otherwise it is "hypothetical", that is not adequately describable by actual experiments. With this attitude I completely agree.[13]

Your definition of experiments, by using point-sets, is perfectly satisfactory to me. I thought however, that it might be good to say explicitly, that a computation may be part of an "observation".

As to $a \cup b$ and $a \cap b$ for non-simultaneously observable a, b, I think that we should proceed like this: The primitive notion is $a \leq b$ ("a implies b"). More precisely: Let $a, b, c, \ldots$ be "experimental propositions", as you defined them in your letter of Jan. 17. Define $a \leq b$ ("a implies b") as you described it in your letter. Then

(1) $a \leq a$

(2) $a \leq b$, $b \leq c$ imply $a \leq c$

[12]Mariette Kövesi, von Neumann's wife. The Editor.

[13]Von Neumann and Birkhoff are discussing quantum logic in their exchange of letters; see Section 6 of the Introductory Comments. The Editor.

are obvious. Now *define* $a \sim b$ by

(3) $a \sim b$ means $a \leq b$, $b \leq a$.

By (1),(2) this (3) is reflexive, symmetric and transitive, thus an "equivalence". Call the "experimental propositions" identified with the help of this "equivalence" "physical qualities", as you suggested it. Now $a \sim b$ means $a = b$ (for "physical qualities"), and writing $a < b$ for $a \leq b$ but $a \not\sim b$, then (1), (2), (3) gives

(4) $a \not< a$
(5) $a < b$, $b < c$ imply $a < c$.

Now the existence of a c for which

(6) $c \leq x$ is equivalent to $a \leq x$, $b \leq x$,

and of a d for which

(7) $x \leq d$ is equivalent to $x \leq a$, $x \leq b$,

must be *postulated*. c, d are unique by (4), (5), ($\sim$ means now equality!), and we can define

(6') $a \cup b = c$ from (6)
(7') $a \cap b = d$ from (7')

The existence of $a \cup b = c$ and $a \cap b = d$ are postulates of the same sort as the (Dedekind) modular axiom: They originate from our knowledge about actual quantum theory. I think, that this should be our actual axiomatic treatment: That is, that $a \leq b$ should be the "undefined relation" and not $a \cup b$, $a \cap b$. Do you agree with this?

The comm.[utative] and assoc.[iative] laws for $a \cup b$, $a \cap b$ can now be proved, the existence of an inverse must be postulated, but is physically obvious.

I would emphasize, as you mention, that for simultaneously decidable a, b – but only for these – $a \cup b$, $a \cap b$ can be formed directly. (Physically!)

Using the operator-description,
$a \cup b$, $a \cap b$ can be formed, if the
physically significant operators form a *ring*. (← I believe this).
This, I think should be assumed anyhow,
even if one does not require that
all operators are phys.[ically] significant. (← but I am rather doubting lately this.)

But we need probably not insist on this point too much.

The division algebra $\mathcal{Z}$ which we obtain might have a finite characteristic. All we know, is, that it possesses an involutorial antiautomorphism $\mathcal{I}$ and elements $\alpha_1, \ldots, \alpha_n$ with $\mathcal{I}\alpha_i = \alpha_i$ and

$$\sum_{i=1}^{n} x_i \alpha_i \mathcal{I} x_i = 0 \qquad \rightarrow \qquad x_1 = \ldots = x_n = 0$$

Let $\mathcal{Z}$ be the algebra of all rational functions of the free variables $Z_1, Z_2, \ldots$, with integer coefficients (mod 2), and put

$$\mathcal{I} = \textit{identity}, \qquad \alpha_i = Z_i$$

This fulfills our requirements and has characteristic 2. Of course it is "unreasonable" in many ways – but we have not defined yet what we mean by "reasonable", beyond our present conditions. (One should require: Existence of transition-probabilities, of additive linear aggregates, etc.)

I am hoping, that we will have plenty of time to discuss everything in detail in Cambridge, including interpretation-questions in quantum mechanics. I am looking forward to this trip with much expectation.

With best greetings until then

Cordially

John

P.S. Mariette's skiing was a lot better than my cruising. So I feel very inferior on this score.
P.S.' I will bring your letter and proofs of Murray's and my paper[14] with me.

Original in Papers of Garrett Birkhoff, Harvard University Archives, Accession 13493, language of original: English

---

Princeton

November 6, [1935 ?]

Dear Garrett,

I too felt the urge to discuss a few more points concerning our paper.[15] I think that it will be possible to find a neutral formulation, which is compatible with our criticisms, and I will send you a tentative mscr. of this type in a few days.

The remarks, which you made, are, I think very much to the point. I will try to describe what I think about the matter (1.-3.). Then I will pass to some observations on the subject (4.-8.) I have made since you left. They may have an essential bearing on our problem, too.

1. I would describe the logical entities, the mathematical equivalents of which are the closed-linear-subspaces of the space $\mathcal{H}$ (or their operators, the projections), and which you called "Qualities," as follows:

(*) Let S be a certain experiment, followed by certain computations, which finally lead to one of the two statements "yes" or "no," to be denoted by a variable x, equal to 1 resp. 0 correspondingly.
Let a system Γ be given – I mean *one* specimen of Γ, *not an ensemble* of Γ's! The statement "Γ possesses the quality S" means: "I prophesize,

---

[14]F.J. Murray, J. von Neumann: "On Rings of Operators", Collected Works, Vol. III. No. 7. The Editor.

[15]G. Birkhoff, J. von Neumann: "The Logic of Quantum Mechanics" Collected Works, Vol. IV. No. 7. See Section 6 of the Introductory Comments. The Editor.

that if S is performed on Γ, the result will certainly be $x = 1$, and Γ will be unaffected by the experiment."

(I) This "prophesy" can be checked even on one system. This is obvious for its first part. (That is: A result $x = 0$ would disprove it. A result $x = 1$ is only a partial confirmation – because $x = 1$ might have had a probability $> 0, < 1$. But in reality, all "empirical confirmations" are such: Unsafe in the affirmative sense, while stringent in the negative one.)

(II) For the second part it is true, too, if one uses certain considerations about the connection of entropy and measurement. (I detailed them in my book on Quantum Mechanics[16]. They amount to this: A measurement on a system Γ is possible without entropy-compensations if and only if it does not change the state of Γ – that is: if it merely ascertains the presence of a state of Γ which existed already before the measurement. This is of course a very exceptional occurrence.)

But it is naturally much easier to check all parts of the "prophesy" on an ensemble of many identical systems Γ: $\Gamma_1, \ldots, \Gamma_N$, in the obvious way.

One has to stress: Although it is more convenient to test (*) on an ensemble $\Gamma_1, \ldots, \Gamma_N$, it is a notion which is defined with the help of a unique Γ. This is its justification: If the ensemble $\Gamma_1, \ldots, \Gamma_N$ had been introduced *a limine*, then there would be no excuse for our not considering more general "qualities" (corresponding to sets not closed-linear-manifolds), as you pointed out.

So the lattice which we consider consists of those statements, which one can make "with certainty" in Quantum Mechanics. We ignore the much vaster domain of "statistical" prophesies.

Now the definition of "negation" is immediate: "Not S" consists of all measurements and computations of S, plus a final replacement of x by $1 - x$.

The definition of "implication" is as before: $S \to T$ means, that whenever the prophesy S can be made, then T can be made too.

"And" and "or" are defined by means of the "implication," as you described it.

As a matter of practical policy: In an actual paper I would rather insist on (I) than on (II), because all matters concerning entropy are somewhat controversial. But I think, that the entropy-argument is OK between friends.

2. Nevertheless I agree with you, that it is good to point at the definition of qualities "constructively," that is by saying: "The values of the physical quantities (i.e. self-adjoint operators) $\alpha_1, \ldots, \alpha_N$ are contained in the (numerical) intervals $I_1, \ldots, I_n$ resp."

3. I am somewhat scared to consider all physical quantities = bounded self-adjoint operators as a lattice.

The definition of $\alpha \leq \beta$ by "$\beta - \alpha$ is definite and α, β commute", as you propose it, is not transitive ($\alpha \leq \beta$, $\beta \leq \gamma$ do not secure the commutativity of α, γ - put e.g. $\beta = 1$); while a definition by "$\beta - \alpha$ is definite" alone would, as you observe, not secure in general the existence of "meets" and "joins".

My remarks are these:

[16] J. von Neumann: *Mathematische Grundlagen der Quantenmechanik*, Berlin, Springer, 1932. First English edition: *Mathematical Foundations of Quantum Mechanics*, New York, Dover, 1943. The Editor.

4. In any *linear space* H the *linear subspaces* $K, L, M, \ldots$ form a B-lattice[17] $\mathcal{L}$ with the

"meet" $K \cap L$: intersection of K and L in the sense of set theory
"join" $K \cup L$: linear sum of K and L, i.e. set of all $f+g$, $f \in K$, $g \in L$

(Proof obvious.) But in a *metric-linear space* $\overline{H}$ the lattice $\overline{\mathcal{L}}$ of all *closed-linear subspaces* $\overline{K}, \overline{L}, \overline{M}, \ldots$, for which the "join" is defined as

"join" $\overline{K} \vee \overline{L}$: closure of the linear sum of K and L, i.e. the set of all condensation points of the $f+g$, $f \in K$, $g \in L$

while the "meet" is as above, is *not necessarily a B-lattice! This is in particular the case in Hilbert space.* ($\overline{K} \cup \overline{L}$ and $\overline{K} \vee \overline{L}$ are identical if $\overline{K}, \overline{L}$ are both closed and *orthogonal* to each other, but not for any two closed $\overline{K}, \overline{L}$!)

In fact, it is possible to find three closed-linear subspaces $\overline{K}, \overline{L}, \overline{M}$ of Hilbert space $\overline{H}$, for which

$$\overline{K} \subsetneqq \overline{M}, \quad (\overline{K} \vee \overline{L}) \cap \overline{M} \supsetneqq \overline{K} \vee (\overline{L} \cap \overline{M})$$

Example: Let Hilbert space $\overline{H}$ be the set of all trios $[f, g, h]$, where $f, g, h \in \overline{H_0}$, $\overline{H_0}$ being a Hilbert-space. (Define

$$([f,g,h],[f',g',h']) = (f,f') + (g,g') + (h,h')$$

I.e.: $\overline{H}$ is the direct sum $\overline{H_0} + \overline{H_0} + \overline{H_0}$). Choose three Hermitean operators A, B, C in $\overline{H_0}$ with

(1.) $A^2 + B^2 + C^2 = I$
(2.) A^{-1}, C^{-1} exist
(3.) $Af + Bg = 0$ imply $f = g = 0$.

(This is done as follows: Using the results of my paper Journ. für Math., Vol. 161,[18] particularly pp. 230-234, choose to[19] unbounded, self-adjoint operators X, Y with

$$\text{Domain}X \cdot \text{Domain}Y = 0$$

Then $X^2 + 2I$ and $Y^2 + 2I$ are self-adjoint, too, and

$$A = (X^2 + 2I)^{-1}, \qquad B = (Y^2 + 2I)^{-1}$$

are bounded, Hermitean, and

$$\|A\| \leq \frac{1}{2}, \qquad \|B\| \leq \frac{1}{2}$$

Thus $I - A^2 - B^2$ is definite, put $C = \sqrt{I - A^2 - B^2}$. So (**1.**), (**2.**) are satisfied. As to (**3.**), $Af + Bg = 0$ means: $h = Af = -Bg$ belongs to the domains of both $A^{-1} = X^2 + 2I$ and of $B^{-1} = Y^2 + 2I$, so to those of X and of Y. Therefore $h = 0$, $f = A^{-1}h = 0$, $g = B^{-1}h = 0$.

Consider now the maping

$$f \rightleftharpoons (Af, Bf, Cf)$$

[17] Modular lattice in today's terminology. The Editor.

[18] J. von Neumann: "Zur Theorie der unbeschränkter Matrizen", Collected Works, Vol. II. No. 3. The Editor

[19] Should be "two". The Editor.

It is a one-to-one isomorphism of $\overline{H_0}$ with a subset $\overline{L}$ of $\overline{H}$, owing to (**1.**). So $\overline{L}$ is a Hilbert-space along with $\overline{H_0}$, and therefore a closed-linear subset of $\overline{H}$.

Now let $\overline{K}$ be the set of all $(0,0,h)$, and $\overline{M}$ the set of all $(0,g,h)$. Clearly

$$(\alpha) \qquad \overline{K} \subsetneqq \overline{M}$$

$\overline{K} \vee \overline{L}$ is the closure of the set of all $(Af, Bf, Cf+h)$, that is, of all (Af, Bf, h). Its orthogonal complement consists of all (f', g', h') for which

$$(Af, f') + (Bf, g') + (Cf, h') = 0$$

identically (in f), that is

$$(f, Af' + Bg') + (f, Ch') = 0$$

Thus $Af' + Bg' = 0$, $Ch' = 0$, and so by (**3.**) and (**2.**) $f' = g' = h' = 0$. Thus $\overline{K} \vee \overline{L} = \overline{H}$, and

$$(\beta) \qquad (\overline{K} \vee \overline{L}) \cap \overline{M} = \overline{M}$$

$\overline{L} \cap \overline{M}$ consists of all (Af, Bf, Cf) with $Af = 0$, so by (**2.**) $f = 0$. Thus $\overline{L} \cap \overline{M} = \emptyset$, and

$$(\gamma) \qquad \overline{K} \vee (\overline{L} \cap \overline{M}) = \overline{K}$$

Together $(\alpha), (\beta), (\gamma)$ contradict the B-lattice postulate.

(Examples could be constructed which make no use of operator theory, but I think that this example shows more clearly "what it's all about": It is the existence of "pathological" operators – like X, Y above – in Hilbert space, which destroys the B-lattice character.)

Thus we have the following situation:

(I) *Either* we define the "join" by $\cup$ (as a honest linear sum), then the lattice is B, but we must admitt all (not-necessarily-closed-) linear subspaces,

(II) *or* we define the "join" by $\vee$ (closure of the linear sum), then the B-character is lost.

5. Let us first consider the alternative (**I**). The orthogonal complement K' still has the property $K' \cup L' = (K \cap L)'$, but $K' \cap L' = (K \cup L)'$ and $K'' = K$ are lost. We have $K \cap K' = 0$, while $K \cup K'$ is everywhere dense, but not necessarily I. There is a funny relationship between K and its "closure" K''. (For instance: All probabilities in the state K are equal to those in the state K'', but "meets" ($K \cap L$ and $K'' \cap L''$, even for closed L's) may differ.)

The situation is strongly reminiscent of the "excluded middle" troubles, although I did not yet compare all details with those of the class-calculus in "intuitionistic" logics.

After all it is so in normal logics, too, that these troubles arise as soon as you pass to infinite systems, although I must admitt, that the difficulties there are more "optional" then[20] here.

It has to be said, finally, that even in this case (**I**) complements exist, i.e., that for every K there exists K^*'s for which $K \cup K^* = I$, $K \cap K^* = 0$, but one needs the Hamel-basis-construction to get them.

[20]Probably spelling error, should be "than". The Editor.

6. Alternative **(II)** seems to exclude Hilbert space, if one sticks to B-lattices. Still one may observe this:

Consider a ring $\mathcal{R}$ of operators in Hilbert space. The idempotents of $\mathcal{R}$ form a lattice $\mathcal{L}_{\mathcal{R}}$. One sees easily, that $\mathcal{L}_{\mathcal{R}}$ is irreducible (= no direct sum), if and only if the center of $\mathcal{R}$ consists of the αI (α=complex number) only, i.e. if $\mathcal{R}$ is a ring of the sort which Murray and I considered[21]. (We called them "factors".) $\mathcal{L}_{\mathcal{R}}$ contains $0, I$ and a complement which dualises $\cup$ and $\cap$. (Now $\cup$ corresponds to what I called $\vee$, case **(II)**.) One may ask: When is $\mathcal{L}_{\mathcal{R}}$ a B-lattice?[22] The answer is (this is not difficult to prove): If and only if the ring $\mathcal{R}$ is finite in the classification Murray and I gave. I.e.: $\mathcal{R}$ must be isomorphic:

(1) either to the full matrix-ring of a finite-dimensional Euclidean space (say n-dimensional, $n = 1, 2, \ldots$),
(2) to one of those of our rings, in which each idempotent has a "dimensionality", the range of which consists of all real numbers $\geq 0, \leq 1$, and which is uniquely determined by its formal properties.

We called (1) "Case I_n" and (2) "Case II_1".

Thus for operator-lattices the B-lattice axiom

$$a \leq b \quad \rightarrow \quad (a \cup b) \cap c = a \cup (b \cap c)$$

leads directly to the cases $I_1, I_2, \ldots$ *and* II_1!

This makes me strongly inclined, therefore, to take the ring of all bounded operators of Hilbert space ("Case I_∞" in our notation) less seriously, and Case II_1 more seriously, when thinking of an ultimate basis of quantum mechanics.

7. Mathematically – and physically, too – it seems to be desirable, to try to make a general theory of dimension in *complemented, irreducible B-lattices*, without requiring "finite chain conditions". I am convinced, that by adding a moderate amount of continuity-conditions, the existence of a numerical dimensionality could be proved, which

(1) is uniquely determined by its formal properties,
(2) and after a suitable normalisation has either the range $d = 1, 2, \ldots, n$ ($n = 1, 2, \ldots$, finite!) or $d \geq 0, \leq 1$.

I have already obtained some results in this direction, which connect the notion of dimension in a very funny way with the perspectivities and projectivities in projective geometry.

It will perhaps amuse you if I give some details of this. Here they are:

8. Keeping the analogy of Euclidean spaces in mind, it seems reasonable to introduce in a *complemented, irreducible* B-*lattice* $\mathcal{L}$ (no "chain conditions"!) a provisional notion of equidimensionality $a \approx b$: $a, b \in \mathcal{L}$:

$a \approx b$ if and only if there exists an $x \in \mathcal{L}$ with

$$\begin{aligned} a + x &= b + x = I \\ ax &= bx = 0 \end{aligned}$$

---

[21] F.J. Murray, J. von Neumann: "On Rings of Operators", Collected Works, Vol. III. No. 7. See Section 3 of the Introductory Comments for von Neumann's work on operator algebras. The Editor.

[22] Recall that B-lattice means modular lattice. The Editor.

I.e. if and only if a, b have a common complement x.[23]

I can prove (easily):

Theorem I.: $a \approx b$ is equivalent to any of these conditions:

α) There exists an x with

$$\begin{aligned} a + x &= b + x \\ ax &= bx \end{aligned}$$

β) There exists an x with

$$\begin{aligned} a + x &= b + x \\ ax &= bx = 0 \end{aligned}$$

γ) There exists an x with

$$\begin{aligned} a + x &= b + x = I \\ ax &= bx = 0 \end{aligned}$$

δ) There exists an x with

$$\begin{aligned} a + x &= b + x = a + b \\ ax &= bx = ab \end{aligned}$$

Theorem II: The condition β) for a, b, x is equivalent to this one:

ϵ) The conditions

(1) $$u \leq a \quad , \quad v \leq b$$

(2) $$u + x = v + x$$

determine v uniquely if u is given, and conversely.

Now in projective geometry it is easy to make a picture of ϵ), and of (1)-(2) in particular:

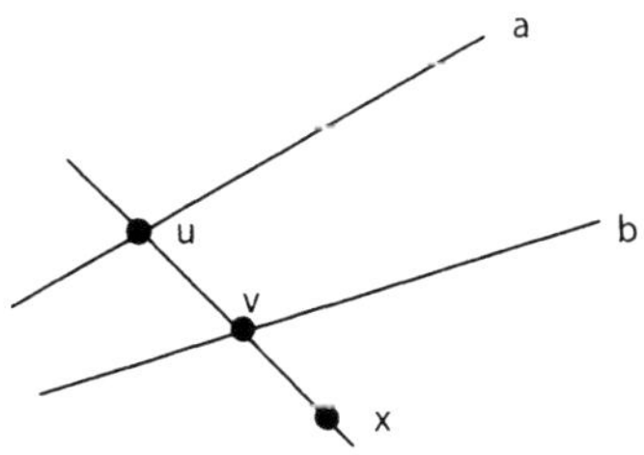

[23]Von Neumann's footnote: *A* common complement. The complement a', for which

$$\begin{gathered} a'' = a \\ (a \cup b)' = a' \cap b' \\ (a \cap b)' = a' \cup b' \\ a \cup a' = 1 \\ a \cap a' = 0 \end{gathered}$$

is another matter. I think that it is needed, too, but only later.

(1)-(2) means that u, v, x are collinear, and thus ϵ) describes a *perspectivity*. Thus $a \approx b$ means: *It is possible to set of a perspectivity between a and b.*

Of course $\mathcal{L}$ is much more general, than this picture: there may be no irreducible elements (i.e. such a's, $a \neq 0$ for which $x \leq a$ implies $x = 0$ or a), that is, no point, etc. But it is still remarkable, that the reasonable definition of equidimensionality is so closely linked to such fundamental notion of geometry, as perspectivity.

The main defect of $a \approx b$ is, that it may not be transitive (?). Then it is obviously necessary to define the final notion of equidimensionality, $a \sim b$, like this:

$a, b \in \mathcal{L}$: $a \sim b$ if a finite chain

$$a = a_0, a_1, a_2, \ldots, a_{n-1}, a_n = b$$

of elements of $\mathcal{L}$ with $a_{i-1} \approx a_i$ for $i = 1, \ldots, n$ exist.

This $a \sim b$ corresponds in the same way to *projectivities*, as $a \approx b$ does to *perspectivities.*

I am ow working out a systematic theory of dimensionalities, based on this notion. Although it is not quite complete yet, I think that I see the way, which carries in this direction, to numerical dimensionalities satisfying 1. and 2. I think, that I will soon be able to write you more about it.

We will certainly come to Cambridge this winter, we are only waiting for the return of Mrs. Stone to Cambridge, to make the final arrangements. Please give our best wishes to your parents, we will be very glad to see them again, we both regret that we had no chance to meet them for quite a considerable length of time.

We hope that you too will renew your visit to Princeton before long.

Please excuse the barbaric length of this letter.

With the best greetings from all of us

as ever

John

P.S. Many thanks for the reprint of your interesting paper on "Integration in Banach spaces" which arrived just now. I would like to ask you several thinks[24] about it – but my letter is much too long anyhow. J.

Original in Papers of Garrett Birkhoff, Harvard University Archives, Accession 13493, language of original: English

---

[24]Misspelled word: things. The Editor.

Princeton

Nov. 13., Wednesday, [1935 ?]

Dear Garrett,

Many thanks for your letter. Your idea of requiring $a \leq c \to a \cup (b \cap c) = (a \cup b) \cap c$ in Hilbert space for the finite a, b, c only is very interesting, but will it permit to differentiate between Hilbert-space and other Banach-spaces?[25]

I would like to make a confession which may seem immoral: I do not believe absolutely in Hilbert space any more. After all Hilbert-space (as far as quantum-mechanical things are concerned) was obtained by generalizing Euclidean space, footing on the principle of "conserving the validity of all formal rules". This is very clear, if you consider the axiomatic-geometric definition of Hilbert-space, where one simply takes Weyl's axioms for a unitary-Euclidean-space, drops the condition on the existence of a finite linear basis, and replaces it by a minimum of topological assumptions (completeness + separability). Thus Hilbert-space is the straightforward generalization of Euclidean space, if one considers the *vectors* as the essential notions.

Now we[26] begin to believe, that it is not the *vectors* which matter but the *lattice of all linear (closed) subspaces*. Because:

(1) The vectors ought to represent the physical *states*, but they do it redundantly, up to a complex factor, only.
(2) And besides the *states* are merely a derived notion, the primitive (phenomenologically given) notion being the *qualities*, which correspond to the *linear closed subspaces*.

But if we wish to generalize the lattice of all linear closed subspaces from a Euclidean space to infinitely many dimensions, then one does not obtain Hilbert space, but that configuration, which Murray and I called "case II_1." (The lattice of all linear closed subspaces of Hilbert-space is our "case I_∞".) And this is chiefly due to the presence of the rule[27]

$$a \leq c \to a \cup (b \cap c) = (a \cup b) \cap c$$

This "formal rule" would be lost, by passing to Hilbert space!

So I am very much in favor to stick in our paper to (finite dimensional-) Euclidean-spaces and to describe merely the difficulties in Hilbert-space, adding to it a cautious indication of the possibility of an atom of a shade of doubt relatively to that venerable object.

I have completed (and written down in imbroglio) the proofs about which I wrote you last time. I am stating the results in what follows, as it may interest you 1) because it is a generalization of your results on the relationship between the modular (Dedekind-)axiom and the existence of dimension, 2) it illustrates the point I was making against Hilbert-space above.

Consider the following axioms:

[25] Von Neumann and Birkhoff are discussing quantum logic; see Section 6 of the Introductory Comments. The Editor.

[26] With F.J. Murray, von Neumann's coauthor. The Editor.

[27] This condition is known as modularity. The Editor.

Let us consider a system L of elements $a, b, \dots, y, z$ with the following properties:

I. L is a *partially ordered set.* That is, there exists a relation $a < b$ (or $b > a$) in L such that

I_1. Never $a < a$.

I_2. $a < b$, $b < c$ imply $a < c$.

II. L is *continuous in the Dedekind sense.* That is:

II_1. For every set $S \subseteq L$ there exists an $a \in L$, such that

(i) $x \in L$[28] implies $a \leq x$.

(ii) If $x \in L$[29] implies $a' \leq x$, then $a' \leq a$.

II_2. For every set $S \subseteq L$ there exists an $a \in L$, such that

(i) $x \in L$[30] implies $a \geq x$.

(ii) If $x \in L$[31] implies $a' \geq x$, then $a' \geq a$.

Remark 1. The a of II_1 and II_2 are uniquely determined by S, owing to I_2. Denote them by $\mathcal{I}(S)$ and $\mathcal{S}(S)$. They are clearly the greatest lower bound resp. the least upper bound of S.

Remark 2. II_1 and II_2 are "dual", but they are redundant: Assuming I., each one of them implies the other.

Remark 3. If $S = (a, b)$, write

$$\mathcal{I}(S) = a \cap b, \qquad \mathcal{S}(S) = a \cup b$$

One proves easily:

$$a \cap b = b \cap a \qquad\qquad (a \cap b) \cap c = a \cap (b \cap c)$$

$$a \cup b = b \cup a \qquad\qquad (a \cup b) \cup c = a \cup (b \cup c)$$

$$a \cap b = a \quad \text{is equivalent to} \quad a \cup b = b$$

and both are equivalent to

$$a \leq b$$

Thus L is an S-lattice.

Remark 4. Write $\mathcal{I}(L) = 0$, $\mathcal{S}(L) = I$. One proves easily

$$a \cap 0 = 0 \quad , \quad a \cup 0 = a$$

$$a \cap I = a \quad , \quad a \cup I = I$$

$$0 \leq \ a \ \leq I$$

Remark 5. Let Ω be an infinite aleph of Cantor. Consider a sequence a_α, defined for all ordinals of Cantor $\alpha < \Omega$. Let S be the set of the a_α, $\alpha < \Omega$. If $\alpha < \beta$ implies $a_\alpha \leq a_\beta$, write $\lim_{\alpha \to \Omega} a_\alpha = \mathcal{S}(S)$, if $\alpha < \beta$ implies $a_\alpha \geq a_\beta$, write $\lim_{\alpha \to \Omega} a_\alpha = \mathcal{I}(S)$.

One proves easily

(*i*) If $\alpha < \beta$ implies $a_\alpha \leq a_\beta$: $(\lim_{\alpha \to \Omega} a_\alpha) \cup b = \lim_{\alpha \to \Omega} (a_\alpha \cup b)$

(*ii*) If $\alpha < \beta$ implies $a_\alpha \geq a_\beta$: $(\lim_{\alpha \to \Omega} a_\alpha) \cap b = \lim_{\alpha \to \Omega} (a_\alpha \cap b)$

---

[28] Should be $x \in S$. The Editor.

[29] Should be $x \in S$. The Editor.

[30] Should be $x \in S$. The Editor.

[31] Should be $x \in S$. The Editor.

Princeton, Nov. 13., Wednesday.

Dear Garrett,

many thanks for your letter. Your idea of requiring $a \leq c \to a \cup (b \cap c) = (a \cup b) \cap c$ in Hilbert-space for the finite a, b, c only is very interesting, but will it permitt to differentiate between Hilbert-space and other Banach-spaces?

I would like to make a confession, which may seem immoral: I do not believe absolutely in Hilbert-space any more. After all Hilbert-space (as far as quantum-mechanical things are concerned) was obtained by generalising Euclidean space, footing on the principle of "conserving the validity of all formal rules". This is very clear, if you consider the axiomatic-geometric definition of Hilbert-space, where one simply takes Weyls axioms for a unitary-Euclidean-space, drops the condition on the existence of a finite linear basis, and replaces it by a minimum of topological assumptions (completness + separability). Thus Hilbert-space is the straightforward generalisation of Euclidean space, if one considers the *vectors* as the essential notions.

Now we begin to believe, that it is not the *vectors* which matter, but the *lattice of all linear (closed) subspaces*. Because:

1) The vectors ought to represent the physical *states*, but they do it redundantly, up to a complex factor, only.
2) And besides the *states* are merely a derived notion, the primitive (phenomenologically given) notion being the *qualities*, which correspond to the *linear closed subspaces*.

But if we wish to generalise the lattice of all linear closed subspaces from a Euclidean space to infinitely many dimensions, then one does not obtain Hilbert space, but that configuration, which Murray and I

The first page of John von Neumann's original seven-page handwritten letter to Garret Birkhoff on November 13 around 1935 while at Princeton.

III. Addition and multiplication in L are *continuous* for the $\lim_{\alpha\to\Omega}$ as defined above. That is

$$\text{III}_1 \qquad \text{If } \alpha < \beta \text{ implies } a_\alpha \leq a_\beta \quad : \quad (\lim_{\alpha\to\Omega} a_\alpha) \cap b = \lim_{\alpha\to\Omega} (a_\alpha \cap b)$$

$$\text{III}_2 \qquad \text{If } \alpha < \beta \text{ implies } a_\alpha \geq a_\beta \quad : \quad (\lim_{\alpha\to\Omega} a_\alpha) \cup b = \lim_{\alpha\to\Omega} (a_\alpha \cup b)$$

Remark 6. The complete statement of continuity consists, of course, of (i),(ii) in Remark 5., plus III_1, III_2.

IV. L is *modular.* That is:

IV_1. $a \leq b$ implies $a \cup (b \cap c) = (a \cup b) \cap c$

V. L is *complemented.* That is:

V_1. For every a there exists at least one x with

$$a \cap x = 0 \qquad a \cup x = I$$

VI. L is *irreducible.* That is:

VI_1. If for an a V_1. has a unique solution x, then $a = 0$ or I.

Remark 7. This can be shown (using Axiom I.,II.,V. alone) to be equivalent to this:

VI_1^*. L is *no direct sum.* That is:

If a, b have these two properties:

(i) $ab = 0$[32]

(ii) the $u + v$[33], $u \leq a$, $v \leq b$ exhaust all L, then

$$a = I, \quad b = 0, \quad \text{or} \quad a = 0, \quad b = I$$

(Of course these remarks are well-known to you. I put them in only for the sake of completeness.)

Observe that this system is a direct generalization of your system. The essential difference is, that I have III. in place of the "finite chain conditions":

$\text{III}^*/\text{III}_1^*$. If $\alpha < \beta$ implies $a_\alpha < a_\beta$ (a_α defined for all $\alpha < \Omega$), then Ω is finite.

$\text{III}^*/\text{III}_2^*$. If $\alpha < \beta$ implies $a_\alpha > a_\beta$ (a_α defined for all $\alpha < \Omega$), then Ω is finite.

All other differences are either formal, or unessential on the basis of our present knowledge. (The greatest one is, that I require "unrestricted" addition and multiplication, that is II. for all sets $S \subseteq L$. But if II. was replaced by II^*: "Require II. for *finite* sets $S \subseteq L$ only", this would be exactly equivalent to your system. And II^*. plus III^*. imply II.)

Now make these definitions:

Definition 1. $D(a)$ is a *dimension-function,* if

α) $D(a)$ is defined for all $a \in L$, and has numerical values

β) $D(0) = 0$, $D(I) = 1$, always $0 \leq D(a) \leq 1$.

γ) $D(a \cap b) + D(a \cup b) = D(a) + D(b)$.

---

[32] It should read: $a \cap b = 0$. The Editor.

[33] It should read: $u \cup v$. The Editor.

Definition 2. $D(a)$ is a *proper dimension function*, if

α) $D(a)$ is defined for all $a \in L$, and has numerical values
β) $D(0) = 0, \quad D(I) = 1$, and for $a \neq 0, I \quad 0 < D(a) < 1$
γ) $D(a \cap b) + D(a \cup b) = D(a) + D(b)$.
δ) If a_i is defined for $i = 1, 2, \ldots$, and if $a_1 \leq a_2 \leq \ldots$ or $a_1 \geq a_2 \geq \ldots$, then

$$D(\lim_{i \to \infty} a_i) = \lim_{i \to \infty} D(a_i)$$

My essential results are:

Theorem 1. If L fulfills I.-VI., then the notions of dimension-function and of proper dimension-function are equivalent, and there exists one and only one such function.

Theorem 2. If L fulfills I., II., V. (that is if it is a complemented lattice with unrestricted addition and multiplication and without any finiteness-restrictions), then the validity of III., IV., VI. (that is of all I.-VI.) is equivalent to either of these conditions:

(i) There exists a unique proper dimension-function in L.
(ii) VI. holds (irreducibility), and there exists at least one proper dimension-function in L.

Assume now, that L fulfills I.-VI., and that $D(a)$ is its (proper) dimension-function.

Definition 3. Define:

(i) $a \sim b$ if a, b have a common complement, that is, if an x with

$$a \cap x = b \cap x = 0, \qquad a \cup x = b \cup x = I$$

exists.
(ii) $a \prec b$ (that is $b \succ a$) if an a' with $a \sim a' < b$ exists.

Theorem 3.

(i) $a \sim b$ is reflexive, symmetric, transitive.
(ii) $a \prec b$ is non-reflexive, non-symmetric, transitive.
(iii) Always one and only one of three cases $a \sim b$, $a \prec b$, $a \succ b$ exists.
In other words $a \sim b$, $a \prec b$, $a \succ b$ describes an equivalence and a complete ordering of its equivalence classes in L.

Theorem 4. $a(\succ, \sim, \prec)b$ is equivalent to $D(a)(>, =, <)D(b)$ resp.

Theorem 5. The range Δ of $D(a)$, $a \in L$, is one of the following sets:

Case D_n: The set of all $\eta = 0, \frac{1}{n}, \frac{2}{n}, \ldots 1$ for a certain $n = 1, 2, \ldots$.
Case D_∞: The set of all $\eta \geq 0, \leq 1$

This allows *ex post*, that III. is in reality void for non-enumerably infinite Ω's.

The "Cases $D_1, D_2, \ldots$" coincide, of course, with the projective geometries determined by you. So they all exist.

The "Case D_∞" exists too: If R is a ring of operators in Hilbert-space, which in Murray's and my classification is of "class II_1"[34], then it is easy to verify (with the help of Theorem 2. (ii)), that it fulfills I.-VI., and that it is a "Case D_∞".

I am now trying, to set up a connection between an L of "Case D_∞" and some hypercomplex number-system, imitating the classical procedure in projective

[34]See Section 3 of the Introductory Comments for the Murray-von Neumann classification theory. The Editor.

geometry (= "Case D_n", $n = 1, 2, \ldots$), hoping that they[35] hypercomplex system may be more restricted here, then[36] it was there.

What do you think of these things? I have now stopped worrying about them, and I am *really* writing the logics-paper.[37]

With many greetings, and hoping to see you soon in Cambridge.

Yours as ever

John

P.S. I forgot to say, what you have probably noticed anyhow, that an L of "case D_∞" may be looked at (considering your results on the "cases D_m", $m = 1, 2, \ldots$) as a "projective geometry without points". This aspect of the matter, to make geometry without using the "minimum configurations" = the points, seems quite funny. J.

Original in Papers of Garrett Birkhoff, Harvard University Archives, Accession 13493, language of original: English

---

Princeton

Nov. 21, 1935

Dear Garrett,

I am sending you enclosed the "preliminary manuscript" of our paper[38]. It has the outward appearance of a manuscript, but do not trust it, because it is incomplete in several respects.

I attempted to write it down, *as if* it were a logical whole, but I did it merely because this seemed the simplest way to me, to indicate what I wanted to include in it. It is now up to you, to add anything you wish to include in it, or to omitt any part, which you think is superfluous, or where you think that I went to too great length.

On page 21, § 15, I omitted the derivation of the antiisomorphism $\mathcal{I}$ and of the definite form $\phi(\xi_1, \ldots, \xi_N) = \sum_{i=1}^{N} \oplus\xi_i \otimes \alpha_i \otimes \mathcal{I}\xi_i$. If I remember correctly, you pointed out to me, that my original derivation of these notions (from the negation a') could be simplified essentially.

The list of axioms for a Boolean Algebra, which I gave on pages 12-13, § 10, is redundant and probably inelegant. But I wanted to write down something definite. As you know the ins and outs of this subjects much better than I do, please replace the list by a more reasonable one. The same may be true for the lists of axioms for lattices on pages 18-19, § 13.

---

[35]Should be "their". The Editor.

[36]Should be "than". The Editor.

[37]Reference to the G. Birkhoff, J. von Neumann: "The Logic of Quantum Mechanics" paper, Collected Works, Vol IV. No. 7; see Section 6 of the Introductory Comments. The Editor.

[38]G. Birkhoff, J. von Neumann: "The Logic of Quantum Mechanics", Collected Works, Vol. IV. No. 7; see Section 6 of the Introductory Comments. The Editor.

The footnotes I put in are very incomplete, and open to many objections. Here I would like to remark:

(i) I know that I quote probably too much on quantum mechanics, and certainly too much of myself. I did it merely, because again this was the simplest way to indicate, to which ideas I wanted to refer. Please use your judgement in reducing this type of footnotes to a reasonable size, which will certainly be $\sim 50\%$ of the present one.
(ii) I filled in no footnotes, which refer to literature on Boolean Algebras and Lattices, as you know it much better, than I do.
(iii) Footnotes 42,43,46 are marked $*$, I will add some quotations here, later.

Looking at the paper now I see, that I forgot to say this, which should be said somewhere in the first §: That while common logics did apply to quantum mechanics, if the notion of simultaneous measurability is introduced as an auxiliary notion, we wished to construct a logical system, which applies directly to quantum mechanics – without any extraneous secondary notions like simultaneous measurability. And in order to have such a consequent, one-piece system of logics, we must change the classical class calculus of logics.

These are all self-criticisms I can think of at this moment. No, I can think of one more: I am using the phrases "We believe", "It seems to us", quite freely on controversial subjects. Please mark those places, where you disagree with the opinions voiced. You will undoubtedly find other criticisms. At any rate my responsibility decreases now again, although somewhat late. But I hope, that you will agree, that late is better than never.

We are coming to Cambridge on Jan. 24. I am very sorry that December was unworkable, but as we leave for California on Dec., 14., there was too little time available.

Is there no chance of seeing you again before January? There would be many things I would be so glad to discuss with you, particularly about the continuous dimensionality in infinite lattices, which occupies me chiefly now.

I am very interested to hear, that you proved the metrisability of all 1-Hausdorff-countability-axiom-fulfilling-groups. How does this surprising result arise?

Hoping to hear from you soon.

I am yours as ever

John

Original in Papers of Garrett Birkhoff, Harvard University Archives, Accession 13493, language of original: English

The Institute for Advanced Study
School of Mathematics
Fine Hall
Princeton, New Jersey

Nov. 27, [1935 ?]

Dear Garrett,

Many thanks for your godly intentions with our paper[39], I hope that you will be much better in living up to them, than I was.

The definition of a "statement," as I gave it, was more complicated, than necessary, but I thought that it is worth while to make it so lengthy for this reason: By separating the "measurement" and the "computation", we make it clear, that the "negation of a " is the same measurement as "a", merely followed by a different computation. The forms F and F' given by you are simpler, but they have this defect: F' has not the form of F, and I think, that one should begin by giving a general pattern, which covers all "statements". Of course we could say, that the general form of a "statement" is this:

F*: The measurement M on the system Γ gives a result ξ with certainty in the set (of numbers) Σ. But Σ is merely a more abstract way to describe the "computations."

But I will leave it to you to adjust these matters.

I think, that your suggestion of saying more about "logical equivalence" is a very good one.

Your general remarks, I think, are very true: I, too, think, that our paper will not be very exhaustive or conclusive, but that we should not attempt to make it such: The subject is obviously only at the beginning of a development, and we want to suggest the direction of this development much more, than to reach "final" results. I, for one, do not even believe, that the right formal frame for quantum mechanics is already found.

I must report with somewhat mixed feelings that I can build up a continuous-dimensional projective geometry on any (not necessarily commutative, but associative) division algebra. (I hoped, as you know, that this might only be possible for some of them, perhaps for real, complex and quaternionic numbers only.) As the construction is quite simple, I give it in full. At the same time I can answer your question, as to how I attach a division algebra to a continuous-dimensional lattice.

The isomorphism question is not so well settled: I do not know, whether all continuous dimensional projective geometries with the same division algebra are isomorphic or not. (I give the mathematical business on a separate leaf.)

Both Mariette[40] and I are looking forward with great pleasure to our trip to Cambridge. I am very glad to dine with you at the "Society of Fellows" dinner on January 27. We both will enjoy seeing your parents again greatly, the Stones - who are, in fact, our bosses resp. *maitres de plaisir* in Cambridge - mentioned it to us, too.

[39] G. Birkhoff, J. von Neumann: "The Logic of Quantum Mechanics", Collected Works, Vol. IV. No. 7; see Section 6 of the Introductory Comments. The Editor.

[40] Mariette Kövesi, von Neumann's wife. The Editor.

Is there any hope of your coming here before? We are going to Cleveland for Thanksgiving, to stay with some friends: Leaving (by car) in some hours, and returning on Monday. On Dec. 11 we leave for California, returning around Jan. 14. But we will be here from Dec. 2 to Dec. 10 inclusively. I know, that I should not pester you too much with this suggestion, as you have already traveled all the way to Princeton, while I have not yet done the same thing with the opposite sign – but it would be very good, to see you before Jan. 24. And we could discuss the lattices with continuous dimensions much more completely, than in writing.

Many greetings from all of us, cordially,

John

P.S. It seems that I will not be able to include the "mathematical appendix" because I have to leave for Cleveland... So I will send it to you from somewhere in the mountains of Pennsylvania or from Cleveland!

Original in Papers of Garrett Birkhoff, Harvard University Archives, Accession 13493, language of original: English

---

December 10, 1941

Professor Garrett Birkhoff
45 Fayerweather Street
Cambridge, Massachusetts

Dear Garrett:

Many thanks for your letter and the manuscript on "Lattice ordered groups". Clifford's opinion, to which you refer, has also been received.

I am reading your paper now – just out of curiosity, because it is of course accepted by the Annals. It is really a pleasure to read it, both esthetically and objectively, and I will write you about it later (I see your skeptical smile, but I will *really* write you about it later).

Your remark about my lately acquired filthy habits is justified,[41] but – although it bodes no good – may I inquire what was the significant question put to you in Cincinnati concerning the same? The thing is getting worse because I have decided to publish the work on games in the form of a book on economic theory in collaboration with the Princeton economist Otto Morgenstern. Besides, I really begin to suffer from an acute shortage of time in consequence of war work.

But I'll try to reform. As part of my reform drive, I actually completed a paper with Halmos, by using the neat device of having him write it altogether.

I can appreciate your sorrow on having to lecture on continuous groups. It must be one of the most terrible subjects to prepare on. Since it is sweet to thrive upon the misfortune of others, I should like to know whether there is any chance

[41] In a letter to von Neumann, Birkhoff remarks that von Neumann has acquired the filthy habit of not publishing his results. The Editor.

of your turning out notes of these lectures? I should be very much interested in having them.

It's hard to write about everyday events, or anything local for that matter, since the situation is changing rather explosively now. In spite of the serious developments of the last days and of what is probably ahead in the immediate future, I can't deny that I feel considerably lighter, – at least all hesitations and theological arguments are over.

What are your plans for Christmas? Do you come to Jersey? Ours are nill. We shall presumably stay here throughout the vacation. Best regards from house to house,

As ever,

John

P.S. I have much sympathy with you, and especially with Ruth on account of the *Blitzkrieg*[42] of your maid. I fear that increased industrial production will lead to more phenomena of this type. Congratulations on your success in Plan A, in spite of the $< 1/2$ Counsellors!

Original in Von Neumann Papers, Library of Congress, language of original: English

---

April 7, 1947

Professor Garrett Birkhoff
1979 Monte Vista
Pasadena
California

Dear Garrett:

I would be very glad to talk on computing machines at the meeting of the Mathematical Association at Yale, on September 1-2. I suppose there is plenty of time for us to discuss the duration, character and details of this talk, including the division of the subject between Aiken and myself.

Regarding the algebra of measurable sets modulo sets of measure zero: The subject of their characterization is touched upon in my 1935-1936 Princeton lecture notes on continuous geometries. I think, however, that the complete characterization occurred mainly in conversations which we had with each other in the late 1930's. I did write a paper which dealt with this algebra but primarily from the point of view of selecting representatives from each class of measurable sets modulo the sets of measure zero so that the algebraical relations between the classes become exactly valid between their representant sets. This paper appeared in Vol. 165 (1931) of the "Journal fur Reine und Angewandte Mathematik", pages 109-115.[43] In the last part of the Continuous Geometry notes, I gave an invariant property

[42] "Ligthning war" – German. The Editor.

[43] J. von Neumann: "Algebraische Representanten der Funktionen 'bis auf eine Menge vom Masse Null' ", Collected Works, Vol. II. No. 6. The Editor.

which distinguishes the algebra of the classes modulo sets of measure zero from those modulo the sets of first category.

How is life in sunny Southern California?

With best regards from house to house.

Yours as ever,

John von Neumann

Original in Von Neumann Papers, Library of Congress, language of original: English

Letter to W.J.E. Blaschke

Princeton, N.J.

January 28, 1935

Dear Professor Blaschke:

Although not a German, I am very much indebted to German science and especially to many representatives of mathematics and physics in every part of Germany. I had received my scientific education in the German speaking part of the World and have spent part of my scientific career in German Universities – a part, which remains for me unforgettable for ever. I take special pride in having been active also in the University of Hamburg.

Nevertheless I cannot reconcile it with my conscience to remain a member of the German Mathematical Society[44] any longer, after another international member, Mr. H. Bohr, was condemned by the 1934 Assembly for his having made a political statement abroad, the resolution was made public, and Mr. H. Bohr has drawn the consequences on his part.

Therefore I must ask you to have my name deleted from the membership list.

It is my hope that my paths and those of the *D.M.V.*, whose true interests I still believe to be serving, are not separating for ever.

I am asking you to believe that I remain,

Truly yours,

Original in Von Neumann Papers, Library of Congress, language of original: German. Translated by M. Rédei.

[44] *Deutsche Mathematiker Vereinigung (DMV).* The Editor.

Letter to R.S. Burington

January 19, 1951

Dr. Richard S. Burington
5200 N. Carlin Spring Road
Arlington, Virginia

Dear Dr. Burington:

This letter contains my views on the subject of the request in your letter of January 12, which we also discussed at the conference with yourself and Drs. R.J. Seeger and H. Polachek in Princeton on January 8.

At our conference I emphasized the circumstances which may limit the validity of my judgement on the matter in question. I understand that you are entirely aware and appreciative of these things, and that you want me to give you my views nevertheless. I am doing this in what follows and leave the evaluation to you.

I.

I have no doubt that the Bureau of Ordnance and the activities that it controls have an immediate, as well as a lasting need for an automatic, high speed computing machine, and specifically for one of the most advanced design. This is true for any largest-scale organization whose complex of activities requires the solution of a wide variety of problems in applied mathematics and physics, and it applies, for the general reasons, as well as for a number of specific and particular ones, to the Bureau of Ordnance. I shall enumerate some particularly outstanding subjects which are integral parts of the Bureau of Ordnance activities, and indicate how high speed calculation will contribute essentially to the Bureau work in these subjects.

1. Aerodynamics

Calculations of (subsonic as well as supersonic) air flows in more than one spatial dimension definitely require high speed computing. Ordinary problems in two spatial dimensions, which include time dependence, already tax the capacity of machines like MARK III., SSEC, ENIAC[45] to the limit. Machines with greater capabilities become necessary when problems of the above mentioned kind are affected with further complications, e.g. certain forms of turbulence, interactions between hydrodynamics and heat conduction processes or chemical reactions, etc. The difficulties are even more serious in problems with three spatial dimensions, even if these offer no other complications and are stationary. These last mentioned problems call for the most advanced machine designs and probably, in addition, for a great deal of mathematical inventivity. All of these problems are of decisive

[45]See Section 8 of the Introductory Comments. The Editor.

importance in the design of most of the important high speed aerodynamic devices like airplanes, missiles, rocket motors, supersonic nozzles, etc.

Another complex of hydrodynamics problems that are important for the Bureau of Ordnance and its subordinate activities, and which is calling for high speed computing machines, involves the formation, propagation, decay, and interaction phenomena of blast waves. Even the problems of the formation and decay of a spherical blast wave in a homogeneous medium have never been quite satisfactorily treated with devices that are less advanced than, say, MARK III, SSEC, ENIAC. In fact, a program concerning these problems, on one of the above mentioned machines would be desirable. As soon as one comes to more involved forms, like the propagation of a blast wave in a not completely uniform atmosphere, more advanced devices than the above mentioned ones are necessary.

2. Hydrodynamics

Multidimensional flows in an incompressible medium (e.g. water, when the incompressible idealization is appropriate) bring about the same situations that were mentioned above for multidimensional flows in air.

The propagation, decay and interaction of pressure waves in water present difficulties which are similar to those mentioned above in air, but they offer some additional complications of the theory, which make the use of very advanced high speed computing machines even more necessary. In this connection, it may suffice to mention the role of the gas bubble that forms in an underwater explosion, and the inhomogeneities and boundary conditions that are necessarily introduced by the existence of the surface and of the bottom of the water mass.

3. Elasticity and Plasticity

Multidimensional calculations in these fields are especially called for in determining the vulnerability of various structures, that is, the interactions between blast waves in water and air, and the attacked solid structures, etc.

4 Explosives

Theoretical questions of deflagration and detonation, in conjunction with the effects of shape of the explosive, require very complicated calculations which also justify the use of very advanced machine computing. Shape effects in the above sense arise in a variety of ways: Even the necessarily finite extension of any explosive charge has to be viewed as causing such effects, and, in addition to this, composite explosive systems can only be understood on the basis of the exact spatial relationships of their parts, which are also sources of shape effects. Finally, there are the effects of shaped charges proper of various kinds, and in this field a great deal of theoretical investigation, beyond the patterns now in use, is certainly called for.

5. Missile Design

The aerodynamic properties of missiles call for investigations of this type, but these may be taken to be covered by the remarks concerning aerodynamics further above.

Another aspect of missile design which requires very high speed computing, equally peremptorily, is this: Missile design would be greatly advanced if extensive simulation by calculation were possible. In other words, if for any given set of aerodynamic missile properties, control and steering element characteristics, communications system and noise level characteristics, the performance of such an hypothetical missile could be calculated in a variety of relevant situations. The same applies to pre-calculations of weapon effects on the basis of given fuse characteristics, fragmentation or blast characteristics, and target mobility and vulnerability characteristics. Such problem setups can lead into extremely intricate analytical and combinatorial discussions, also involving a wide variety of mathematically very difficult questions concerning the targets tactics and countermeasures. All existing simulations handle only a very small fraction of these problems and these only with rather inferior precision. There is no doubt that they will be adequately treated by digital computers only, and among these only by those of the most advanced characteristics.

The savings and the acceleration that will be achieved in the missile field, when such techniques can be routinely used, are evident: Most missile designs can then be tested by mathematical simulation on computing machines, and only a few critical and especially promising ones, selected on the basis of the computational simulations, will have to be carried on into the hardware stage.

6. Atomic Weapons and Motors

The blast properties offer one more aspect of the complex discussed under aerodynamics further above.

Radiation properties call for calculations of shielding and of other related effects which can only be handled digitally, and with existing machines only on a rather modest scale. There is no doubt that this subject is of the greatest importance and calls for a very fast automatic computing.

Similar considerations apply to the integration of atomic components into larger weapons systems.

II.

The characteristics of the proposed IBM machine are very bold, but not at all impossible. The speed characteristics exceed by a considerable amount those of all machines about which I am informed, whose completion in the nearer future is at all likely.

Specifically: The acoustic memory machines have multiplication times of several milliseconds. The Institute for Advanced Study machine (electrostatic memory) has a multiplication time of half a millisecond, or possibly a little less. Other machines which are to follow this design (Illinois, Ordvac, Los Alamos, Argonne, Rand) will have the same, or, in some cases, possibly even somewhat longer multiplication times. The multiplication time of WHIRLWIND lies, depending on how many of the subordinate operations are counted in, between 20 and 40 microseconds, however, the precision of WHIRLWIND amounts to about three times fewer places than that of the proposed IBM machine and, therefore, its multiplication time should be multiplied by a factor lying somewhere between five and nine before a comparison is made. In view of this, the proposed multiplication speed of 100 microseconds, including certain subsidiary operations and the manipulation of a floating decimal point, represents a marked advance beyond the field.

The characteristics proposed for the electrostatic memory are less extreme but also remarkable and desirable. The same remark applies to the proposed input-output system, although here a different distribution of emphases should be considered (cf. further below).

To sum it up, the proposed characteristics describe a machine which should be definitely viewed as the next step beyond the machines that are now under construction, or that may now exist in more or less completed form. I am not at all surprised that the IBM Company estimates completion in two years, in fact I would expect that it will take somewhat longer to produce such a machine and to make it operable. To put it in another way: This proposal does probably not represent the quickest or the cheapest way to acquire a very high speed computing machine. It corresponds to a program of proceeding to the next stage beyond the present one and to obtain by a considerable effort a machine which is likely to have its peak usefulness toward the middle 1950's.

The estimated expenditure, for development and construction, does not seem to me unreasonable per se, in view of the boldness of the program. The relevant considerations must be those of the evaluation of duration and feasibility.

In this last respect I would suggest that the representatives of the Bureau inform themselves in detail about the state of the component developments on which the proposed machine will be based. I understand that the IBM Company already has extensive experience with certain critical components, while others may yet have to be developed. It is very important for the Bureau to assess what the reliability and reproducibility of the existing components are, and what the nature of the problems of combining these and of integrating them into larger systems will be. The assessment of prices, datelines and comparisons with alternative developments should ultimately depend on these.

III.

If the construction of such a machine is decided upon, certain phases of the proposal concerning mathematical arrangements, programming, balance of various components should be considered in great detail and reviewed critically.

Specifically: I think that the following questions should be considered very extensively and carefully, and on the basis of specific problems and codes, of the known statistical relationships of various parts of a calculation in typical mathematical and allied problems, and of the needs and limitations of the users:

(1) Relative merits of the decimal and binary systems.

(2) Relative merits of the floating decimal (or binary) point and of other possible ways to handle the problems and dangers of varying number size.

(3) Evaluation of the merits of the proposed very fast input-output system in comparison with those of a large secondary memory (which should presumably be in the form of a magnetic drum). The latter may achieve those functions, which are expected to be achieved by the high speed input-output system, considerably more effectively. In this case, the justified speed level for the input-output system may prove to be quite different from what is generally believed to be the case for such machines.

(4) The order system to be used in controlling the operations of the machine calls for a careful study and evaluation, from the viewpoints enumerated above, and also from that one of the experience with other machines and machine designs and their relationship to the IBM proposal.

IV.

All considerations of the above type are incomplete without further ones concerning the right type of training and personnel programs required in conjunction with such a machine. Without them an efficient use of such a machine, or even a much less ambitious one, is impossible. These problems are even more critical than e.g. the much discussed ones of mere machine maintenance. I am sure that they can be solved, but they require considerable and continued attention.

I am,

Very truly yours,

John von Neumann

Original in Von Neumann Papers, Library of Congress, language of original: English

Letters to V. Bush

November 15, 1949

Dr. Vannevar Bush
Carnegie Institution in Washington
1530 P Street, Northwest
Washington 5, D.C.

Dear Dr. Bush,

Please excuse my delay in answering your letter of October 31st and in thanking you for having sent me a pre-publication copy of your book[46]. I understood from your letter that you expected to be on the Pacific Coast during the first two weeks of November, so I assume that my reply is now properly timed to reach you with little delay after your return.

First of all I want to tell you how much I appreciated having had an occasion to read your book at this early date. I have read it and I found it most interesting. To talk about its technical parts, and especially, those which deal with military matters: I am convinced that they fill an important and dangerous lacuna which exists in the public's, as well as in many experts' minds. I mean the over-estimation of the role of strategic bombing, the high flying, heavy bomber, and the fission bomb. I feel that it is exceedingly important, and that it will have a salutary effect coming from an absolutely informed and absolutely impartial expert like yourself, to have this clear-cut stand against the belief that any approximation to a single-operation, quasi-pushbutton, war is possible. Your detailed discussion of how diversified war and the tools and organizations required for its prosecution will still have to be, makes very instructive reading, and I only hope that the lesson will be learned by anyone whom it concerns. I also feel that your discussion of the non-technical aspects of war and political organization round out the whole very appropriately.

On the matters that concern the Harvard Panel, I can report this: I have not yet received a statement from Professor Aiken. It is perhaps best if I wait another four or five days, and then proceed to write you a summary of my impressions. For the present, however, I would like to add my comments to those which you made in your letter of October 31, concerning one particular phase of the Harvard affairs – the Computing Laboratory:

I need not tell you that I completely share your views that a computing laboratory is an important addition to any applied mathematics or applied physics group,

[46]Reference is most likely to V. Bush: *Modern Arms and Free Men: A Discussion of the Role of Science in Preserving Democracy*, Simon and Schuster, New York, 1949. The Editor.

and therefore, to any engineering school conceived on a high level. I am equally convinced that the description, "important addition", is on the way to becoming an understatement. I think that such a laboratory either is now, or will soon be, an absolutely integral part in an establishment in applied mathematics or physics. The existence of such an organization at Harvard seem to me, therefore, to be a valuable asset which ought to be developed.

Regarding the computer situation in the country at large, I find myself in agreement with your general statement: There is reason for concern. I would qualify somewhat the nature of this concern. I, too, feel that the number of computer projects which are now being prosecuted is somewhat higher than optimum. It seems to me, however, that the time at which it will be possible to make a final assessment of every one of them, favorably or unfavorably, is not very far. I would be surprised if one could not differentiate unambiguously the white sheep from the black by the latter part of 1950.

I think, too, that among the projects, examples of a variety of sins are easy to find. Prominent among these are excessive spending, lack of visualization of the problem as a whole, getting bogged down in disproportioned component gadgetry, etc. etc. The latter trait worries me in particular a great deal: It seems to me that very few people have a clear idea of how a computing machine functions as a whole, and how, in addition, the entire computing machine is merely one component of a greater whole, namely, of the unity formed by the computing machine, the mathematical problems that go with it, and the type of planning which is called for by both. There exists, it seems to me, in many projects a quite catastrophic misjudgement of the proper balance between these. It would be easy to name examples of machines which have arithmetical speeds out of proportion with their precisions, or with their memory access times, logical capacities out of proportion with their memory capacities, in which certain special operations are over-instrumented to the detriment of other equally important ones, etc. etc.

Another worrisome symptom is the lack of understanding that one frequently encounters with respect to the concept of what constitutes an "all purpose" machine – what amounts of information and of ignorance about the problems to be handled go into its make-up. That is, claims, that a "one-purpose" machine, or an "all purpose" machine with certain very special and very extreme characteristics is needed for a particular group of applications, are made much too easily on the basis of altogether inadequate analyses of the essentials of the situation.

The parallel problem, that once the machines are built, it will be difficult to find problems for them is a quite insidious one. I agree with your diagnosis to the extent that I feel convinced that many organizations are now asking for computing machines, although they will be completely unable to use them, and have only the most amorphous ideas as to how those might be useful for them. On the other hand, I am equally convinced that it would not be difficult to keep a large number of properly designed computing machines usefully occupied if qualified people went about it in the proper way.

What is greatly needed, it seems to me, is a general clarification of ideas and a great deal of education and training. By the way, I have always felt that the more or less informed public opinion, as well as the not very broadly informed experts, frequently hold completely unsound ideas as to where the difficulties of high-speed computing and the attendant planning do lie and do not lie.

A thorough analysis and discussion of this and various parallel subjects would, I think, be very useful. In fact, I think that this will be absolutely necessary if the country is to get a fair return for the effort which is now going into these machines.

Returning to Harvard affairs, I agree with you that we have to consider the situation there in the light of the national trends. I agree with you, that the committee at Harvard, which is to allocate machine-time, is excellent in its membership, and that it was a sound move to set up such a committee. My impression is, however, like yours, that it does not in fact function as a committee, but that Aiken rules his laboratory as an autocrat. This, is, of course, to a great extent due to the fact that only a few people have as yet developed the necessary habits and discipline of thinking which a computing machine demands. A more elastic character in Aiken's place, however, should have been able to remedy this condition. It appears to me that Garret Birkhoff and W. Leontief did get useful help from the computing laboratory. Yet others, like H. Emmons, could have profited even more. To be more specific on what seems to me a very relevant, special case: I am convinced that the theory of turbulence, for instance, is a very important field for machine computing and the presence of R. von Mises and H. Emmons at Harvard is a standing challenge to do something in this respect, yet nothing appears to have happened. It is also not quite clear to me to what extent the Navy and the Air Forces obtained at Harvard what they needed. Thus, a typical question in this respect is this: Has Z. Kopal's project at M.I.T. been mechanized to the extent that it might be?

I certainly feel that the central administration at Harvard should take an essential part in determining the policies, the personnel, and the management of the computing laboratory, so as to make it more catholic in its outlook and more flexibly available to the entire university. Aiken's views on many parts of the subject are very rigid, and a certain softening up in this respect would profit everybody.

To sum up, I think that the computer project should be supported. I think that your view, that the support should come primarily from the Government, and that Harvard funds should be used as stimulants and catalysts at a few critical points, is very sound. I think that the participation of the other interested departments could and should be made much more effective. This will certainly require a certain intervention on the part of the central administration. I think that some way should be found by which the computing laboratory is not made absolutely and wholly an expression of Aiken's views, which in certain respects seem to me very dogmatic.

Yet Aiken's presence is a great asset and his contribution is very important. I think that it would be a mistake to move Mark III away from Harvard. With a reasonable organization and a reasonable way of controlling the use of that machine, the Navy, Harvard, and everybody else should be able to get essentially more use from that machine, than if it were moved out of Aiken's institute.

I would greatly appreciate if you would let me know what your impression of any or all of these comments is.

As I mentioned earlier in my letter, I shall give you a more integrated description of my impressions regarding the entire engineering school. They will, of course, be shorter when I come to those parts in which I have had no firsthand experience.

I am,

Cordially yours,

John von Neumann

Original in Von Neumann Papers, Library of Congress, language of original: English

---

December 6, 1949

Dr. Vannevar Bush
Carnegie Institution in Washington
1530 P Street, Northwest
Washington, D.C.

Dear Dr. Bush,

I think that the time has come for me to give you a summary of my impressions in connection with our inquiry at Harvard. In doing this, it seems to me that it will be best if I follow the outline of questions given on the last page of Mr. Raymond's very excellent notes on our panel meeting of October 21, 1948, at Cambridge.

In referring to the 11 questions which were enumerated there, I shall, of course, answer in unequal detail and occasionally not at all – that is, by only setting forth the reasons for not answering at all or for not answering the question exactly as put. This seems to me a justifiable procedure because of the limitations of our mission, as well as of my experience. You have pointed out, and I completely concur with you, that we should not try to give specific administrative advice, only opinions on general principles. The questions, however, reflect the worries of our witnesses and these were frequently directed toward definite administrative action. In addition, my own experience is obviously centered around pure and applied mathematics and certain parts of more or less mathematical physics. I should, therefore, in my contributions emphasize these subjects and not the others. The list of all questions, nevertheless, offers an excellent framework for my observations and I shall, therefore, follow their pattern.

---

1. Write a Charter.

I don't think I should try here to make any suggestions in this respect. If it is at all proper for our panel to attempt to write a charter, it should result from your synthesis of all the opinions of the panel members on detailed questions, and in any case flow from your much more mature experience. What I can contribute to this question will follow from my remarks regarding the other questions.

2. Determine how Graduate courses should be chosen.

The proper answer to this question too can only result from a summary of all opinions given by all of us on every question of detail. In a general way I think the following principle should be observed:

I agree with the philosophy that the Harvard School of Engineering should not emphasize specialization, but should maintain a catholic outlook. There should, therefore, be a concentration on graduate courses, and these should be distributed to cover the major fields of engineering research as uniformly as feasible in view of the time and the faculty personnel available. The desire of the broad approach should certainly control the selections for faculty appointments.

3. Establish method of co-ordination (graduate and undergraduate) with University Departments of Physics and Mathematics.

I agree with the view that the Harvard School of Engineering should not and cannot be a professional school in the ordinary sense of the word. It should, therefore, be run on a higher scholastic level than what is practical for the main part of the educational effort in the traditional professional schools. If this principle is accepted, it should follow that no concessions are to be made in teaching the fundamental subjects: Mathematics, Physics, Chemistry. This would suggest to rely for the teaching of these subjects as much as possible on the respective departments themselves, that is, to forego the convenient expedients of having either of these subjects presented in simplified and mollified "versions for engineers". The addenda which a research engineer needs in these fields, over and above what the general departments teach, should be supplemented in subsequent courses of the School of Engineering.

4. Say how to handle: Electronics, Nuclear Physics, Metallurgy.

G.D.E. suggests the appointment of a Professor in Metallurgy. This appears to me justified and I do not see that more than this action is immediately needed.

A strong, although probably somewhat overspecialized, group in electronics exists at Harvard now. It seems to me that this group should be continued on its present level with a careful policy of replacement when retirements occur. For the appointments that effect this, a very effective mechanism exists at Harvard now. (Cf. the reply to question 8.) I do not think an expansion of the electronics group is called for now or in the immediate future.

It is very difficult to tell when and in what manner a specific development in Nuclear Physics should be undertaken. It would seem to me best to keep the point of view of Nuclear Physics in mind when the appointments in the relevant supporting fields are made. That is, Harvard should acquire, if possible, at least one faculty member in each of several contiguous fields, who has some personal experience with and interest in Nuclear Engineering. These contiguous fields should certainly include Applied Mathematics, General Physics, Thermodynamics, Anorganic Chemistry. The main problem of Nuclear Engineering should, it seems to me, only be considered after such a supporting group has been developed.

5. Make recommendations about relations with MIT and Harvard Business School.

The other members of the panel have so much more experience that is relevant to this question that I do not feel competent to make specific recommendations. Your own, as well as Mr. Rowe's, remarks on this subject strike me as very convincing.

6. Say something about fellowships.

In view of the national trend in these matters, and also in view of the fact that a scholastically very high level graduate school is called for, a generous fellowship policy seems indicated to me.

7. Establish what kind of students the school should attract and what should be done to attract them.

The answer is partly contained in the replies to the other questions. I, too, feel that the Harvard School of Engineering should train research engineers and academic teachers of engineering rather than engineers for industry at large. The type of student to be attracted should therefore be a research-minded and/or the unconventionally-minded one. A sound fellowship policy and a broadly conceived graduate instruction, together with the general reputation of Harvard, should suffice to attract the right students.

8. Recommend a mechanism for passing on proposed appointments and for getting the right men.

The present Harvard policy of "ad hoc committee" appears to me to be an excellent one. This procedure has its virtues in existing, well-established departments, but it is even more called for in a part of the university which is in a state of flux, and of reorganization, and in many relevant respects of formation. The ad hoc committees should contain appropriately selected members of the contiguous departments, and outsiders who are competent to judge the implementation of the policy outlined above.

9. Resolve the present conflict between GDE and ESAP.

With the present conflict between GDE and ESAP, it seems to me clearly desirable to replace the rather nominal combination which exists now by a real merger. From this a new department should emerge, that will be different from both that exist now, but more like ESAP than anything else. I realize that this involves grave questions of personalities. This will call for delays and possibly transient compromises. What delays and what compromises will have to be accepted in this respect is a matter for specific and detailed administrative savoir faire, on which we can hardly give any general and abstract advice.

10. Advise how best to bring together the University Mathematics Department, Computation Laboratory, Applications of Mathematics in other Departments, both inside and outside Engineering.

With respect to the Computing Laboratory, I have described my views in my letter to you on November 15. I would like to add the following remarks: As I mentioned in that letter, the Computing Laboratory seems to me a very important component of the program. It should be supported and strengthened. Most of its expenses could and should be met by government support. The basic needs (especially salaries of the key members of the staff, reasonably broadly construed) should be met by Harvard. It may even be wise to apply this to some important, but not excessively expensive, parts of the equipment. Harvard money may also be needed to "catalyze" other support.

The MARK III computer should, if possible, be kept at the Laboratory, with an appropriate arrangement for using a suitable fraction of its time for government purposes. Personnel and the outlook of the personnel should be broadened.

The Departments of Mathematics, Physics and Chemistry, and to a lesser extent several other departments, should be educated to make intelligent and increasingly independent use of the facilities of the Computing Laboratory. In the case of the Mathematics Department, this is likely to happen spontaneously. In the case of

Departments of Physics and Chemistry a good deal of help and cooperation, both from the Computing Laboratory and the Mathematics Department's side, will be called for. This will be even more true for other departments. This should be kept in mind in some of the appointments.

I greatly admire Aiken's work, but I do not concur with all of his views. Specifically, I think that the question of planning and coding for a machine is more a matter of education than for mechanization. Aiken has made an important contribution to this type of education, but it will have to be carried on on a much broader basis, with a more inclusive program in Applied Mathematics. In this respect collaboration with MIT may be indicated. I do, furthermore, disagree with the view that the problems of developing, designing, and constructing more advanced computing machines are already mature enough to turn them over to industry. Industry, and especially industrial research laboratories, will play an increasing role in this respect. But it seems to me that academic research, and even research and development, will nevertheless have to play a quite vital role in this field for a considerable time to come.

A well-integrated program in Applied Mathematics seems to me absolutely called for. I think that it would be desirable to concentrate it into one department. This might be a new Department of Applied Mathematics, or an extension of the existing Department of Mathematics. If the former solution is chosen, the two departments should be rather closely connected. In what way the Applied Mathematics Department or the applied mathematical part of the Mathematics Department, should straddle between the general Mathematics Department and the School of Engineering will be a matter for specific administrative decision. I am sure that a satisfactory solution for this problem can be found, and it seems to me that the actual mode of implementation need only be taken up after the main questions of principle have been settled.

The Computing Laboratory would probably be best off as a part of the Department of Applied Mathematics. The exact settlement will, of course, have to depend upon the basis of the relationship of the Pure Mathematics and the Applied Mathematics groups.

The teaching of statistics should be taken up in a well-integrated way in the Department of Applied Mathematics. A major faculty appointment in statistics would therefore seem to be justified.

11. Recommend how money should be allocated, as between Staff and new appointments, Buildings, Equipment, Fellowships.

Obviously, our panel can only contribute partially to answering this question, and I personally can contribute even much less.

Generally speaking, it seems to me that the amount available for buildings is adequately circumscribed by the consideration that only accumulated interest from the Mckay Fund can be used for buildings.

Fellowships, while very important, will presumably require considerably less funds than the other items.

Thus, the main problem appears to me to lie in the assignment of quotas to staff and new appointments vs. equipment. The only relevant observation I can make on these is that I would expect that a large part of the equipment can be obtained from other sources, primarily, from government contracts. This is certainly true in

Computing and Applied Mathematics. I expect that it is to a considerable extent true in the central fields of Engineering, too.

The above observations do not cover certain existing and particularly successful groups within the School of Engineering. The Soil Mechanics Group is an important instance of this. Nothing I have said above is, of course, intended to mean that the successful work of such groups should be interfered with. Existing values of this type can be conserved without seriously distorting the trend that is considered desirable in the central fields, and it would be a grave mistake not to conserve them.

I would appreciate learning from you to what extent my observations are to the point, and helpful to you. I shall be glad to amplify any part where you may consider this desirable. I am leaving Princeton on December 8 for about six weeks in the Midwest and Southwest, expecting to be back about February 1. My mail, however, will reach me with little delay via Princeton.

I am,

Cordially yours,

John von Neumann

Original in Von Neumann Papers, Library of Congress, language of original: English

---

Tulsa, Oklahoma

January 25, 1950

Dr. V. Bush
Carnegie Institution in Washington
1530 P Street, Northwest
Washington, D.C.

Dear Dr. Bush,

I have just received your telegram regarding the report of the Harvard Panel. I am sorry that I did not write to you earlier – I have been away from Princeton for some time (partly on vacation in Mexico, partly in Los Alamos, and since, a few days here) and this slowed down my actions.

I have read the draft of your report, which you are circulating for comments and proposals for modifications. I find myself in substantial agreement with what you say, and I would be very glad if the panel's report went essentially as this draft now stands.

I find, in particular, that your observations on the digital computer reflect my views too, with considerable precision. It might, perhaps, be worth while to add a few lines on training. You are pointing out, I think very rightly, that the computer organisation should not be run only as a service to the other departments, with those departments acting as naive users of ready-made packages, but that they should be gradually educated to think for themselves in using the computer. They should learn to collaborate in developing the computer's philosophy, the study of the desirability of its characteristics, and to contribute their intrinsic knowledge of their respective fields to the gradual definition of the milieu in which the computing machine has to live, with respect to which its "efficiency" has to be assessed, and which therefore, in fine, should set the criteria of "typical performance" on which the componentry-development is to orient itself. You point out that this will be neither easy nor efficient at first. There is a constant temptation to take the opposite course (that of straight service) – but this is nevertheless the direction in which the great rewards of the future lie. I would add to this some further remarks about the need of outright instruction in the computer field, in coding, planning, assessing problems, etc. – especially in view of the overall, national situation. Aiken is, of course, aware of this latter circumstance, and he is effecting a program of computer-instruction, but the need of expanding it further should be kept in mind. In discussing the Applied Mathematics group you point out that it will have its natural connections with the Computer group, and may well be its parent organisation. From the point of view of instruction, it is the Applied Mathematics group that could round out the computer instruction in the sense indicated above.

I agree wholeheartedly with your observations on Applied Mathematics. I do not see that anything more should be added to them – I think that you have succeeded to point out with remarkable brevity and lucidity the essentials of the situation.

I am also in agreement with your final remarks on the Administration, and equally with your Conclusion.

I am,

Cordially yours,

John von Neumann

Original in Von Neumann Papers, Library of Congress, language of original: English

Letter to R. Carnap

Berlin W. 10,
Hohenzollernstrasse 23

June 7, 1931

Dear Mr. Carnap,

Many thanks for your letter and your suggestions.

In my contribution[47] I have referred to the detailed bibliography in Weyl's handbook article without giving any further literature. I have done this mainly because I did not want to give any value judgements by means of my choice. There are some programmatic publications by Hilbert in which he claims that certain things have been proven or almost proven while this is in fact not even approximately the case (continuum problem and so on). Therefore I would like neither to quote these nor to correct or ignore them, and I therefore believe that the best course of action is to include a general reference to the handbook article.

Concerning your suggestion on mailing the off-prints together I would like to say the following:

I have known from Gödel (whom you also know?) about his results in Königsberg and later in correspondence. These theorems (which have since then appeared) show that certain logical and mathematical statements, such as for instance the consistency of analysis, are unprovable in certain formal-logical systems. In my opinion, however, they also show that it is also impossible contentually, for I am of the opinion that all contentual inferences are reproducible in *one* specific formal system (which I do not want to describe more precisely but for which Gödel's theorems hold). Thus I am today of the opinion that

(1) Gödel has shown the unrealizability of Hilbert's program.[48]
(2) There is no more reason to reject intuitionism (if one disregards the aesthetic issue, which in practice will also for me be the decisive factor).

Therefore I consider the state of the foundational discussion in Königsberg to be outdated, for Gödel's fundamental discoveries have brought the question to a completely different level. (I know that Gödel is much more careful in the evaluation of his results, but in my opinion on *this* point he does not see the connections correctly). I have several times discussed with Reichenbach whether it makes any

[47]Here and below von Neumann refers to the lecture he delivered during the famous conference on foundations of mathematics in Königsberg September 5-7, 1930; see Section 2 of the Introductory Comments. The Editor.

[48]Von Neumann's footnote: I would like to emphasize: *nothing* in Hilbert's aims is *false*. Could they be carried out then it would follow from them absolutely what he claims. But they cannot be carried out, this I know only since September 1930.

sense, under these circumstances, to publish my lecture. Had I had to give the lecture four weeks later it would have sounded fundamentally different. We agreed finally that it should be written down as the description of a certain outdated state of things.

As you will be able to gather from these lines I would not otherwise send out a single off-print of my lecture. I do not want, however, to interfere with a joint action and I agree to your proposal and to the list you suggested.

With best regards,

John von Neumann

Original in Rudolf Carnap Nachlass: RC 029-08-01, Archives of Scientific Philosophy, Special Collections, Hillman Library, University of Pittsburgh, Pittsburgh, U.S.A. Language of original: German. Quoted by permission of the University of Pittsburgh. *Both the original and an English translation (by P. Mancosu) is published in P. Mancosu: "Between Vienna and Berlin: The immediate reception of Gödel's incompleteness theorems",* History and Philosophy of Logic **20** (1999) 33-45. *The present English version is identical (except for minor changes), with the translation by P. Mancosu.*

Letter to W. Cattell

April 8, 1940

Dr. Ware Cattell, Managing Editor
The Scientific Monthly
Smithsonian Institution Building
Washington, D.C.

Dear Dr. Cattell:

Many thanks for your kind letter of March 28 and its enclosures (the article by Gamow and the criticisms by Professor Moulton). It is very kind of you to encourage me to give my opinion about this matter.

I am no expert on non-technical exposition, but Professor Gamow's article strikes me as a precise, if brief, description of quantum mechanics. In any case it is scientifically absolutely sound: that is, it gives merely a concise and untechnical exposition of matters which are considered by the great majority of physicists interested in this subject, as non-controversial. Professor Moulton's criticisms are directed not so much against this article as against the present system of quantum mechanics, or at least against the generally accepted interpretation of the same. I don't think that an expository article of this type is the appropriate background for such a discussion, if such a discussion is at all desired.

It is of course, up to the editors of the Scientific Monthly to decide whether they consider quantum mechanics as a sufficiently well established part of theoretical physics to be presented to the general scientific public. I am sure, however, that there is as much agreement in the positive sense concerning the subject among theoretical physicists as there ever can be about a modern theory.

Before closing I should like to mention that the "uncertainties" discussed by Professor Gamow are generally accepted to be of an entirely different type from those pointed out by Professor Moulton. The former are believed to be absolute limitations (if the present theory is accepted) while the latter are only due to the imperfections of our senses, more prefect ones than which are at least conceivable. The situation is comparable to the absolute maximum velocity in the special theory of relativity (light velocity) compared to the much lower practical limits of velocity of macroscopic objects. I cannot agree with Professor Moulton's rejection of "conceptual experiments". They have played an important and generally accepted role, e.g. in classical thermodynamics, and Bohr and Heisenberg have merely applied this procedure to atomic physics.

For all these reasons I think that Professor Gamow's article would only lose value as a picture of our present views of atomic physics, by modifications or qualifications.

Sincerely yours,

John von Neumann

P.S. I am returning enclosed the material you kindly sent me.

Original in Von Neumann Papers, Library of Congress, language of original: English

Letter to T.M. Cherry

November 27, 1950

Professsor T.M. Cherry
Athenaeum
551 South Hill Avenue
Pasadena 5, California

Dear Professor Cherry:

The mathematical phenomena, to which the remark that you mention refers, are primarily the following ones:

α. In the theory of harmonic functions, it is well known that the highest degree of regularity obtains in two dimensions, while in three dimensions (and a fortiori in more than three dimensions) potential functions can be a good deal more singular than in two. The immediate technical reason for this is the change in the character of the singularity of Green's function of this problem. The inner reason is, that volume of the sphere decreases (as the radius tends to zero) the faster the higher dimension, and, therefore, an isolated singularity is somehow less "noticeable" in its effects in higher dimension, and, accordingly, less likely to be excluded by the general specifications of the problem.

β. If I am not mistaken, biharmonic functions behave similarly. Here a higher degree of regularity holds up to some dimension higher than two (I forget which), and then this problem, too, begins to allow pathological possibilities for higher dimensions, like the potential equation does from three dimensions up.

γ. The same is true for the n-th power of the Laplacian operator (i.e. n-harmonic functions), the critical dimension in the above sense being an increasing function of n.

δ. It is well known that the theory of surface (i.e. 2-dimensional area in the 3-dimensional space) is more difficult than the theory of length (1-dimensional area in 2-dimensional space). Thus, the isoperimetric problem has more pathological possibilities, and, generally speaking, worse complexities in the first-mentioned case than in the second one – i.e. everything is worse in the higher dimensional case.

ϵ. The existence and uniqueness of the solutions of the Navier-Stokes equations (viscous, non-conductive, incomprehensible fluid), has been investigated by J. Leray in his thesis (Journal de Math. Pures et Appliquees, 1933-34). Leray found, essentially, that existence and uniqueness could be established in two dimensions, but in three dimensions he ran into great difficulties. If, in defining solutions, he insisted on as much regularity as is usually taken for granted in physics (existence of all necessary derivatives, etc.), he could prove uniqueness but not existence. If, on the other hand, he relaxed the regularity requirements in a manner which is plausible by the standards of modern real-variable function theory (existence of the highest order necessary derivatives in the "almost always" or in the "en measure convergence" sense only), he could prove the existence, but not the uniqueness. As far as is known at this moment, it is by no means excluded that existence and uniqueness in the three dimensional case cannot be had together under any dispensation, and it has even been suggested that these phenomena (if real) may have something to do with the existence in Turbulence in three (but not in two) dimensions.

I do not plan to publish my 1949 report on Turbulence[49] until I have very thoroughly revised it. A number of the references to the literature are incomplete, and some of the interpretations are not in the form in which I would like to state them after more study and reflection.

Please let me know, if the above contain, reasonably completely, what you are interested in.

I am,

Yours very sincerely,

John von Neumann

Original in Von Neumann Papers, Library of Congress, language of original: English

[49] Von Neumann: "Recent theories of turbulence. (Report made to Office of Naval Research)" Collected Works Vol. VI. No. 32. The Editor.

Letter to H. Cirker

October 3, 1949

Mr. H. Cirker
President
Dover Publications
1780 Broadway
New York 19, New York

Dear Mr. Cirker,

Please excuse my delay in writing to you as I promised in my letter of September 3rd, replying to yours of August 23rd. The operation of settling down in Princeton for a normal fall and winter, and to dispose of the work accumulated during my long absence during the summer, proved more time-consuming than I had expected.

In your letter of August 23rd, you informed me that the Alien Property Custodian has asked you to submit an application for the publication of the translation and revision of QUANTENMECHANIK,[50] and that in order to obtain a 2% royalty arrangement they request "a complete and detailed description of the scope, general plan, manner of presentation of material, the material itself, etc. in the new manuscript as contrasted with the scope, general plan, etc., of the original work." You suggested accordingly that I send this material.

In trying to assess what it is that the Alien Property Custodian wants, I am comparing your description of the probably desirable material, as you gave it in your detailed letter to me of April 1, and the statements that Mr. E.P. Tilleux of the Alien Property Custodian's office made to me when I saw him in Washington on May 25, and that I described in my letter to you of May 30. As you will see from that letter, Mr. Tilleux was quite encouraging, and he stated that the fact that about 60% of the new book would be precise translation (Mr. Beyer's contribution 340+50=390 *double-spaced* typed pages) and about 40% reformulation and revision (my contribution, 126 *single-spaced* typed pages), was quite sufficient. (The arithmetic is as follows: 390 double-spaced pages are equivalent to 195 single-spaced pages. 195 [Beyer]+126 [myself]=321. Hence of the equivalent of a total of 321 pages the equivalent of 195 is old – 60.7% – and 126 is new – 39.3%.) He stated that even an 80% old vs. 20% new ratio would justify the 2% royalty arrangement.

As a general comment to this I would add: The general plan and layout of the new work, as reflected in the table of contents, is the same as that of the old. The text had to be extensively rewritten, because a literal translation from German to

[50] J. von Neumann: *Mathematishe Grundlagen der Quantenmechanik*, Springer, Berlin, 1932. See Section 5 of the Introductory Comments. The Editor.

English is entirely out of question in the field of this book. The subject-matter is partly physical-mathematical, partly, however, a very involved conceptual critique of the logical foundations of various disciplines (theory of probability, thermodynamics, classical mechanics, classical statistical mechanics, quantum mechanics). This philosophical-epistemological discussion has to be continuously tied in and quite critically synchronised with the parallel mathematical-physical discussion. It is, by the way, one of the essential justifications of the book, which gives it a content not covered in other treatises, written by physicists or by mathematicians, on quantum mechanics. This peculiar scope requires a very specific and sensitive use of the language, and a true translation is therefore out of the question. The fact that I practically had to rewrite Dr. Beyer's translation confirms this. The process of writing the new book was really a two-stage one: First Dr. Beyer provided the basis, the framework, by writing a translation in the ordinary sense, then I had to do the rewriting, which increased the work in the approximate ration of $\frac{126}{195}$ (cf. above), that is by about 64.6%. Note that the net work of rewriting took me from January 25, 1949 to May 20, 1949, that is, 4 months net, and with the preparations that went into it, etc., about 6 months. The writing of the original book, with the greater difficulties of the original work, took about 1 year in 1930/31. This, too, confirms the above assessment of the importance of the new work.

A copy of my additions and modifications to Dr. Beyer's translation can be furnished to the Alien Property Custodian, if he so desires.

In addition to this, the new book will contain, as we discussed, an index and a bibliography, which were not parts of the old one.

Do you think, that these statements, properly formulated for submission to the Alien Property Custodian, will be sufficient? Or shall we go into the matter further and consider the whole approach further? Mr. Tilleux's remarks indicated that the above is what he wants. Is it safe to go ahead on this basis, or may we spoil something? Might you perhaps get Mr. Tilleux's views on this subject again? I am,

Sincerely yours,

John von Neumann

Original in Von Neumann Papers, Library of Congress, language of original: English

Letter to F.W. Crocker

September 1, 1955

Mrs. Frank W. Crocker
345 Mahatan Street
Westwood, Massachusetts

Dear Mrs. Crocker:

This is in reply to your letter of August 17, 1955, in which you request that I join in the World Affairs Council of Boston program, "Diplomats Off the Record", to make some remarks on "atomic energy as it relates to the future treaty making and peace".

It is most kind of you to consider me for your fine program, but I find that I must, for this occasion, beg off for two reasons. Of course, in the Atomic Energy Commission, we have learned never actually to "go off the record" in public statements, because our statements must either be fully on the record or made on a classified basis for audiences of security cleared persons only. Therefore, even the appearance of an "off-the-record" position must be avoided to prevent any possible public impression that we are not properly careful of classification and security. That is one reason I feel I could not accept an opportunity to speak on a program having a title such as yours has.

My second reason is that the only remarks I could give you on the topic would be personal and in no sense official, since my official responsibilities do not extend into treaty making or similar diplomatic specialties and your audience, expecting more, would doubtless be disappointed.

May I suggest that your program would be better served if the proposed discussion was by a representative of the State Department rather than by me.

I hasten to add that my reluctance to meet with your Council is caused only by the two factors noted above, and that I would be happy, with a more suitable format and more familiar topic, sometime to talk to the organization.

I appreciate having been asked to meet with the Council and, as well, your kind offer of the hospitality of your home.

Sincerely yours,

John von Neumann

Original in Von Neumann Papers, Library of Congress, language of original: English

Letter to M.R. Davie

May 3, 1946

Mr. Maurice R. Davie
133 Hall of Graduate Studies
Yale University
New Haven, Conn.

Dear Mr. Davie:

Please excuse my delay in answering your letter of April 24th. I was away from Princeton and returned this morning.

I first came to the United States in January 1930 and, since conditions in Europe at that time were, in the main, normal, I would not consider myself a refugee scientist.

Sincerely yours,

John von Neumann

Original in Von Neumann Papers, Library of Congress, language of original: English

Letter to W.E. Deming

January 23, 1945

Dr. W. Edwards Deming
Adviser in Sampling
Bureau of the Budget
Executive Office of the President
Washington, D.C.

Dear Ed:

Many thanks for your two letters of January 16. I will certainly be glad to cooperate with the work of the group on statistics. I also think that we ought to have Hotelling. Probably the best procedure will be if I wait until we have accumulated some more names, and then circularize those interested and write invitations. What you say about the subject in general is very gratifying.

I read your review of Morgenstern's and my book[51] with interest and much gratification. Since you asked for comments, I should like to mention a few minor points:

Page 1, line 7 from the bottom: It is not necessarily true for the solution in a 2-person game by mixed strategies, that each player should have, if he plays right, a 50-50 probability of winning – or a zero expectation value. This is the special case of a "fair" game, and in general only when the game is "symmetric". In general our theory gives for a 2-person game a value v which can be positive or negative as well as zero.

Page 1, last 3 lines: It is certainly true that we have not found any method so far to determine all solutions of an n-person game. On the other hand, I would perhaps make a somewhat stronger statement than that we have a few general principles only, since we have formulated general necessary and sufficient conditions for solutions. Also, all solutions for 3-person games as well as for some other classes of games are determined.

Page 4, last 3 lines: I realize that you make it clear at this point that you are giving an interpretation which may not agree with our views. I also agree with you that the role which coalitions play in our theory makes one inclined to draw conclusions of the type you indicate. On the other hand, we were very careful to assert nothing specific about such political and controversial subject as trusts and cartels, with as few real and general

[51] O. Morgenstern, J. von Neumann: *Theory of Games and Economic Behavior*, Princeton University Press, 1945; see Section 9 of the Introductory Comments. The Editor.

> results as we have now. Of course I realize that it is for you to judge what conclusions you wish to draw, as the reviewer.

I would like to add that we hope to go into this subject in detail later, and get mathematical results about monopoly, duopoly, cartels, etc., but this is only a program so far.

With best regards, I am,

Yours sincerely,

John von Neumann

Original in Von Neumann Papers, Library of Congress, language of original: English

Letter to J.L. Destouches

December 18, 1951

Professor J.L. Destouches
Professeur a la Sorbonne
4 Rue Thenard
Paris 5^0, FRANCE

My dear Colleague:

I have just received the typescript of your paper, "Sur l'interpretation ...", and of Mrs. Destouches-Fevrier's paper "Sur le charactere ...", as well as a copy of Mrs. Destouches-Février's book,[52] "La structure ...". I want to thank you most warmly for this attention, and also express my thanks and my respects to Mrs. Destouches-Février.

I need not tell you that the subject interests me greatly, and I have read both papers and I am now beginning to read the book. I have the impression, that I should read the book carefully before I make any really relevant and substantive comments on your approach – it expresses a definite philosophical system and all of it hangs together. Any observations that I can now make are therefore provisional and very superficial – and for these circumstances I am most apologetic.

Much in your approach is conceived in a sense with which I find myself in agreement, but I need enlightenment on some points. The main point of this nature is the following one: The "corpuscular system" S, that you consider, is clearly a quantum mechanical one. Adding an observing apparatus A, or rather its representant system W_A to S, the composite system $S + W_A$ obtains. Do you view this as a non-quantum-mechanical, i.e. a classical system? I always assumed that $S + W_A$, too, had to be viewed as quantum-mechanical, since it, too, could be observed – say with some further apparatus B, with the representant system W_B, and yielding the doubly composite system $S + W_A + W_B$. I assumed that the relevant thing was to show, that the observer's classical way of talking about the actually quantum mechanical fate of S under observation was compatible in the framework of $S + W_A$ and of $S + W_A + W_B$, providing that B was observing the observation of A. I also considered chains

$$S + \ldots + W_X + W_Y + W_Z + W_A + W_B$$

in which the dichotomy between "observed" and "observer" could be placed anywhere: at Y/Z, or at Z/A, or at A/B, or anywhere before or after – all these approaches being equivalent. Do you propose a classical view of these things?

[52]Destouches, Paulette Février: *La structure des théories physiques*, Paris, Presses Universitaires de France, 1951. The Editor.

That the classical circumlocutions of $S+W_A$ and of $S+W_B$, for non-simultaneously measurable A and B, are irreconcilable, is clear, but this need not preclude viewing $S+W_A$ (as well as $S+W_B$) as quantum-mechanical systems. Is this view acceptable to you?

I hope we can discuss these things further at your convenience. The only really satisfactory procedure would, of course, be an oral discussion – the nuances that are involved are sometimes too delicate for an adequate rendering in writing. I hope that we will have an occasion to talk about these matters in a not too distant future.

In the main I remain, with the expression of my best personal regards,

Yours very truly,

John von Neumann

Original in Von Neumann Papers, Library of Congress, language of original: English

Letter to P.A.M. Dirac

The Institute for Advanced Study
School of Mathematics
Fine Hall Princeton, New Jersey

Jan. 27, 1934

Dear Dirac,

Many thanks for your kind letter. I would be delighted to spend a year in Cambridge, although it is too bad of course, to miss you on both sides of the Ocean.

But my absence for a year here would be only practicable with considerable difficulties. As the activities of the Institute have just begun 1933, a certain continuity in courses delivered (if any are delivered) etc. is highly desirable – and at any rate desired by the administration. But I could perfectly well get away for the later part of the year, so that I could come to Cambridge for the spring term. Would this, or a similar, arrangement be agreeable to you?

Hoping to hear from you soon, I am with the best wishes,

Yours truly

John von Neumann

P.S. I have lately thought of a quantum mechanical problem which connects indirectly with your theory of the spinning electron in De Sitter's space and which throws some light on its solutions, too. It is this:[53]

1. It should be perhaps desirable, to have operators X, Y, Z which have discrete (point) spectra, in order to avoid the difficulties connected with the point electron (in electrodynamics). The obvious argument against "quantized" X, Y, Z is, that the rotational symmetry of space would necessitate the same behavior for every $\alpha X + \beta Y + \gamma Z$ (α, β, γ real numbers, $\alpha^2 + \beta^2 + \gamma^2 = 1$), which is impossible if X, Y, Z commute.

[53]See Sections 5.3 and 5.4 of the Introductory Comments. The Editor.

2. For non commuting X, Y, Z however this is possible: f.i. [for instance] the spin components, or the operators

$$\begin{array}{ccc} i(x\frac{\partial}{\partial y} & - & y\frac{\partial}{\partial x}) \\ i(y\frac{\partial}{\partial z} & - & z\frac{\partial}{\partial y}) \\ i(z\frac{\partial}{\partial x} & - & x\frac{\partial}{\partial z}) \end{array}$$

behave like this. Still these particular operators have not the other properties required for coordinates, and their relativistic invariance cannot be secured.

3. The relativistic invariance would amount to this: find 4 operators X, Y, Z, T, such that every linear aggregate $\alpha X + \beta Y + \gamma Z + \delta T$ ($\alpha, \beta, \gamma, \delta$ real numbers, $\alpha^2 + \beta^2 + \gamma^2 - \delta^2 = 1$) has the same discrete point spectrum, whereas T (and every linear aggregate $\alpha X + \beta Y + \gamma Z + \delta T$, $\alpha^2 + \beta^2 + \gamma^2 - \delta^2 = -1$) may or may not behave this way. (I would prefer a continuous spectrum for "time".) $X = i(y\frac{\partial}{\partial z} - z\frac{\partial}{\partial y}), \ldots, \ldots$ seem not to allow such an extension.

3.[54] In your theory of the relativistic spinning electron in De Sitters' space you have a homogeneous wave function $\psi(x_1, x_2, x_3, x_4, x_5)$, and your differential equation; that is puting

$$E = \sum_{n=1}^{5} x_n \frac{\partial}{\partial x_n}$$

$$D = \frac{1}{2}\sum_{n,m=1}^{5} \gamma_m\gamma_n(x_m\frac{\partial}{\partial x_n} - x_n\frac{\partial}{\partial x_m})$$

$\gamma_1, \ldots, \gamma_5$ matrices of degree 4, $\gamma_m\gamma_n + \gamma_n\gamma_m = 2\delta_{mn}1$, you use these equations:

$$E\psi = 0, \qquad D\psi = c\psi \tag{1}$$

(c a complex constant). You further consider the operators

$$P_{mn} = x_m\frac{\partial}{\partial x_n} - x_n\frac{\partial}{\partial x_m} + \frac{1}{2}\gamma_m\gamma_n \qquad (m \neq n)$$

which commute with E, D. x_1, x_2, x_3, x_4 and $\frac{x_5}{i}$ are real, $\sum_{n=1}^{5} x_n^2 > 0$.

You show that all P_{nm} with $m, n \neq 4$ have discrete spectra, and

$$[P_{mn}, P_{pq}] = 0 \text{ for } m, n \neq p, q \tag{2}$$

$$[P_{mn}, P_{nq}] = P_{mq} \tag{3}$$

4. This shows, that

$$X = iP_{14}, \quad Y = iP_{24}, \quad Z = iP_{34}, \quad T = P_{45} \tag{4}$$

behave just as needed, because the commutation relations (2)-(3) imply, that every $\alpha X + \beta Y + \gamma Z + \delta T$ with $\alpha^2 + \beta^2 + \gamma^2 - \delta^2 = 1$ or -1 can be obtained by a transformation from X or T resp. So by calling your "impulses" "coordinates" and your "energy" "time", one obtains 4 operators, which show at least formally the desired behavior.

5. The spectrum of T, that is of P_{45} is of course of importance. One might expect that it omits a neighborhood of 0 (as the "energy" should always $> M$ or

[54] The numbering is not consequent in the letter: this should be 4. The Editor.

$< -M$, $M=$ rest mass) – which of course is an undesirable behavior for "time". Now *it is not so.* To see this, I used a method of solving (1) which runs like this:

6. I put

$$x_1 = \rho \sin\alpha \cos\beta \cos\gamma \tag{5}$$
$$x_2 = \rho \sin\alpha \cos\beta \sin\gamma \tag{6}$$
$$x_3 = \rho \sin\alpha \sin\beta \tag{7}$$
$$x_4 = \rho \cos\alpha \cosh t \tag{8}$$
$$\frac{x_5}{i} = \rho \sin\alpha \sinh t \tag{9}$$

(ρ is merely a homogeneity-factor, as the normalization $\sum_{n=1}^{5} x_n^2 = 1$ is not used. α stands for something like the distance from the center of the "static part" of the Sitters' space under consideration; β, γ are polar coordinates; t is "static time". This of course covers only the "static part" $x_4 > |\frac{x_5}{i}|$, but one can show easily, that every solution which is found in it, can be extended over the whole de Sitter space.) Now by quantizing simultaneously the (commuting) operators P_{12}, $P_{12}^2 + P_{13}^2 + P_{23}^2$, and P_{45} – that is by the usual method of "separation of nodes", one finds:

α. That in order that the problem be Hermitean, c must have a real part (as you and Wigner found in 1931), namely $c = 2 + iM$ (M real).

β. That the rotational quantum numbers m, k

$$k = \pm 1, \pm 2, \ldots,$$
$$m = -k + \frac{1}{2}, -k + \frac{3}{2}, \ldots, k - \frac{1}{2}$$

have to be introduced in the usual way, and that the wave function depends on the polar angular coordinates β, γ in the usual "spin-spheric-polynomial"-manner.

γ. That after all this has been done, the radial proper-value problem is

$$(\rho_1 \frac{\partial}{\partial r} + \rho_2 \frac{M}{\cosh r} + \rho_3 \frac{k}{\sinh r})\phi = \mathcal{E}\phi \tag{10}$$

Herein $r = \ln\tan\left(\frac{\pi}{4} + \frac{\alpha}{2}\right)$

$$\psi = \Delta e^{-\frac{1}{2}\gamma_1\gamma_4\alpha} e^{-\frac{1}{2}\gamma_1\gamma_3\beta} e^{-\frac{1}{2}\gamma_1\gamma_2\gamma} e^{\frac{i}{2}\gamma_4\gamma_5 t} e^{i\mathcal{E}t}\phi$$

Δ is the normalization factor made necessary by the change of volume element, $\rho_1\rho_2\rho_3$ are three γ_m-products with $\rho_m\rho_n + \rho_n\rho_m = 2\delta_{mn}1$, and $\mathcal{E}$ is the energy parameter (proper value of P_{45}).

7. Your special relativistic spinning electron equation gives for

$$x = r\cos\beta\cos\gamma$$
$$y = r\cos\beta\sin\gamma$$
$$z = r\sin\beta$$
$$\psi = \Delta e^{-\frac{1}{2}\gamma_1\gamma_3\beta} e^{-\frac{1}{2}\gamma_1\gamma_2\gamma} e^{i\mathcal{E}t}\phi$$

similarly (light velocity $=c=1$, rest mass $= M$)

$$(\rho_1 \frac{\partial}{\partial r} + \rho_2 M + \rho_3 \frac{k}{r})\phi = \mathcal{E}\phi \tag{11}$$

Thus (10) has (11) as limiting case (if r is small, $\cosh r \sim 1$, $\sinh r \sim r$), but it differs from it essentially, f.[or]i.[nstance] because M is replaced by $\frac{M}{\cosh r}$, which $\to 0$ for $r \to \infty$. From this last circumstance it follows easily, that the continuous spectrum of (10) extends *over all* $\mathcal{E} \geq 0$, $\mathcal{E} < 0$! It is not difficult to see, what this means physically: The wave functions belonging to $\mathcal{E}$'s with $|\mathcal{E}| < M$, are appreciably $\neq 0$ only in far away regions of space (where $|\mathcal{E}| \geq \frac{M}{\cosh r}$), that is near the "apparent mass horizont", where highly abnormal "apparent" gravitational conditions (great "red shifts", etc.) seem to make such miracles possible!

8. Your equation (1) can be somewhat generalized, by introducing

$$\begin{aligned} R &= \sqrt{\sum_{n=1}^{5} x_n^2} \\ X &= \sum_{n=1}^{5} \gamma_n x_n \\ P &= \sum_{n=1}^{5} \gamma_n \frac{\partial}{\partial x_n} \end{aligned}$$

R, X, P too commute with all P_{mn}'s, and formulating

$$E\psi = 0 \qquad (\mathcal{A}D + \mathcal{B}\frac{X}{R} + \mathcal{C}RP + \mathcal{D})\psi = 0 \tag{12}$$

($\mathcal{A}, B, C, D$ numbers.) All these equations can be transformed, by using

$$\mathcal{D} = XP - E, \qquad X^2 = R^2$$

into

$$E\psi = 0, \qquad RP\psi = (\mathcal{A}_1 \frac{X}{R} + \mathcal{B}_1)\psi \tag{13}$$

($\mathcal{A}_1, \mathcal{B}_1$ numbers), but the result (10), etc. remains essentially the same!

What is your opinion about all this, and particularly about the existence of solutions of (10) (that is of (1), or (12), of (13) with $|\mathcal{E}| < M$)? The X, Y, Z, T operators in (4) (in 4.) are a little strange for coordinate operators, but I still can see no reason which would preclude their use for this purpose.

With the best wishes

Yours truly

John von Neumann

Original in Von Neumann Papers, Library of Congress, language of original: English

Letters to J. Dixmier

November 24, 1953

Professor J. Dixmier
Universite de Dijon
Faculte des Sciences Mathematiques
Dijon, FRANCE

My Dear Colleague,

I am exceedingly apologetic about my performance in connection with your paper, "Sous-anneaux abeliens maximaux dans les facteures de type fini". The paper, together with your letter dated May 25, 1953, reached me properly in the course of the summer. However, I was away from Princeton, travelling for several months, and so I did not have my old papers and notes around to answer it then and there. This was my demise, because in the course of my subsequent travels and distractions, it got mixed up with other papers and I forgot to act on it later on. Your second letter of October 31, 1953, luckily reminded me of this, and I am only sorry that you did not write me earlier. I hope that you will not be too angry with me for being so disorganized and inconsiderate.

Coming to the subject matter of your letter, let me say first of all, that I found your paper very interesting and I would be very glad to help you to get it published here in America – I infer from your letter that this would be your first choice. I am, accordingly, submitting it for publication to the *Annals of Mathematics.* Please let me know, if, for any reason, this does not conform with your wishes, in which case I can easily retract it and discuss with you what course you consider preferable.

Regarding your questions, I would say this.[55]

(1) You are, of course, right. My statement in RO IV[56], page 787, lines 12, 11 from below[57], is incorrect as it stands. That is, I should have included in the definition of the examples $(\beta), (\gamma)$ in RO I,[58] page 208, lines 5-8 from above, the observation that the groups mentioned there may be taken either as groups of real numbers or as groups of real numbers (modulo 1). This fits, of course, with the Lemma 13.2.1. in RO I, page 206.

[55]See Section 3 of the Introductory Comments for von Neumann's work on operator algebra theory. The Editor.

[56]F.J. Murray, J. von Neumann: "On rings of operators, IV." Collected Works Vol. III. No. 5. The Editor.

[57]Here (and below) the page numbers von Neumann gives refer to the page numbers of the journals in which the papers were originally published. The Editor.

[58]F.J. Murray, J. von Neumann: "On rings of operators, I." Collected Works Vol. III. No. 2. The Editor.

The reference loc.cit. in RO IV, page 787, should then be limited to the (modulo 1) cases.
The proof of Lemma 5.2.3 in RO IV, page 787, i.e. the proof, that for an abelian group $\mathcal{G}$ the ring M is approximately finite, is, indeed, somewhat lengthy. I have only sketchy notes on it, but I can reconstruct it in detail. May I write you concerning it a little later on?

(2) This, too, is a subject on which I would like to think a little longer, since I have to recall and combine elements on which I have not worked for a long time.

(3) The program to which this refers can be carried out on three consecutive levels, each one in a certain sense is deeper than the preceding one, namely:[59]
First, one can proceed in terms of an abstract algebra. I did this in some notes which I have here, but which I did not publish. I start out with an abstract algebra in which an involutory antiautomorphism A^* and a scalar function with the formal-algebraic properties of the trace $Tr(A)$ are postulated. One can manipulate very much in the style of the later work of I. Gelfand and N. Neumark (*Mat. Sbornik*, 1943), and then construct a Hilbert space whose elements are the elements of this algebra themselves, using the trace to define an inner product in this fashion $(A, B) = Tr(AB^*)$. One then completes this Hilbert space in the metric that one now possesses, reinterprets the elements of the original algebra as operators in this Hilbert space by left multiplication, demonstrates that they are "closed", etc. After this, the whole theory can be tied to that one of RO I. The procedure from here on has a good deal in common with what was done subsequently by I. Kaplansky (*Ann. Math.* v. 56, p. 460, 1952).

Second, one can proceed on a more limited algebraical basis, which analyzes the Hermitean operators only. In this case, one begins with a nonassociative algebra in a manner which was originally suggested by P. Jordan. (Cf. also my paper with P. Jordan, "On Inner Products in Linear, Metric Spaces, "*Ann. Math.*, v. 36, pp. 719-723, 1935.)[60] This approach is worked out in considerable detail in my paper, "On an Algebraic Generalization of the Quantum Mechanical Formalism (Part I)", *Mat. Sbornik*, v. 1. (43), pp. 415-484, 1936.[61] This goes to a point where the main algebraical and topological desiderata are established. I did not publish Part II., but I will be glad to give details on it, if you are interested in this approach. I would suggest that you first look at the published *Mat. Sbornik* paper, since it established the background and the terminology. If you wish, I will be glad to send you a reprint of this paper.

Third, one can proceed on the basis of the algebraical entities that are equivalent to the projection operators, i.e. lattice theory. This lead to my development of "Continuous Geometries". My reason for being so lax about the publication of the work along the two earlier mentioned

[59]See the end of Section 3 of the Introductory Comments. The Editor.

[60]Collected Works Vol. IV. No. 4. The Editor.

[61]J. von Neumann, Collected Works Vol. III. No. 9. The Editor.

approaches was, that I considered this last method as the really satisfactory one, and having worked it out, lost a good deal of my interest in the others.

Needless to say, I will be very glad to receive the reprints from you to which you refer in your letter. To my regret, they have not arrived so far.

I would not like to conclude this letter without apologizing once more for my lack of attention. I would like to assure you, that the subject of your paper and of your letter interests me greatly, and that I would be very happy if we could again discuss it with each other orally and at leisure. I expect to be in Europe this fall, and I hope that we will have an occasion to meet, e.g. at the Congress in Amsterdam.[62] Would you some time consider a visit to this country?

With best personal regards, I am,

Sincerely yours,

John von Neumann

P.S. Simultaneously with this, I am also sending you a reprint of RO II,[63] as requested. (I am sending it to your Dijon address.) JvN

cc: 18 Rue le Brun
Paris (13°), France

Original in Von Neumann Papers, Library of Congress, language of original: English

---

June 18, 1954

Professor J. Dixmier
Universite de Dijon
Faculte des Sciences Mathematiques
Dijon, FRANCE

My dear Colleague,

Thank you for your very good letter of May 29, and for the return of my Mat. Sbornik reprint[64] I am writing to you now because it is only a few days ago that the typescript of your book[65] arrived, and I wanted to combine my reply with my warmest thanks for your sending me that material.

---

[62]The International Congress of Mathematicians took place in Amsterdam between September 2-9, 1954 and von Neumann gave an invited address on "Open problems of mathematics", see the paper [**43**]. The Editor.

[63]F.J. Murray, J. von Neumann: "On rings of operators, II." Collected Works Vol. III. No. 3. The Editor.

[64]Reference is most likely to J. von Neumann: "On an algebraic generalization of the quantum mechanical formalism", Collected Works, Vol. III. No. 9. The Editor.

[65]Reference is most likely to J. Dixmier: *Les algébres d'opérateurs dans l'espace hilbertien (algébres de Von Neumann)*, Paris, Gauthier-Villars, 1956. Dixmier's book was the first monograph on the theory of von Neumann alegbras. The Editor.

Needless to say, I am most happy to have it. I would like to write you later about it, as and when I get into a condition of commenting on it substantively.

Your and Dr. Diedonne's suggestion[66] on nomenclature is very flattering and I greatly appreciate it, since it expresses your judgement on my contribution. I feel somewhat self-conscious about the matter but I might as well admit that I have no objections – and, as I say, I appreciate very greatly this expression of your regard.

Hoping that I will have an occasion to see you this summer in Europe, I am, with best personal regards,

Sincerely yours,

John von Neumann

Original in Von Neumann Papers, Library of Congress, language of original: English

[66]In his letter to J. von Neumann (written in French, copy in Von Neumann Papers, Library of Congress) dated May 29, 1954, Dixmier proposed to call *von Neumann algebras* what were known as "rings of operators". The Editor.

Letter to P.A. Dodd

February 24, 1956

Dean Paul A. Dodd
College of Letters & Science
University of California
Los Angeles 24, California

Dear Paul:

This is in addition to my letters of February 21 to President Sproul, Chancellor Allen, as well as to you.

President Sproul's letter of February 5 mentions that my appointment will become "effective at a date to be agreed upon"[67]. In this connection, I wish to say the following: I am a Member of the U.S. Atomic Energy Commission and any plans that I have to resign from government service are, and must be at this time, highly confidential. Nevertheless, the University of California offer requires that we agree on a date when I might change my status, if for no other reason than for planning purposes.

Therefore, may I suggest that for planning purposes, and in strictest confidence, we think in terms of January 1, 1957. Would you let me know whether such a date would be agreeable to the University of California authorities.

There is one more matter I would like to mention – we have discussed it personally and have found ourselves in complete agreement regarding it, but it seems to me best to settle it explicitly at this time also. In the past I have done a good deal of industrial and government consulting – of course, always in the area of my scientific and intellectual interests. I would expect to continue to do so in the future. Am I right in assuming that such consulting meets with the approval of the University of California and is in conformity with its policies? Are there any special rules which apply, in particular for a person in my specific status? I realize that there would have to be certain upper limits to the time which can be devoted to such activities. I would appreciate your telling me what they are. If you prefer a gentlemen's agreement on this subject, please let me know what its guide lines would be.

I assume that you will consider our correspondence strictly confidential and to be communicated only to those members of the government boards and administrations of the University of California who are properly involved.

[67]See Section 1 of the Introductory Comments for von Neumann's career plans after his job as Atomic Energy Commissioner. The Editor.

Sincerely yours,

John von Neumann

Original in Von Neumann Papers, Library of Congress, language of original: English

Letter to W.M. DuMond

April 17, 1944

Dr. Jesse W. M. DuMond
California Institute of Technology
Pasadena, California

Dear Doctor DuMond:

Please excuse my delay in writing to you. I was away from Princeton for a while. I have written a detailed letter which does not contain the answers to all your questions, but does explain what sort of information we have on the subject you mention, and I would hope provide a basis for our discussion of these matters. Yesterday I received a letter from Mr. Miles J. Martin, Technical Aid of Division 17, of which the following is the pertinent part:

> "If the work that you are doing is classified work under an OSRD[68] contract, I do not have the authority to authorize you to release such information, but will make such a request through official OSRD channels. If, however, your research is not on a classified project or otherwise directly related to the war effort, then I should regard Dr. DuMond's request as a legitimate one.
>
> I should appreciate it if you would be good enough to write me stating whether or not your work is upon a classified project. If such is the case, and if you feel you can legitimately do so, perhaps you will inform me as to the sponsorship under which your work is conducted so that I may make suitable request for the release of the desired information."

Since the projects in question are classified Confidential, I immediately wrote to Mr. Martin informing him who are the authorities who can give the permission.

In this situation I must hold the letter which I have written you until I receive the authorizations and Mr. Martin's reply.

I hope you will not misunderstand my attitude. I am terribly sorry that our collaboration is subject to such delays, and I hope that you will understand that I am not trying to make things more difficult than they are by nature. I hope that it will also be possible for you to obtain authority to discuss such details of your project with me as you consider useful and appropriate; I think it will be of considerable help both ways.

I expect to see Dr. Theodore H. von Karman in Aberdeen tomorrow, and will explain the matter to him too.

[68]Office of Scientific Research and Development. The Editor.

Very sincerely yours,

John von Neumann

Original in Von Neumann Papers, Library of Congress, language of original: English

Letter to R.E. Duncan

December 18, 1947

Mr. Ralph E. Duncan
IBM War History Section
International Business Machines Corporation
590 Madison Avenue
New York 22, New York

Dear Mr. Duncan,

In reply to your letter of December 16th, in connection with the article of William L. Lawrence in the New York Times, May 4, 1947, I can tell you the following things: I did initiate and carry out work during the war on oblique shock reflection. This did lead to the conclusion that large bombs are better detonated at a considerable altitude than on the ground, since this leads to the higher oblique-incidence pressures referred to. The principle that is involved is more precisely described as the "high-burst" principle, rather than by saying that a "miss is better than a hit".

I did receive the Medal for Merit (October, 1946) and the Distinguished Service Award (July, 1946)[69]. The citations are as follows:

"Citation to Accompany the Award of
The Medal for Merit
to
Dr. John von Neumann

DR. JOHN VON NEUMANN, for exceptionally meritorious conduct in the performance of outstanding services to the United States from July 9, 1942 to August 31, 1945. Dr. von Neumann, by his outstanding devotion to duty, technical leadership, untiring cooperativeness, and sustained enthusiasm, was primarily responsible for fundamental research by the United States Navy on the effective use of high explosives, which has resulted in the discovery of a new ordnance principle for offensive action, and which has already been proved to increase the efficiency of air power in the

[69] See Section 1 of the Introductory Comments for von Neumann's work during the war. The Editor.

atomic bomb attacks over Japan. His was a contribution of inestimable value to the war effort of the United States.
HARRY TRUMAN"

"The Secretary of the Navy takes pleasure in presenting the Distinguished Civilian Service Award to Dr. John von Neumann for services set forth in the following

CITATION

For exceptionally outstanding contribution in the field of efficient utilization of high explosives while attached to the Bureau of Ordnance during World War II.

By his technical knowledge of extraordinary breadth, his incisive judgment and outstanding ability, his pre-eminence in the field of research, and his discovery of the "air-burst" principle of increasing the blast effect of a bomb, Dr. von Neumann contributed immeasurably to the understanding of explosion phenomena and to the utilization of explosives throughout the war.

Dr. von Neumann has reflected great credit upon the United States Navy and exemplified its highest traditions of service and devotion to duty. He has distinguished himself in a manner richly deserving of the navy's highest civilian award."
JAMES FORRESTAL"

The calculations which were involved in this work were done at various times between 1942 and the end of the war, and continued after the end of the war. During the war they were done partly by hand and partly by standard IBM machines of the Navy Bureau of Ordnance. Some related work was done by the automatically sequenced computer at Harvard University. Further work on shock waves was done with the electronic computing machine ENIAC at the University of Pennsylvania and the Ballistic Research Laboratory, Aberdeen, Maryland.

Sincerely yours

John von Neumann

Original in Von Neumann Papers, Library of Congress, language of original: English

Letter to the Editor of the Evening Star

May 29, 1947

The Editor
Washington Evening Star
Washington, D.C.

Dear Sir:

Among the editorials in the Tuesday, May 6, 1947, issue of the 'Evening Star' there was one entitled "Near Misses and Hits" which referred to some work that I did during war connected with methods of aerial bombing. The article made certain inferences, which it based in part on this work, and concluded that it had now been shown that near misses were generally preferable to hits, that many bombardiers during the war intuitively were aware of this but that the Army and Navy High Command had influenced them towards aiming for direct hits by distributing "the most coveted medals" accordingly. I would like to correct these statements because I do not think they are historically or scientifically correct and also because – although I am not a resident of Washington – I have always been an interested reader of the 'Evening Star' and would greatly regret it if inexact and incorrectly critical assertions should get into its pages.

It is true that under certain conditions a near miss is preferable to a hit. Thus, this may be true for attacks on ships under certain conditions, and on concrete fortifications under certain conditions, although it is certainly not so when armor-piercing bombs are used, which is likely to be the case for strongly protected structures. In any event, this principle has been known all along, and presumably observed in those cases where it mattered. I would like to add that my work had nothing to do with this.

My work was connected with the issue of large blast bombs, which is a type of ammunition that has to be used in the opposite situation: in attacks on relatively light structures. In this case I did suggest that oblique incidence of the blast wave will usually be preferable to head-on incidence. The military inference from this is not that one should strive to miss the objective but that the bomb should be detonated at a certain height rather than on the ground. This principle was tested both in the laboratory and in the field and was found to be correct, and was used during the war when appropriate conditions arose.

To conclude, I would like to observe that there is not, and cannot be, a general principle stating whether a hit is, or is not, preferable to a near miss. It will depend on the circumstances. A general answer is just as impossible here as an answer to the question "What is the best move in chess" – without any reference to the situation in which it is supposed to take place.

I would like to suggest, in conclusion, that you correct in some form the misunderstandings which your editorial, to which I have reference, may cause. I think that this would be the fairest procedure for everybody concerned and would increase everybody's respect for the precision and objectivity of the 'Evening Star'.

Yours very truly,

John von Neumann

Original in Von Neumann Papers, Library of Congress, language of original: English

Letter to R. Farquharson

November 28, 1955

Mr. Robin Farquharson
The Queen's College
Oxford, England

Dear Mr. Farquharson:

Please excuse my somewhat delayed reply to your letter of October 30, and its attachments – the reprint of your C.R. note and your subsequent typescript.

I am very attracted by your work, which appears to be the first systematic and general effort towards dealing with games whose "payoffs" are general utilities from an ordered domain, and not tied to numerical valuations[70]. I have no doubt that this is an important step forward and I am very interested to see in what elegant way you have taken it.

I certainly want you to publish your work. As to a suitable medium for its publication, I think that my colleague, Professor A. Tucker, of the Mathematics Department, Princeton University (Princeton, New Jersey, USA), can advise you better than I could. Professor Tucker has been conducting a seminar on game theory for several years, and he has also initiated and arranged the publication of several "Annals of Mathematics Studies" collections of papers on game theory. I would suggest that you write to him, and send him a copy of your typescript. Needless to say, I would be very glad to help you with the publication if this becomes necessary, but I think that it will be best to obtain Professor Tucker's advice for this.

Thanking you once more for sending me your papers, and hoping to hear from you again before long, I am

Sincerely yours,

John von Neumann

P.S. Do you want me to return your typescript?

Original in Von Neumann Papers, Library of Congress, language of original: English

[70]See Section 9 of the Introductory Comments for von Neumann's work on game theory. The Editor.

Letter to A. Flexner

September 27, 1939

Dear Doctor Flexner,

I am returning enclosed Gödel's wire which Miss Eichelser showed me this morning.[71] I think that the wire means that his essential difficulty at this moment concerns the U.S. visa, and that if this is removed the German military authorities will release him.

Don't you agree that this is the only way to read his cable?

Under these conditions the situation is considerably better than we originally expected it to be. While it would have been difficult, if not impossible, to intercede with the German authorities, there should be a way to persuade the American consul in Vienna to grant a visa.

As far as I can make out from Gödel's statements and from other cases of similar nature, his difficulty is this: professors' visas are granted only to applicants who go into teaching or research positions in this country and who have had such positions previously for two years in the country they come from. The objection made against Gödel – and against some other refugees – is that the two years teaching in the country of origin have to be immediately preceding their application; whereas Gödel was suspended from his position by the Nazis after the "Anschluss" in 1938. This requirement I think is altogether illogical. Gödel of course had more than two years teaching experience before. He has taught more than two years in the United States – adding several occasions – and it is against all sane policy to penalize a man who has been a professor for many years for having been dismissed from office by the Nazis.

Couldn't you find it possible to intercede with the State Department in this sense?

Further points which may be mentioned advantageously in this connection are as follows:

(1) I know that in many cases *visitors' visas* have been granted to teachers, lecturers, scientists, and these men have been permitted to accept teaching positions in this country. After having taught for two years *in this country*, they were given the *professors' visa*, this previous American position being considered as the equivalent of the two years teaching in the country of origin. All these cases were of men and women of scientific distinction vastly inferior to that of Gödel.

[71]For comments on von Neumann's evaluation of Gödel's work see Section 2 of the Introductory Comments. The Editor.

(2) The claim may be made with perfect justification that Gödel is unreplaceable for our educational program. Indeed Gödel is absolutely irreplaceable; he is the only mathematician alive about whom I would dare to make this statement. He represents a very important branch of mathematics, formal logics, in which he outranks everybody else to a much higher degree than usually happens in any other branch of mathematics. Indeed, the entire modern development of formal logics concerning "undecidable questions", the solution of the famous "continuum hypothesis", and quite unexpected connections between this field and other parts of mathematics, are his entirely individual contribution. Besides, the ouvre of his scientific achievements is obviously still in steep ascent, and more is to be expected from him in the future. I am convinced that salvaging him from the wreck of Europe is one of the great single contributions anyone could make to science at this moment.

(3) I know several cases where the requirement of "two years teaching immediately preceding application" was finally waived by United States consuls.

(4) Gödel actually possessed a United States immigration visa, and only forfeited this claim by returning to Austria in 1936 for two years on account of sickness. His return didn't even imply an intention to repatriate himself in Austria; it was conditioned purely by considerations of health.

(5) After his dismissal by the German Government, Gödel has been in this country and taught in Princeton and Notre Dame for one year. It is hard to understand why an obvious vacation trip to Europe, the main motive of which was to meet his wife, should trap him outside the pale.

I hope you will forgive me for bothering you with this long letter, but we all feel that the matter is of highest importance.

Yours sincerely,

John von Neumann

Original in Von Neumann Papers, Library of Congress, language of original: English

Letter to R.O. Fornaguera

December 10, 1947

Professor Ramon Ortiz Fornaguera
Residencia Consejo Investigaciones
Pinar, 21
Madrid, Spain

Dear Sir:

I have just received your very kind letter of December 1st, and I hasten to answer it.

I have had some contacts with the Alien Property Custodian regarding the copyright of my book[72], but they haven't yet been quite conclusive. I also found it somewhat difficult to explain to him in what way his office could influence your translation project in Spain. I would be much obliged to you if you could, without too much of a burden to yourself, explain to me in another letter exactly what it is that I could do at this end. The Alien Property Custodian, I assume, has only vested J. Springer's property in the United States. There may or may not be an international agreement between Spain and the Allies concerning German property, but you will probably find it easier to ascertain that matter in Madrid. In any event, I shall be very glad to help you in connection with any specific suggestions you may make. I need not tell you that I shall be most honored and pleased if a Spanish edition of my book on "Mathematische Grundlagen der Quantenmechanik" will appear.

I am very much obliged to you for the kind interest in my book, and I am much obliged also for the suggestions you make in connection with the misprints or errors in it. I shall appreciate it if you will send me your list of misprints.

Regarding the difficulty on page 229 which you mention, it seems to me it can be removed by correcting what I think was merely a misprinting. Formula (2) on page 229 should have contained not ξ_m but $\overline{\xi_m}$. I think if you re-read the text with this correction in view (it may involve a few parallel adjustments in the subsequent text), you will find that everything is self-consistent and correct.

Your questions on the nature of mathematical physics and theoretical physics are interesting but a little difficult to answer with precision in my own mind. I have always drawn a somewhat vague line of demarcation between the two subjects, but it was really more a difference in distribution of emphasis. I think that in theoretical physics the main emphasis is on the connection with experimental physics and those

[72] J. von Neumann: *Mathematische Grundlagen der Quantenmechanik*, Springer, Berlin, 1932. For comments on von Neumann's work on the mathematical foundations of quantum mechanics see Section 5 of the Introductory Comments. The Editor.

methodological processes which lead to new theories and new formulations, whereas mathematical physics deals with the actual solution and mathematical execution of a theory which is assumed to be correct per se, or assumed to be correct for the sake of the discussion.

In other words, I would say that theoretical physics deals rather with the formation and mathematical physics rather with the exploitation of physical theories. However, when a new theory has to be evaluated and compared with experience, both aspects mix.

With highest regards, I am, sir,

Yours very sincerely,

John von Neumann

Original in Von Neumann Papers, Library of Congress, language of original: English

Letter to N.H. Goldsmith

October 5, 1948

Dr. N.H. Goldsmith
Bulletin of the Atomic Scientists
1126 East 39th Street
Chicago 37, Illinois

Dear Dr. Goldsmith,

I have been away from Princeton throughout the latter part of the summer, and I only returned here a few days ago. Upon my return I found on my desk a circular letter of Dr. Oppenheimer's which, as I see, came to me in his absence through you. It deals with a Board that is to act both as sponsor of and advisor to the Bulletin of Atomic Scientists. Since Dr. Oppenheimer is still in Europe and I see from your note that a letter from me before his return is desirable, I think that it is best if I now write to you on this subject. I expect that I shall discuss the matter with Dr. Oppenheimer after his return, but I would like to tell you, first of all, that I found the numbers of the Bulletin which you sent me very well made and interesting, and this applies equally to other numbers which I have had occasion to see previously. As a matter of principle, however, I would prefer not to join the Board, since I have throughout the last years avoided all participation in public activities which are not of a primarily technical nature, and it seems to me that the objectives of the Bulletin are defined somewhat more broadly than would be the case for a technical publication. I think that this statement is correct now – I am certain that it was correct with respect to the definition of objectives which the Bulletin had at its inception, and I suppose that these objectives have not changed essentially.

I realize that all these things are imponderables, but all public activities are primarily affected by such imponderables.

I am

Sincerely yours,

John von Neumann

Original in Von Neumann Papers, Library of Congress, language of original: English

Letter to W.H. Gottschalk and H. Rademacher

November 30, 1955

Professor W.H. Gottschalk
and Professor Hans Rademacher
Department of Mathematics
University of Pennsylvania
Philadelphia 4, Pennsylvania

Dear Professor W.H Gottschalk and
Dear Professor Hans Rademacher:

I have just received your letter of November 23, and I hasten to reply to it. By the way, I have no objection against your quoting the contents of my letter to the responsible group of University personnel to which you refer.

You are asking me about my opinion on the desirability to tie a major applied mathematics effort in the University of Pennsylvania into the Department of Mathematics, and also about the role that it would be best for the Department of Mathematics to play in connection with the establishment and operation of a Regional Computing Center at the University. You point out, that it is a statement of principles upon which the disposition of these problems should be based, rather than any specific recommendation for the actual disposition, that you are inquiring about.

I am in full accord with your view that an outsider like myself can best contribute by a statement of general principles he believes in, rather than by specific recommendations. I am not qualified to make the latter since I do not know the organizational and personnel structure of the University of Pennsylvania, and its various divisions and departments, well enough.

The principle that I would like to recommend is this. You ought to determine and evaluate how much effort, interest and growth potential there exists in your Department of Mathematics, with respect to applied mathematics, and particularly to computational phases. In principle, I think, all these applied mathematical activities could be performed by a mathematics department better than by anyone else. The fact that the Mathematics Department represents many other mathematical interests, aside from the applied ones, guarantees that the enterprise will be steered from a sufficiently broad and differentiated viewpoint, and will remain flexible and responsive to changes in technique, which may occur in the future. However, I think that the Mathematics Department must, from the start, possess the characteristics to which I referred above, otherwise the marriage of these activities cannot be fruitful. If the Mathematics Department does not contain able people, with interests of the sorts described above, and if it cannot, without impairing its main functions in pure mathematics, expand in the applied direction, then applied mathematics and computational activities are best developed in a separate set up. I repeat: I believe that the best way to do these things is in the Mathematics Department, but only if there is a spontaneous desire and capability in the Mathematics Department for assuming the responsibilities that are involved. Otherwise, it would seem to me, that one has to resort to a set of second best solutions, that is, starting a separate Department of Applied Mathematics, or of proceeding in conjunction with some other contiguous Department (Statistics, Electrical Engineering). May I add that the examination of the Mathematics Department, as to its desire and capability, ought to be primarily a self-examination, but as such, it should be very searching and rigorous. In other words, the Mathematics Department should express its willingness to assume the responsibilities in question only if the answer to both questions of desire and capability to undertake the task is affirmative without any ambiguity and without any possibility of substantive doubt.

I hope that I have not misunderstood the nature of your inquiry by answering in this rather general way. If you wish my views on any more specific aspect of the matter, please let me know. I will always be extremely glad to be at your disposal.

Sincerely yours,

John von Neumann

Original in Von Neumann Papers, Library of Congress, language of original: English

Letters to K. Gödel

Berlin, Hohenzollernstrasse 23, (Tiergarten)

November 20, 1930

Dear Mr. Gödel,

I have recently concerned myself again with logic, using the methods you have employed so successfully in order to exhibit undecidable properties. In doing so I achieved a result that seems to me to be remarkable. Namely, I was able to show that the consistency of mathematics is unprovable[73] .

This is more precisely as follows:

In a formal system that contains arithmetic it is possible to express, following your considerations, that the formula $1 = 2$ cannot be the endformula of a proof starting with the axioms of this system – in fact, this formulation is a formula of the formal system under consideration. Let it be called W.

In a contradictory system any formula is provable, thus also W. If the consistency [of the system] is established intuitionistically, then it is possible, through a "translation" of the contentual intuitionistic considerations into the formal [system], to prove W also. (On account of your result one might possibly doubt such a "translatability". But I believe that in the present case it must obtain, and I would very much like to learn your view on this point.) Thus with unprovable W the system is consistent, but the consistency is unprovable.

I showed now: W is always unprovable in consistent systems, i.e., a putative effective proof of W could certainly be transformed into a contradiction.

I would be *very* much interested to hear your views on this, in particular on the "translatability issue". If you are interested, I will send you the proof details, as soon as they are written up ready to be printed and communicated, which will soon be the case – or as soon as we can welcome you in Berlin.

When is your paper going to appear, and when are proof sheets going to be available? This is also of technical interest to me, as I would like to follow you as closely as possible both substantively and notationally, and on the other hand I would like to publish as soon as possible.

And again: aren't we going to see you soon in Berlin? E. Schmidt, to whom I communicated your result as you had presented it in Königsberg, was delighted by it. He considers it, as I do, to be the greatest logical discovery in a long time.

[73]See Section 2 of the Introductory Comments for von Neumann's and Gödel's work and views on foundations of mathematics. The Editor.

This article originally appeared in Kurt Godel Collected Works, Volume V (2003 AMJ) edited by Feferman, S. et al., Oxford University Press (OUP). Click here for the OUP web page.

Hoping for an answer by return mail and to see you again very soon, I remain yours sincerely,

J. v. Neumann

Original in Gödel Archive, Manuscripts Division, Department of Rare Books and Special Collections, Princeton University Library; language of original: German. Both the original German version and the English translation (by W. Sieg), identical with the text published here, was published in K. Gödel: Collected Works, Vol. V. Correspondence H-Z, S. Feferman, J. W. Dawson, Jr., W. Goldfarb, C. Parsons, W. Sieg (eds.) (Oxford University Press, New York, 2003) pp. 337-338.

Berlin

November 29, 1930

Dear Mr. Gödel,

Many thanks for your letter and your reprint[74]. As you have established the theorem on the unprovability of consistency as a natural continuation and deepening of your earlier results, I clearly won't publish on this subject.

On the basis of your communications I think I can reproduce your line of thought, and therefore I can tell you that I employed a somewhat different method. You proved $W \to A$, I showed independently the unprovability of W, in fact with a different argument, that also mimics antinomies.

Has anything new happened since then? Are you able, for example, to decide or to conjecture whether mathematics is incomplete or ω-inconsistent?

I believe that every intuitionistic consideration can be formally copied, because the "arbitrarily nested" recursions of Bernays-Hilbert are equivalent to ordinary transfinite recursions up to appropriate ordinals of the second number class. This is a process that can be formally captured, unless there is an intuitionistically definable ordinal of the second number class that could not be defined formally – which is in my view unthinkable. Intuitionism clearly has no finite axiom system, but that does not prevent its being a part of classical mathematics that does have one.

Thus, I think that your result has solved negatively the foundational question: there is no rigorous justification for classical mathematics. What sense to attribute to our hope, according to which it is de facto consistent, I do not know – but in my view that does not change the completed fact.

I am looking forward to your proof sheets with great interest. I am very sorry that there is so little prospect of being able to welcome you here.

Hoping to hear from you again soon,

---

[74]See Section 2 of the Introductory Comments for background notes on the exchange of letters between von Neumann and Gödel. The Editor.

I remain with best wishes, yours sincerely

Johann v. Neumann

Original in Gödel Archive, Manuscripts Division, Department of Rare Books and Special Collections, Princeton University Library; language of original: German. Both the original German version and the English translation (by W. Sieg), identical with the text published here, was published in K. Gödel: Collected Works, Vol. V. Correspondence H-Z, S. Feferman, J. W. Dawson, Jr., W. Goldfarb, C. Parsons, W. Sieg (eds.) (Oxford University Press, New York, 2003) pp. 340-341.

Berlin W. 10
Hohenzollerstrasse 23

January 12, 1931

Dear Mr. Gödel,

Many thanks for your two letters and the proof sheets. Your remark on ω-consistency was very interesting to me. Incidentally, the other day I developed a method that always allows a finite decision for the effective provability question concerning propositions that are built up solely by means of the concepts "not", "or" (thus also "and", "follows", etc.), [and] "provable" (starting from the identical truth-consistency is for example such a proposition). Perhaps, if it interests you, or you have not yet thought of it, then I can write it up.

Concerning ω-consistency I am actually reassured, because it is implied by the consistency of the next type.

I absolutely disagree with your view on the formalizability of intuitionism. Certainly, for every formal system there is, as you proved, another formal one that is (already in arithmetic and the lower functional calculus) stronger. But intuitionism is not affected by that at all.

To be more precise: let us denote by A the arithmetical axiom system that contains number variables, but neither function nor set variables, and uses freely the quantifiers $((x), (Ex))$ for the number variables. If also first-order function variables are available (functions of just one variable, for example) together with their quantifiers $((f), (Ef))$, then this system may be called M. Finally, let for example my set theoretic axiom system be called Z.

Clearly I cannot prove that every intuitionistically correct *construction of arithmetic* is formalizable in A or M or even in Z – for intuitionism is undefined and undefinable. But is it not a fact, that not a single construction of the kind mentioned is known that cannot be formalized in A, and that no living logician is in the position of naming such [a construction]? Or am I wrong, and you know an effective intuitionistic arithmetic construction whose formalization in A creates difficulties? If that, to my utmost surprise, should be the case, then the formalization should certainly work in M or Z!

I would be very grateful if you would tell me whether you are really conjecturing the existence of such examples, or whether you even know some?

Your paper is very nice; I am quite delighted, how briefly and elegantly you carried out the difficult and lengthy "enumeration" of formulas. However, I believe that the proof of the unprovability of consistency can be shortened, i.e., that the general formal repetition of all considerations, as you propose, can be avoided.

It is possible to argue roughly as follows:

(1) Let a be any recursive proposition. Then

$$a \to B(a)$$

can be shown (where B stands for provable, in your sense). (Isn't that approximately your Theorem 5?)[75]

(2) If $b = (Ex)a$, we can conclude

$$b \to B(b)$$

from

$$a \to B(a)$$

(3) As every $B(a)$ is of this form, we have

$$B(a) \to B(B(a)),$$

for arbitrary a.

(4) You constructed an a with

$$* \qquad \bar{a} \sim B(a)$$

According to (3) we have

$$** \qquad \bar{a} \to B(\bar{a})$$

Let O be absurdity and W consistency; then we have according to * and ** that

$$\bar{a} \sim B(a) \& B(\bar{a}) \sim B(a \& \bar{a}) \sim B(O) \sim \bar{W},$$

[and consequently] $W \sim a$.

Thus, W is exactly as unprovable as a.

(I am sure you can fill in the gaps in my presentation.)

With best regards,

Yours sincerely,

Johann v. Neumann

Original in Gödel Archive, Manuscripts Division, Department of Rare Books and Special Collections, Princeton University Library; language of original: German. Both the original German version and the English translation (by W. Sieg), identical with the text published here, was published in K. Gödel: Collected Works, Vol. V. Correspondence H-Z, S. Feferman, J. W. Dawson, Jr., W. Goldfarb, C. Parsons, W. Sieg (eds.) (Oxford University Press, New York, 2003) pp. 342-345.

---

[75] Of K. Gödel: *Collected Works*, Vol. I. Publications 1929-1936, S. Feferman, J. W. Dawson, Jr., S.C. Kleene, G. H. Moore, R. Solovay, J. van Heijenort (eds.) (Oxford University Press, New York, 1986) p. 170. Footnote of the editors of Gödel's Collected Works, Vol. V. Correspondence H-Z; S. Feferman, J. W. Dawson, Jr., W. Goldfarb, C. Parsons, W. Sieg (eds.) (Oxford University Press, New York, 2003).

Budapest, V, Arany János utca 16

February 14, 1933

Dear Mr. Gödel,

I just returned to Europe from Princeton, and I would be very glad if during my European stay there would be an opportunity to meet you. At the end of September I will go back to Princeton, where in the future I will spend two terms per year, as I received an offer from the new Bamberger-Flexner "Institute for Advanced Study", and I accepted.

Perhaps you would like to know more than you have gathered from the brief correspondence up to now about the conditions in Princeton and about the structure of this new Institute, of which you will fortunately be a member next year. Therefore, I write to you first of all, and also on Veblen's request, to tell you that I am happy to provide you with any information you may be interested in.

Perhaps we can also meet sometime and somewhere – this summer I will probably still lecture in Berlin (for the last time), and I will repeatedly be in or travel through Vienna. What is your program?

With best regards, which I ask you to extend also to Menger and Hahn,

Yours sincerely,

Johann von Neumann

Original in Gödel Archive, Manuscripts Division, Department of Rare Books and Special Collections, Princeton University Library; language of original: German. Both the original German version and the English translation (by W. Sieg), identical with the text published here, was published in K. Gödel: Collected Works, Vol. V. Correspondence H-Z, S. Feferman, J. W. Dawson, Jr., W. Goldfarb, C. Parsons, W. Sieg (eds.) (Oxford University Press, New York, 2003) p. 345.

Letter to G. Haberler

October 31, 1947

Professor Gottfried Haberler
Economics Department
Harvard University
Cambridge, Massachusetts

Dear Professor Haberler:

Oscar Morgenstern showed me your very kind letter in which you suggest that I might review Dr. Samuelson's book[76] on the mathematical foundations of economics. Let me first of all apologize for the delay in answering. It was due solely to a conflict between the temptation to do the review and the obvious difficulties of carrying this out properly. After a certain amount of introspection I feel that I had better not accept this responsibility[77]. Samuelson's book is very interesting and very detailed, and clearly represents the result of a great deal of work. It would be improper to accept the reviewing in any different spirit from this. Besides, the whole subject of the use of mathematical methods in economics is one which I wouldn't like to deal with in print except after a very careful study of the corpora delicti and of my corresponding formulations. The methodological questions which are involved are very delicate and it is very easy with respect to them to sin by overstatement as well as understatement.

On the other hand, I am very badly overloaded with work with no relief in sight for at least another year, so I think that it would be a mistake to accept such a serious additional responsibility.

I am sure that you will understand my position, and thanking you once more for having thought of me in this connection, I am

Sincerely yours,

John von Neumann

Original in Von Neumann Papers, Library of Congress, language of original: English

[76]Most likely: P.A. Samuelson: *Foundations of Economic Analysis*, (Cambridge, Mass. Harvard University Press, 1947). The Editor.

[77]For von Neumann's work on game theory and economics and for possible further reasons why he declined to review Samuelson's work, see Section 9 of the Introductory Comments. The Editor.

Letters to I. Halperin

The Institute for Advanced Study
School of Mathematics, Fine Hall
Princeton, New Jersey

June 1 [1939 ?]

My dear Mr. Halperin,

This is to tell you that I have established the existence of factors of class (III_∞) at last[78]. The components of what I called "ring $C^{\#}$" in the various incomplete direct products $\Pi^{\mathcal{D}}_{\otimes n=1,2,\ldots}(\mathcal{H}_{n,1}\mathcal{H}_{n,2})$ (cf. my paper on Infinite direct products of Operators, Compositio Math. 1938, pp. 71-77[79]), which are, as you know, characteristical by a sequence of constants $\alpha_1, \alpha_2, \ldots$, all of them $\geq 0, \leq 1$, do it: In the case which I mentioned to you i.e. always, when for some $\delta > 0$ infinitely many $n = 1, 2, \ldots$ with $\delta \leq \alpha_n \leq 1 - \delta$ exist.

In fact, this is only a special case of a more general result, namely:

Let S be a set, μ a Lebesgue-measure in S, G an enumerable group. Every $a \in G$ generates a mapping of S on itself: $x \rightleftharpoons xa$. I assume, of course, that $x1 \equiv x$ and $x(ab) \equiv (xa)b$.

I do *not* assume, as in my paper with Murray,[80] that always $\mu(Ma) = \mu(M)$ (for $M \subset S$, $a \in G$), but only that

(1) $$\mu(M) = 0 \text{ implies } \mu(Ma) = 0 \quad (\text{for } M \subset S, a \in G)$$

Hence, by the differentiation-theorem of Lebesgue-Nikodym, there exists a function $\omega_a(x)$ $(a \in G, x \in S)$, which is always > 0, and such that generally

(2) $$\int_S f(xa)\omega_a(x)\mathrm{d}\mu x = \int_S f(x)\mathrm{d}\mu x$$

Hence the operator

(3) $$U_{a_0} f(x) = \sqrt{\omega_{a_0}(x)} f(xa_0)$$

is unitary in the L_2 over S.

I assume, as in my paper with Murray that G has these further properties:

---

[78]See Section 3 of the Introductory Comments for background notes on operator algebra theory and especially for the classification of von Neumann algebras. The Editor.

[79]J. von Neumann" "On infinite direct products" Collected Works Vol. III. No. 6. The Editor.

[80]F.J. Murray, J. von Neumann: "On rings of operators" Collected Works, Vol. III. No. 2. The Editor.

A): G is ergodic in S i.e.:

If for a measurable set $M \subset S$ we have[81] for every a

$$\mu((M + Ma) - (M.Ma)) = 0 \tag{4}$$

then either $\mu(M) = 0$ or $\mu(S - M) = 0$

B): For every $a \neq 1$ the set N_a of all $x \in S$ with $x = xa$ has $\mu(N_a) = 0$.

Now form the L_2 over $S \times G$, and define in it the operators[82]

$$\begin{array}{llll}
(5) & \overline{U}_{a_0}F(x,a) & \equiv & \sqrt{\omega_{a_0}(x)}F(xa_0, aa_0) \qquad (a_0 \in G) \\
(6) & \overline{V}_{a_0}F(x,a) & \equiv & F(x, a_0^{-1}a) \qquad (a_0 \in G) \\
(7) & \overline{W}F(x,a) & \equiv & \sqrt{\omega_{a-1}(x)}F(xa^{-1}, a^{-1}) \qquad (a_0 \in G) \\
(8) & \overline{L}_{\phi(x)}F(x,a) & \equiv & \phi(x)F(x,a) \qquad \phi(x) \text{ complex bounded} \\
(9) & \overline{M}_{\phi(x)}F(x,a) & \equiv & (xa^{-1})F(x,a) \qquad \phi(x) \text{ complex bounded}
\end{array}$$

$\overline{U}_{a_0}, \overline{V}_{a_0}, \overline{W}$ are unitary, $\overline{W}^2 = I$, $\overline{L}_{\phi(x)}, \overline{M}_{\phi(x)}$ are bounded. Put

$$\begin{array}{lll}
\mathcal{I} & = & \text{set of all } \overline{U}_{a_0} \text{ and } \overline{L}_{\phi(x)} \\
\mathcal{J} & = & \text{set of all } \overline{V}_{a_0} \text{ and } \overline{M}_{\phi(x)}
\end{array}$$

Put[83]

$$\begin{array}{llll}
(10) & \mathcal{M} & = & R(\mathcal{I}) \\
(11) & \mathcal{N} & = & R(\mathcal{J})
\end{array}$$

then also[84]

$$\begin{array}{llll}
(12) & \mathcal{M} & = & \mathcal{J}' = \mathcal{N}' \\
(13) & \mathcal{N} & = & \mathcal{I}' = \mathcal{M}
\end{array}$$

These $\mathcal{M}, \mathcal{M}'$ are factors.

Now I define further: μ is equivalent to ν, if and only if this is true: If ν is another Lebesgue measure in S, then $\mu(M) = 0$ is equivalent to $\nu(M) = 0$. ν is *invariant*, if this is true: $\nu(Ma) = \nu(M)$ for all $M \subset S$, $a \in G$. It is easy to verify:

(1) $F \rightleftharpoons WF$ is an involutory anti-isomorphism of $\mathcal{M}, \mathcal{M}'$.

(2) $\mathcal{M}, \mathcal{M}'$ remain isomorphic to themselves, if μ is replaced by any equivalent ν.

I define: G is *measurable* if and only if μ is equivalent to an invariant ν. (In our previous discussions both measurable and non measurable G's have frequently figured. Examples of both kinds are easy to construct.)

Now this is true:

I) *G measurable.*

---

[81] The equation (4) below reads in proper set theoretic notation: $\mu((M \cup Ma) \setminus (M \cap Ma)) = 0$. The Editor.

[82] ϕ is missing from the right hand side of equation (9), the correct formula should be $\overline{M}_{\phi(x)}F(x,a) \equiv \phi(xa^{-1})F(x,a)$. The Editor.

[83] $R(\mathcal{I})$ and $R(\mathcal{J})$ in eqs. (11) and (10) denote the von Neumann algebras generated by $\mathcal{I}$ and $\mathcal{J}$ respectively. The Editor.

[84] A comma is missing from the right hand side of equation (13), the correct equation is $\mathcal{N} = \mathcal{I}' = \mathcal{M}'$, c.f. Lemma 3.6.5 in J. von Neumann: "On Rings of operators III", Collected Works, Vol. III. No. 4. The Editor.

Since replacement of μ by an equivalent ν does not change $\mathcal{M}, \mathcal{M}'$ up to isomorphism, we may assume in this case, that μ itself is invariant. Then this coincides with the examples discussed in my paper with Murray. Hence in this case $\mathcal{M}, \mathcal{M}'$ belong to the classes (I_n) $(n = 1, 2, \ldots)$, (I_∞), (II_1), (II_∞), and they give indeed specific examples for each one of these classes.

II) *G non measurable.*

In this case I proved now, that $\mathcal{M}, \mathcal{M}'$ are of class (III_∞).

The proof is rather involved. I have written it up, the manuscript is already in its final form, and is being typed. The paper will be about 50 printed pages, I wrote somewhat broadly but I don't think, that it would be compressed to less than 35-40 pages.

I have now even some hopes and ideas about the isomorphism-situation in factors of class (II_1). I think, that I am making headway here, too – but this may just be a fit of optimism, due to the euphoria about killing of (III_∞)!

What news do you have?

Best greetings

As ever

John von Neumann

P.S. It seems, that I will stay on this side all summer while my wife will probably go to Europe for 2 month's or so. What are your plans?

Original in Von Neumann Papers, Library of Congress, language of original: English.

The Institute for Advanced Study
School of Mathematics, Fine Hall
Princeton, New Jersey

December 17, [1939 ?]

My dear Mr. Halperin,

Please excuse me for answering your letter of Oct. 31 only now. I was working with a certain amount of "fury" at a lemma on finite matrices, in connection with the problem of existence of "non-approximately-finite" operator rings of class (II_1)[85] and I did not want to write until I knew more about it. I have solved the question (I mean that lemma) now.[86]

[85]See Section 3 of the Introductory Comments for notes on the classification theory and types of von Neumann algebras. The Editor.

[86]The lemma stated in this letter is the precursor of Definition 6.1.1 and Lemma 6.1.2 in F.J. Murray, J. von Neumann: "On rings of operators IV" Collected Works Vol. III. No. 5. The Lemma 6.1.2 states that approximate finiteness of a type II_1 factor implies a so-called "property Γ", a property closely related to 6–8 described in this letter. It is this Lemma 6.1.2 that Murray and von Neumann use to prove the existence of not approximately finite, type II_1 factors (c.f. Lemma 6.2.1, 6.2.2 and Theorem XV' in "On rings of operators IV"). The Editor.

The lemma is one, about which we have talked before, it is somewhat complicated to state, but simple in its motive and in its basic idea. It may be stated as follows:

(1) Let M_n be the space of all n-th order complex matrices $a = (a_{ij})$, $i, j = 1, \ldots, n$.

(2) Define for $a = (a_{ij}) \in M_n$

$$|a| = \sqrt{\frac{1}{n}\sum_{ij}|a_{ij}|^2}$$

$$\|a\| = \sqrt{Max\frac{\sum_i |y_i|^2}{\sum_i |x_i|^2}} \qquad (y_i = \sum_j a_{ij}x_j)$$

(3) Denote the set of all $a \in M_n$ with $\|a\| \leq 1$ by E_n.

(4) If $n = ml$, then imbed M_m into M_n by identifying the element $a = (a_{\mu\nu}) \in M_m$ with the element $\breve{a} = (\breve{a}_{ij}) \in M_n$, where[87]

$$\breve{a}_{ij}\begin{cases} = a_{\mu\nu} & \text{for } \rho = \sigma \\ = 0 & \text{for } \rho \neq \sigma \end{cases} \left| \begin{array}{lll} i, j = 1, \ldots, n, & \mu, \nu = 1, \ldots, m, & \rho, \sigma = 1, \ldots l \\ i = m(\rho - 1) + \mu & j = m(\sigma - 1) + \nu & \end{array}\right.$$

(5) The imbedding (4) of M_m into M_n is obviously isomorphic for all matrix operations as well as for $|\ldots|$ and $\|\ldots\|$.

(6) Let M_n^m be the set of all $u^{-1}au$, where $u \in M_n$ is unitary, and $a \in M_m \subseteq M_n$.

(7) Let $M_n^m(\epsilon)$ be the set of all $b \in M_n$, for which a $c \in M_n^m$ with $|b - c| \leq \epsilon$ can be found.

(8) *Question*: Given an $\epsilon > 0$, does there exists an $m_0 = m_0(\epsilon)$, such that for $m \geq m_0$ and $n = ml$ always

$$E_n \subseteq M_n^m(\epsilon)$$

As you will perhaps remember, the answer to the Question (8) is *yes*, if the matrices under consideration (in M_n as well as in M_m) are restricted to those of *Hermitean* character. The same is the case if they are required to be *unitary*. And even (this embraces the two above cases as special ones) if they are only required to be regular.

Nevertheless we suspected, that the answer to the unqualified Question (8) is *no*. But – in spite of trying it constantly for the last two years – the matter could not be settled.

It is clear, that finding an answer to the Question (8) is a plausible step in orienting oneself about the possibility of "non-approximately-finite" operator rings of class (II_1). I must admitt, however, that I do not know as yet, what the precise effect of either answer to Question (8) would be on that problem.[88]

My present result is, that the general answer to Question (8) is *no*. I succeeded computing the volumes of the sets involved (they are domains in M_n, and M_n may be looked at as a ln^2 dimensional real Euclidean space), and I proved, that

[87] The original letter is typed but the formula below in it contains hand-written insertions that are not obvious and need to be interpreted. The interpretation given here is based on the corresponding formula 5.4 in J. von Neumann: "Approximative properties of matrices of high finite order", Collected Works Vol. IV. No. 24. The Editor.

[88] See footnote 86. The Editor.

(9)

$$\frac{\mathrm{Vol}(E_n \cdot M_n^m(\epsilon))}{\mathrm{Vol}(E_n)} \to 0 \quad \text{for } n \to \infty$$

uniformly in l, if $n = ml$, $l = 2, 3, \ldots$ provided that ϵ is sufficiently small. Indeed, it suffices to have

(10)

$$\epsilon \leq \frac{1}{110,000}$$

although this upper limit for ϵ is probably much too low. (Observe, that (9) holds for $l = 2, 3, \ldots$ For the denial of (8) it would be sufficient to have $E_n \cdot M_n^m(\epsilon) \neq E_n$, i.e. $E_n \not\subseteq M_n^m(\epsilon)$, for $l \geq l_0 = l_0(m)$, where $l_0 \to \infty$ for $m \to \infty$ [of course $m \geq m_0 = m_0(\epsilon)$, ϵ field].)

The above result is *not* satisfactory for the following reason: It proves the existence of an $a \in E_n - E_n \cdot M_n^m(\epsilon)$ by a volume consideration, hence without constructing such an a explicitely. (I do not mean this from the rigorous "intuitionistic" point of view: The sets involved are of such a nature ["polyhedra" with a finite number of analytic "faces"], that one could "construct" such an a explicitely – but this construction would be so involved and so indirect, that it would probably not be illuminating at all.) Yet the knowledge of such an a, with as many algebraic details as possible, would be of great importance: It would probably show, what algebraic properties make a quantity a unfit for approximation by matrices of any given order m. And this would probably lead to ideas how to construct "non-approximately-finite" rings of class (II_1).

On the other hand I think, that the "volumetric" method which decided (8), can also be used to solve other algebraical questions about matrices of given and of very high order (m and n) – and I hope that this will lead to those algebraic insights, which are needed to construct "non-approximately-finite" rings of class (II_1). I have obtained some results in this direction and I hope that I will have more soon.

Would you let me know what you are doing now and what your plans are? Is it possible and likely that you will visit the U.S. in a near future?

Hoping to hear from you soon at any rate,

I am cordially yours

John von Neumann
26 Westscott Road
Princeton (N.Y.)
Phone 2186

Original in Von Neumann Papers, Library of Congress, language of original: English

The Institute for Advanced Study
School of Mathematics, Fine Hall
Princeton, New Jersey

Febr. 22. 1940

My dear Mr. Halperin,

I want to report to you, that Murray and I have finally succeeded in showing that a certain operator ring (factor) of class (II_1) is not "approximately finite"[89]. The construction and the proof are not so bad, especially considering that it took us 5 years to find it – but the real difficulty was, as it looks now, to find the right example.

The example is constructed with the help of a general technique for the construction of class (II_1) factors, which is simpler than our original method to build such factors (Ann. of Math., 1936[90]) – although it is closely related to it. Since this technique is quite amusing in itself – because it is closely connected with certain procedures in the representation theory of finite groups – I will describe it first in full generality. This is it:

Let G be a finite or enumerably infinite group. Form a unitary space H_G, in which the elements of G are used as indices of the coordinates: The general element $\xi \in H_G$ is a $\xi = \{x_a ; a \in G\}$, x_a a complex number, defined for each $a \in G$, such that $\sum_{a\in G} |x_a|^2$ is finite. So for $\xi = \{x_a \; ; \; a \in G\}, \eta = \{y_a \; ; \; a \in G\}$

$$(\xi, \eta) = \sum_{a\in G} x_a \bar{y}_a$$

If we denote the element

$$\xi_b = \{\delta_{a,b} \; ; \; a \in G\} \qquad\qquad \delta_{a,b} \begin{cases} = 1 & \text{for} \quad a = b \\ = 0 & \text{for} \quad a \neq b \end{cases}$$

of H_G simply by b, then we see:

The b, $b \in G$, form a complete, normalised orthogonal set in H_G. For

$$\xi = \{x_a \; ; \; a \in G\}$$

the expansion in this orthogonal system is

$$\xi = \sum_{a\in G} x_a \cdot a \tag{1}$$

The mapping W defined by

$$W\{x_a \; ; \; a \in G\} = \{x_{a^{-1}} \; ; \; a \in G\}$$

i.e.

$$Wa = a^{-1}$$

is clearly linear, isometric, and involutory in H_G, – i.e. it defines a unitary operator W in H_G, with $W^2 = 1$.

---

[89]See Section 3 of the Introductory Comments for notes on the theory and classification of von Neumann algebras. The Editor.

[90]F.J. Murray, J. von Neumann: "On rings of operators", Collected works, Vol. III. No. 2. The Editor.

We also define two operators L_b and R_b for every $b \in G$:

$$L_b\{x_a\ ;\ a \in G\} = \{x_{b^{-1}a}\ ;\ a \in G\}$$

i.e.

$$L_b a = ba$$

and

$$R_b\{x_a\ ;\ a \in G\} = \{x_{ab}\ ;\ a \in G\}$$

i.e.

$$R_b a = ab^{-1}$$

Clearly W transforms R_b into L_b and vice versa. Put

$$\mathcal{I} = \text{ Set } (L_a\ ;\ a \in G), \qquad \mathcal{J} = \text{ Set } (R_a\ ;\ a \in G)$$

Remembering that

$$\begin{aligned} L_a L_b &= L_{ab}, & L_a^* &= L_{a^{-1}}, & L_I &= 1 \\ R_a R_b &= Rab, & R_a^* &= R_{a^{-1}}, & R_I &= 1 \end{aligned}$$
$$L_a R_b = R_b L_a$$

one proves easily:

If $A \in \mathcal{I}'(\mathcal{J}')$, then put (remember, that $1 \in H_G$)

(2) $$A1 = \sum_{a \in G} z_a a \qquad (\sum_{a\in G} |z_a|^2 \text{ finite}).$$

Then

(3) $$A = \sum_{a \in G} z_{a^{-1}} R_a \qquad (\sum_{a \in G} z_a L_a)$$

in the sense of strong operator convergence. In particular then

$$A \sum_{a \in G} x_a a = \sum_{a \in G} y_a a$$

means

$$y_a = \sum_{b \in G} z_{a^{-1}b} x_b \qquad (\sum_{b \in G} z_{ab^{-1}} x_b).$$

So (2), (3) associate every $A \in \mathcal{I}'(\mathcal{J}')$ with a

$$\xi = \sum_a z_a a \in H_G$$

The ξ which correspond to such A's form a subset of H_G, to be denoted by $H_G^R(H_G^L)$. One verifies easily, that $H_G^R[H_G^L]$ is a linear set, everywhere dense in H_G – it is $=$ H_G if and only if G is finite.

(3) also shows that[91]

(4) $$\mathcal{J}' = \mathcal{R}(\mathcal{J}), \qquad \mathcal{J}' = \mathcal{R}(\mathcal{I})$$

Put[92]

(5) $$\mathcal{M}_G = \mathcal{I}' = \mathcal{R}(\mathcal{J}), \qquad \text{then} \qquad \mathcal{M}_G' = \mathcal{J}' = \mathcal{R}(\mathcal{I})$$

---

[91]Equation (4) below contains misprints: it should read $\mathcal{J}' = \mathcal{R}(\mathcal{I})$ and $\mathcal{I}' = \mathcal{R}(\mathcal{J})$; cf. Lemma 5.3.5 *in* F.J. Murray, J. von Neumann: "On Rings of Operators IV., Collected Works Vol. III. No. 5. (also see eq. (5) and footnote 92 below). The Editor.

[92]Equation (5) below is misprinted: it should read $\mathcal{M}_G = \mathcal{J}' = \mathcal{R}(\mathcal{I})$ and $\mathcal{M}_G' = \mathcal{I}' = \mathcal{R}(\mathcal{J})$. The Editor.

$\mathcal{M}_G$ is not a factor in general, e.g.:

(1) If G is Abelian then (and only then) $\mathcal{M}_G$ is also Abelian – in fact $\mathcal{M}_G = \mathcal{M}'_G$.

(2) If G is finite, then

$$C = \text{Center of } \mathcal{M}_G = \text{ Center of } \mathcal{M}'_G = \mathcal{M}_G \cdot \mathcal{M}'_G$$

consists of those $A = \sum_{a \in G} z_a R_a \quad (A = \sum_{a \in G} z_a L_a)$ for which

$$z_a = z_b \quad \text{whenever } b = uau^{-1} \quad \text{for any } \ u \in G \tag{6}$$

Hence it contains elements $\neq z1$.

But just 2. makes this clear: C can always be characterised by (6) in 2. Hence if G is infinite, then G will consist of the $z1$ only if this is true:

(##): For every $a \in G$, $a \neq 1$, there are infinitely many different uau^{-1}, $u \in G$.

(Combine (6) with the finiteness of $\sum_{a \in G} |z_a|^2$!)

So we see: $\mathcal{M}_G$ $(\mathcal{M}'_G)$ is a factor if and only if G fulfills condition (##).

It is also clear, that

$$t(A) = z_1 \qquad (\text{for } A = \sum_{a \in G} z_a R_a [\sum_{a \in G} z_a L_a]) \tag{7}$$

possesses all properties of a trace for $\mathcal{M}$ $(\mathcal{M}')$.[93] Hence when $\mathcal{M}_G$ $(\mathcal{M}'_G)$ is a factor (cf. above), then (7) gives *the* trace of $\mathcal{M}$ $(\mathcal{M}')$.[94] Now one verifies immediately, that the class of $\mathcal{M}_G$ $(\mathcal{M}'_G)$ is (II_1).

The joke is, that for a *finite* G this $\mathcal{M}_G$ is the so called *regular representation*, which is reducible, i.e. not at all a factor. (Cf. 2. above). In fact it contains *all* irreducible representation of G. In spite of this, when G is infinite and (##) is satisfied, then the center of $\mathcal{M}_G$ disappears, and $\mathcal{M}_G$ becomes a factor.

Now we proved, that $\mathcal{M}_G$ is *not* approximately finite for certain G's, e.g. for this one:

$G =$ The free group with two generators u, v, of orders 2,3 respectively: $u^2 = v^3 = 1$.

We did not prove, however, that this $\mathcal{M}_G$ is not isomorphic to a *subring* of some approximately finite class (II_1) factor.

In fact, I found two groups G_1, G_2 for which this can be shown.

$\mathcal{M}_{G_1}, \mathcal{M}_{G_2}$ are both class (II_1) factors, each one isomorphic to a subring of the other, but they are not isomorphic to each other.

For these reasons, and owing to certain algebraical analogies, I suspect that this is true:

H Every class (II_1) factor is isomorphic to a subring of an approximately finite class (II_1) factor.

H is equivalent to this:

HH Any two class (II_1) factors are isomorphic to subrings of each other. (But not always isomorphic to each other !)

---

[93]Clearly $\mathcal{M}$ and $\mathcal{M}'$ are meant to be $\mathcal{M}_G$ and $\mathcal{M}'_G$. The Editor.

[94]See footnote 93. The Editor.

Getting my finite-order-matrices paper typed, formulae written in, etc. lasted much longer than I expected. I will have a spare manuscript in 7-10 days, and I will send it to you then.

What are your news? Is there any chance of our getting together in the near future?

Hoping to hear from you soon,

I am yours

John von Neumann

Original in Von Neumann Papers, Library of Congress, language of original: English.

Letter to G.B. Harrison

November 20, 1945

Dean George B. Harrison
Massachusetts Institute of Technology
Cambridge 39, Massachusetts

Dear Dean Harrison:

Three months ago you made me a very tempting proposal of a professorship of mathematics, connected with a later assumption of the chairmanship of the Mathematics Department. You also indicated, and so did President Compton and Vice President Killian, that the Institute would continue its present projects and undertake new projects in the field of automatic, high-speed computing, and that in particular a new project of this type, aiming at a fully electronic device, could be initiated by the Mathematics Department and the new Vacuum Tube Laboratory of the Institute. I expressed to you, as well as to Professors Wiener and Phillips, who took a great interest in these matters, my great appreciation of this offer and of this opportunity. I am fully convinced of the great importance, both purely scientific and applied of the new developments in high-speed, automatic, electronic computing. I told you therefore that the primary circumstance determining my decision would be whether efforts at the Institute for Advanced Study in Princeton, which were being made in the same direction (that is, to develop and to construct such a computer), would be successful. Owing to my long association with the Institute for Advanced Study, and owing to the fact that I had always urged the importance of such a project, I felt that the Institute for Advanced Study had the first call on my services if it could offer me a promising opportunity to carry out such a program. I also mentioned at the time, that it would take about three months to reach a conclusion about the possibilities at Princeton.

In your letters to Dr. Aydelotte and to me you showed full understanding of these motives. You also pointed out that in judging the possibilities which might develop in Princeton, it was important to see not only whether adequate financial backing was available, but also whether the necessary manpower and laboratory facilities and the necessary high-grade engineering experience could be found there.

It has now developed that a very excellent opportunity for such work exists at Princeton. The Institute for Advanced Study and the radio Corporation of America (the research laboratory of the latter is, as you know, located at Princeton), with the assistance of Princeton University, have offered me support in carrying out this project. As far as I can see, they are providing adequate backing financially and in manpower. The Radio Corporation will make its exceptionally broad experience in the pertinent fields available to this project. It is agreed between all parties that while this is a community enterprise, the completed computer will be used exclusively as a research tool, at the Institute.

Under these circumstances I feel that it is my first responsibility to stay in Princeton and to direct this project. I must therefore decline your very flattering offer, although I do this with great regret. I do not want, however, to let the opportunity go to express to you my sincere appreciation of the confidence which you have shown me in this matter.

I am

Very truly yours,

John von Neumann

Original in Von Neumann Papers, Library of Congress, language of original: English

Letter to M. de Horvath

September 20, 1955

Mrs. Marianne de Horvath
Hotel Zieglerhof
Dufourstrasse 5
Zurich 8, Switzerland

Dear Mrs. Horvath:

Your letter of September 14 arrived a few days ago, together with your son's paper, and I hasten to reply. First of all, please let me tell you how glad I am to hear from you and to be reminded by your friendly lines of our acquaintance and the friendship of our families in Hungary of over 30 years. I am also very glad to hear from you from Zurich and to know that you and your family are well, and to be reminded of the town in which I passed several very delightful years as a student at E.T.H.[95]

I have read your son's paper and also your description of your difficulties in getting an evaluation of it. To my regret, I can only tell you that the matter does not strike me as a hopeful one. The views about the possibility of reaching the zero of absolute temperature and the phenomena that will occur when that point is reached are not supported by anything theoretical or experimental that his paper contains or refers to or that I am familiar with.

I am exceedingly sorry that I cannot tell you anything more favorable or encouraging. Please believe me that my frank expression of an adverse opinion is based on the desire to help your son, whose interest in science is evidently lively and deserving of proper guidance, and on my high regard for you and our old friendship.

Very sincerely yours,

John von Neumann

P.S. I assume that your son will want his paper back, I am therefore, returning it under separate cover.
J.v.N.

Original in Von Neumann Papers, Library of Congress, language of original: English

[95]Eidgenössische Technische Hochschule, the famous Swiss Technical University in Zürich. See Section 1 of the Introductory Comments for von Neumann's student years. The Editor.

Letter to A.S. Householder

February 3, 1948

Dr. Alston S. Householder
Monsanto Chemical Company
Clinton laboratories
P. O. Box 1W
Oak Ridge, Tennessee

Dear Dr. Householder,

Please excuse my delay in answering your letter of January 15th. I have been on the move since January 16th when I left Los Alamos until a few days ago when I finally got back to Princeton.

In our plans for the Monte Carlo problem we intended to use the iterative application of some function $f(x)$ to produce pseudo-random numbers in the following manner: To start with, some 8 or 10 decimal-digit number (depending on the type of computing equipment used) and generating from it a sequence of numbers by iterative applications of $f(x)$. With respect to this procedure, the following observations have to be made.

1. Since we are dealing with digital aggregates which permit only a finite number of combinations, every conceivable function $f(x)$ must after a sufficient number of applications produce a number which has already been produced before, and after that things must repeat periodically. Consequently, no function $f(x)$ can under these conditions produce a sequence which has the essential attributes of a random sequence, if it is continued long enough. What we intended was to produce sequences of something between 1,000 and 10,000 numbers (digital aggregates), which up to this point look reasonably "random." It is possible to rationalize the following attitude: A function $f(x)$ is "efficient" in this respect if it permits producing sequences which are random for practical purposes and the length of which is not much less than $\sqrt{10^n}$ if n-digit aggregates are used. Thus for $n = 8$ or 10, lengths of 1,000 to 10,000, as mentioned above, seem reasonable.

By "reasonably random" for practical purposes, I mean that the numbers thus produced are as equidistributed as one would expect for a random sample of the size in question, and that furthermore, k-th neighbors are correspondingly independent for moderate values of k. We usually worked with $k = 1, \ldots 8$. To be more precise: Our statistical analyses dealt not with the 8 (or 10) digit numbers as such (as wholes), but with their digits separately. We tested the equidistribution of each digit (from No. 1 to No. 8), and the independence of each digit of a number from each digit of its k-th neighbor.

2. There were theoretical reasons to expect that the function $4x(1-x)$ would behave well if only the middle digits are used. We found, however, by making a few test runs (of a length of a few thousand each) that while the equidistribution was reasonably fulfilled, the independence of neighbors was not. We, therefore, finally decided not to use this function.

3 The following function $f(x)$ turned out to be completely satisfactory: $f(x) =$ the middle 8 digits of x^2. (x an 8 digit number.) We tested it with respect to equidistribution as well as independence of neighbors, in sequences of a length of a few thousand. There is every reason to believe that the same for 10-digit numbers would be even better. We also found that the same for 7-digit numbers was entirely satisfactory.

4. In addition to all of this, however, I would like to emphasize the following point: Our main reason for using these arithmetical pseudo-random numbers was that we intend to solve the Monte Carlo problems on an electronic automatic computing machine: the ENIAC. If hand calculation, or calculation on some slower automatic device is contemplated, then the arithmetical-pseudo-random procedure does not seem reasonable to me. It is possible to obtain large quantities of "guaranteed" random numbers recorded in tables or on punch-card (cf. below). The only point is that the ENIAC is much faster in producing a number by some arithmetical function $f(x)$ (in the above sense) than any reading it from a punch-card[96]. In other words, it is preferable to explain to the ENIAC how to produce its random numbers internally and mathematically than to furnish it with a ready-made list of random numbers which it has to read. (I mention for your orientation: the ENIAC can add 2 ten-digit numbers in 1/5 millisecond, and it can multiply them in three milliseconds, while it has to spend 600 milliseconds in reading an 80-digit punch-card.)

If you envisage human computation, then rolling dice or using Tipton's tables of random numbers is preferable. (If you choose to roll dice, it is of course necessary to make sure that the faces are equiprobable. This is never actually the case. It is, however, very simple to introduce combinatorial tricks which eliminate the effect of the "untrue" character of the dice.) If you intend to use some medium-fast reading machine, then feeding random numbers from punch-cards or punched-tape is probably best. I know about a project, supported by the Air Forces (Project RAND, Douglas Aircraft Company, Santa Monica, California), which has produced half a million or more punch-cards with random numbers. I am sure that they will be very glad to let you have some if you want them.

I hope I need not tell you that I am very interested in the Monte Carlo problem and any applications you may make of it. I do believe in particular that the reactor problems in which you are engaged contain quite particularly interesting and important applications of this method. I will always be very glad to help you on this subject to the best of my ability. Please don't hesitate to call on me if you think it will be helpful.

[96]See Section 8 of the Introductory Comments for von Neumann's work on computer development. The Editor.

Sincerely,

John von Neumann

Original in Von Neumann Papers, Library of Congress, language of original: English

Letters to C.C. Hurd

December 3, 1948

Dr. Cuthbert C. Hurd
Carbide and Carbon Chemicals Corporation
Post Office Box P
Oak Ridge, Tennessee

Dear Dr. Hurd,

The work on random numbers which we did at Los Alamos was as follows: We formed a sequence of 10 digit numbers, each one of which was obtained by squaring its predecessor and taking the middle 10 digits of this square. We started with the number consisting of 10 ones (this being zero-th term of the sequence), and building up a sequence of about 3400 successive iterations. In this manner we obtained 3400 punch cards, each one of which contained 10 decimal digits – a total material of 34,000 decimal digits. The statistical test to which this material was submitted was as follows: We formed the matrix a_{ij}, $i, j = 1, \dots 10$, where a_{ij} was the frequency with which the value $i+1$ occurs for the digit number j in the ensemble of the 3400 punch cards. This array was then submitted the χ^2 test for both lines and columns. The result was satisfactory. In other words, the array behaved like an array of the frequencies of independent and equi-probable stochastic variables should behave.

We next calculated the correlation coefficients

$$b_{jkl}, \quad j, k, = 1, \dots, 10, \quad l = 0, 1, \dots, 5$$

Here b_{jkl} is the correlation of the following two stochastic variables: digit j on one punch card and digit k on the punch card, l steps after the first mentioned one. Obviously, j, k, need only be considered when $j < k$, if $l = 0$, while all pairs j, k have to be considered if $l = 1, \dots 5$. This gives a total of 545 correlation coefficients. We calculated the expected size for such a correlation coefficient, say b. It would have been unreasonable to expect that no b_{jkl} should exceed b seriously, since their number was so large – 545. We did, however, check whether the distribution of these 545 numbers looked at all like a normal distribution with dispersion b. This proved to be the case. As you see, this latter test was somewhat "informal", but when the calculations were made I was not immediately aware of any traditional statistical method by which the "significance question" for such a large array of correlation coefficients could be decided.

To sum up, we felt that these two tests together gave a reasonable guarantee of the essential randomness and independence features for the numerical material in question.

I include with this letter a file of my correspondence on this Dr. Preston C. Hammer, the head of the IBM section at Alamos, where this work was done. I think you will have no seeing how the material on the last two pages these tests I gave above.

Some similar work was also done at Los Alamos earlier with 8 digit numbers. We squared and took the middle 8 digits and built up a sequence of, I think, about 1500 numbers. The results with 8 digits were also satisfactory. Similar sequences built with 7 digit numbers were unsatisfactory.

Both groups of numbers, the 10 digit one and the 8 digit one, were used in "Monte Carlo" type problems on the ENIAC, and they seem to be all right.

If there is any information on this subject which is of interest to you, please let me know. I shall be very glad to give you or obtain for you additional information.

I would be glad in principle to consult for Oak Ridge if you judge that this will be useful in your program. Before I answer your question concerning terms, could you tell me what the modus operandi at Oak Ridge is: Do yo make arrangements with consultants on the basis of the rates of government laboratories or along more industrial lines.

I am,

Sincerely yours,

John von Neumann

P.S. I would be much obliged to you if you could return the correspondence with Dr. P.C. Hammer in a week or two. I am likely to need it in Los Alamos where I am going on December 17th. If you return it to me before December 15, please send it to this address. If not, then to Los Alamos, P.O. Box 1663, in my name.

Original in Von Neumann Papers, Library of Congress, language of original: English

---

October 13, 1950

Dr. C.C. Hurd
IBM Corporation
590 Madison Avenue
New York 22, New York

Dear Cuthbert:

The following is a summary of my talk at Endicott. I hope that it is suitable for the purpose for which you need it. If not, please let me know, and we can then try to concoct something better.

SUMMARY

These are some quite general and rather incomplete comments on future uses of very fast computing machines[97].

A major concern which is frequently voiced in connection with such devices, in particular in view of the extremely high speeds which may now be hoped for, is

[97]See Section 8 of the Introductory Comments for notes on von Neumann's work on computer development. The Editor.

that they will do themselves out of business rapidly – that is, that they will out-run the planning and coding that they require and, therefore, run out of work.

I do not believe that this objection will prove to be valid in actual fact. It is quite true that for problems of those sizes which in the past, and even in the nearest past, have been the normal ones for computing machines, planning and coding required much more time than the actual solution of the problem would require on one of the hoped-for, extremely fast future machines. It must be considered, however, that in these cases the problem-size was dictated by the speed of the computing machines then available. In other words, the size adjusted itself essentially automatically so that the problem-solution time became longer, but not prohibitively longer, than the planning and coding time. For faster machines, the same automatic mechanism will exert pressure towards problems of larger size, and the equilibrium between planning and coding time on one hand, and problem-solution time on the other, will again restore itself on a reasonable level once it will have been really understood how to use these faster machines. This will, of course, take some time. There will be a year or two, perhaps, during which extremely fast machines will have to be used relatively inefficiently while we are finding the right type and size problems for them. I do not believe, however, that this period will be a very long one, and it is likely to be a very interesting and fruitful one. In addition, the problem types which lead to these larger sizes can already now be descerned,[98] even before the extreme machine types to which I refer are available.

Another point deserving mention is this. There will probably arise, together with the large size problems which are in "equilibrium" with the speed of the machine, other, smaller "subliminal" problems, which one may want to do on a fast machine, although the planning and programming time is longer than the solution time, simply because it is not worthwhile to build a slower machine for smaller problems, after the faster machine for larger problems is already available. It is, however, not these "subliminal" problems, but those of the "right" size, which justify the existence and the characteristics of the fast machines.

Some problem classes which are likely to be of the "right" size for fast machines are of the following ones: (1) In hydrodynamics, problems involving two and three dimensions. In the important field of turbulence, in particular, three-dimensional problems will have to be primarily considered. (2) Problems involving the more difficult parts of compressible hydrodynamics, especially shock wave formation and interaction. (3) Problems involving the interaction of hydrodynamics with various forms of chemical or nuclear reaction kinetics. (4) Quantum mechanical wave function determinations – when two or more particles are involved and the problem is, therefore, one of a high dimensionality.

In connection with the two last mentioned categories of problems, as well as with various other ones, certain new statistical methods, collectively described as "Monte Carlo Procedures", have recently come to the fore. These require the calculation of large numbers of individual case histories, effected with the use of artificially produced "random numbers". The number of such case histories is necessarily large, because it is then desired to obtain the really relevant physical results by analyzing significantly large samples of those histories. This, again, is a complex of problems that is very hard to treat without fast, automatic means of computation, which justifies the use of machines of extremely high speed.

[98]Misspelled word. Correctly: discerned. The Editor.

I am leaving tonight for Los Alamos, and will be back in Princeton on November 3rd or 4th. I will then telephone to you without delay. Looking forward to seeing you then in Princeton, and with best personal regards,

Yours as ever,

John von Neumann

Original in Von Neumann Papers, Library of Congress, language of original: English

Letter to K. Husimi

September 29, 1937

Mr. Kodi Husimi
Care the Editor
Proceedings of the Physical-Mathematical Society of Japan
Faculty of Science
Tokyo Imperial University
Tokyo, Japan

My dear Mr. Husimi:

I have received with many thanks the reprint of your interesting memoir on "The Foundations of Quantum Mechanics, Part I". I need not tell you that I am much interested in its contents, since it deals with subjects on which I have been working lately. I hope very much that you will inform me of your further results in this field, and that there will be opportunity for us to discuss these questions in correspondence. I am sending you by this mail reprints of some papers of mine which are directly or indirectly connected with this subject. And I am enclosing herewith abstracts of four lectures which I have given lately on "Continuous Geometries" and on the treatment of the foundations of quantum theory with transition probabilities. The "Fourth Lecture" of the series deals with this quantum-mechanical topic. I give there a set of axioms for logics which lead necessarily to a system of quantum theory based on operators in Hilbert space, in about the style in which F.J. Murray and myself obtained the theory of "continuous finite dimensional" operator rings[99]. In obtaining these results it is important that not only the system of strict logics (based on the relations of implication and negation), but also the system of probability logics (based on the numerical function on "transitional probability"), be axiomatized. In this manner also a much more intrinsic connection of logics and of probability (or in the above terminology: of strict logic and of probability logics) is obtained than could be hoped for in the classical theory. (Compare, for instance, Axiom XII in the "Fourth Lecture", which states that every automorphism of strict logics is even one of probability logics. There could of course be no such thing in the classical theory, yet it is true in – and I think very characteristic for – quantum mechanics.) I have a complete manuscript containing detailed proofs of the work announced in the "Fourth Lecture", but I have not published it yet. I intend to lecture on them this year and have the lectures multigraphed.

[99]See Section 6 of the Introductory Comments for von Neumann's work on quantum logic. The Editor.

I shall be very glad to get your comments and reactions, particularly on the contents of this "Fourth Lecture". Thanking you once more for your interesting paper, I am

Very sincerely yours,

John von Neumann

Original in Von Neumann Papers, Library of Congress, language of original: English

Letters to P. Jordan

Urbana-Lincoln Hotel
Urbana, Illinois
Sunday, December 11, 1949

Dear Mr. Jordan,

Many thanks for sending me your reprints – I do not have to tell you how extraordinarily I am interested in them.

First of all, however: it is indeed good to hear from you again, and I am especially glad that we now meet again in the field of "quantum logic"[100].

Naturally, I have read the works on quantum logic and lattice theory you sent me with great interest. You are active in a field in which Garret Birkhoff on one hand and myself on the other have also worked further.

At this moment I am not in Princeton (I will return there in about 6 weeks only), thus, unfortunately, I do not have the older and the more recent literature before me. For this reason, what I am writing you might not be quite precise; I think, however, that it is essentially correct.

First, concerning lattice theory: your investigations seem to be going in a direction similar to that of Garrett Birkhoff. There are probably many overlaps but the results are certainly not identical. Do you know his book "Lattice Theory?" (First edition: about 1940, Second edition: 1948.) Your concepts such as "Sippen", "Darstellungen" and "Arten" seem to be related to his "congruence-classes", "representations" and "defining relations". So are also a few things on more or less "free" lattices with given generators. At this moment I am not sufficiently well-versed in the Scriptures with respect to this subject to be completely sure about the precise connections, similarities, and differences – I think however that your and Birkhoff's ideas, approach and results have a lot in common.

Should you not know Birkhoff's book: is it available in Germany (perhaps in Göttingen)? Shall I write to him on these subjects or would you rather do this yourself?

---

[100]See Section 6 of the Introductory Comments for von Neumann's work on quantum logic. The Editor.

The questions concerning quantum logic are of course closer to my field. You have of course correctly diagnosed the tendency and the gaps in the work on quantum logic by Birkhoff and me. I did further work on the subject after 1936; unfortunately, I have not properly published it, however. The lecture notes "Continuous Geometries" come closest to a publication. These are two volumes; I delivered the lectures in Princeton in 1935/36 and 1936/37. Photostatic copies of the Notes had been made but all copies are gone, unfortunately – I only have got one left. As far as I know, there was a copy available in Germany: Koethe had one (perhaps he still has it – in any case, it seemed still to have existed around 1941). I now want to publish a new edition. May I send you a copy then? I hope the new edition will exist within 4-6 months.

The order in my Proc. Nat. Acad. work[101] and in the Lecture Notes on "Continuous Geometries" deviates from the one desirable for quantum logic in that I introduce the "negation" $\overline{a}$ as late as possible. Indeed: one needs $\overline{a}$ in quantum logic from the very beginning; from the perspective of pure geometry one is interested in getting as far as possible without $\overline{a}$.

In "Continuous Geometries" I have worked with different treatments of distributivity and commutativity properties (the Proc. Nat. Acad. contains hints concerning this), but your approach seems to me the best for the purpose of quantum logic.

I have worked a lot with systems of quantum logic in which the *transition probability* $p(a, b)$ enters axiomatically. From these "Continuous Geometry" can be derived, together with the theorem that the "coordinate field" K has to be that of the real numbers (K_r), or the complex numbers (K_c) or the quaternions (K_q). (In contrast, *each* "Continuous Geometry" – with an arbitrary K – possesses an *apriori* probability $P(a)$: the "dimension".) Thereby what is essential in quantum mechanics follows from the concept of "transition probability".

Nb.: Do you know the work I wrote in Russian for *Matematiceski Zbornik* in about 1937?[102] This is an extension to the infinite dimensional case of the work by Jordan-Wigner-JvN.[103] May I send you a reprint? Do you know the five operator theoretical works written partly by F. J. Murray and myself, and partly by myself alone in 1936-1948 in "Annals of Mathematics" (one in "Mathematical Transactions")?[104]

I had hoped at the beginning that the continuous – but finite dimensional quasi-quantum mechanical systems I had found are not subject to the well known divergence miseries in quantum electrodynamics etc. I did not succeed in proving this but the opposite is also not established. These continuous – but finite dimensional quantum mechanics can be brought into a quite amusing relation with

[101] J. von Neumann: "Continuous geometry", Collected Works Vol. IV. No. 8. The Editor.

[102] J. von Neumann: "On an algebraic generalization of the quantum mechanical formalism", Collected Works Vol. III. No. 9. The Editor.

[103] P. Jordan, J. von Neumann and E. Wigner: "On an algebraic generalization of the quantum mechanical formalism", Collected Works Vol. II. No. 21. The Editor.

[104] These are the so-called "rings of operators papers", see Collected Works, Volume III. The Editor.

"second quantization". In view of the recent crunch in quantum mechanics of elementary particles this possibility merits perhaps some more attention. I would gladly write you about this if it interests you.

In the hope of hearing more from you about your recent interests, and thanking you once again for your reprints, I remain

Sincerely yours

J. von Neumann

Original in Pascual Jordan Nachlass, Nr. 562, Blatt 7r-14r, Staatsbibliothek zu Berlin – Preussischer Kulturbesitz. Languge of original: German. Translated by M. Rédei.

January 12, 1950

Dear Mr. Jordan,

Many thanks for your nice and detailed letter of 12/16/1949.[105] At this moment I am not in Princeton but will be there on 2/1. From there I will send you a list of rings-of-operators papers by Murray and myself. I will of course also send you a copy of the new edition of the "Continuous Geometries" lecture notes as soon as it comes out. I have not published anything on the axiomatics of transition probabilities. At this moment I am somewhat overloaded by work but I will write you about this approach in detail as soon as I can. The convergence question is open but it possibly offers some attractive opportunities.

It is due to my laziness only that Snyder has re-discovered the space-time quantization we discussed in Cologne in 1938. In any case, this is damped by skepsis – the more I thought about the subject the less likely it appeared to me that the solution could lie in this rather complicated direction. The whole thing seemed – and seems – to me an amusing and perhaps also instructive but nevertheless purely formal mathematical tour de force. The development since then and since Snyder's publication is evidence in the same direction. Don't you also think that the trouble of elementary particle physics lies deeper?

I will talk to Birkhoff and I think we will be able without much effort to get for you his book and the other essential lattice-literature.

I have heard something of your work on cosmology. These works – especially the variability of the gravitational constant and the compensation of the rest mass energy in particle creation through negative gravitational energy, and the further relations emerging from this – seem to me very nice and stimulating. I will of course read your "Nature" paper and am looking forward to it with much interest. Any publication or galley proofs that you could send me on this would of course greatly interest me.

I also would be very thankful for further details concerning your work and ideas regarding Cayley-numbers, etc.

[105]Here, and below, dates are written in the convention mm/dd/yr. The Editor.

Please excuse this hastily written letter. In the hope of hearing from you soon, I remain

Sincerely yours

J. von Neumann

Original in Pascual Jordan Nachlass, Nr. 562, Blatt 7r-14r, Staatsbibliothek zu Berlin – Preussischer Kulturbesitz. Languge of original: German. Translated by M. Rédei.

Letters to I. Kaplansky

February 1, 1945

Dr. Irving Kaplansky
401 West 118th Street
New York 27, N.Y.

Dear Doctor Kaplansky,

Many thanks for your letter of January 27 and your manuscript. The results are really very interesting, and they seem to indicate that the concept of the "value" of a rectangular matrix deserves systematic investigation. Your method of reducing the non-quadratic matrices and the reducible quadratic ones, and your recognition of the "completely mixed" case as the "irreducible" one, are particularly fascinating.

I will see to it that the paper is printed in the "Annals".

I think that it would be most important to find a practicable procedure to locate a "superfluous" column or line in a matrix, when your criteria show that there is one.

I have also played with "finite" methods to find solutions, and more specifically with methods which would give at least one solution in a practicable time for matrices of orders ~ 20. I have some results and some surmises on this subject, – I would very much like to discuss it with you sometime.

A. Wald has settled "infinite cases" in considerable generality. His paper will appear in the "Annals"[106] My own preference is for this procedure:[107]

(1) Let $A(B)$ have the – possibly infinite – set of strategies $S(T)$. Then the general, mixed strategy $\xi(\eta)$ of $A(B)$ runs over the set of all – non-negative, total measure $= 1$ – measure functions in $S(T)$: $M_S(M_T)$. However, I prefer to replace M_S, M_T by two convex spaces (fundamental operation: f, g in M and α real > 0, < 1, allow to form $\alpha f + (1-\alpha)g$, this $\alpha f + (1-\alpha)g$ has the obvious arithmetical properties) M, N, without insisting on M, N-s origin. (The desirability of changing the domain of ξ and η, as done in my discussion of Poker, suggests this.) Now let $K(\xi, \eta)$ be a (numerical, bilinear) form for ξ in M, η in N.

(2) M, N have no topologies, or if they have any, I forget them. I now topologize M: Each ξ in M has the coordinates ξ_η, η in M, defined by $\xi_\eta = K(\xi, \eta)$; the topology of M is then the weak topology with respect to these coordinates ξ_η. Similarly for N.

I call $M(N)$ intrinsically compact if it is compact in this topology.

[106] A. Wald: Generalization of a theorem by J. von Neumann concerning zero-sum two person games, Annals of Mathematics, **46** (1945) 281-286. The Editor.

[107] See Section 9 of the "Introductory Comments" for background information about von Neumann and game theory. The Editor.

(3) I want to prove

$$\text{Max}_\xi \ \text{Min}_\eta K(\xi, \eta) = \text{Min}_\eta \ \text{Max}_\xi K(\xi, \eta) \tag{1}$$

If M', N' are subconvexes of M, N, each spanned by a finite number of elements, then

$$\text{Max}_{\xi \text{ in } M'} \ \text{Min}_{\eta \text{ in } N'} K(\xi, \eta) = \text{Min}_{\eta \text{ in } N'} \ \text{Max}_{\xi \text{ in } M'} K(\xi, \eta) \tag{2}$$

follows immediately from my "finite" theorem. Now if M and N are both intrinsically compact, then (1) can be easily inferred from (2) by means of the Borel "covering theorem".

Hoping to hear from you again,

Sincerely yours,

John von Neumann

P.S. I am just about to leave Princeton for four or five weeks, but my mail will reach me through the Institute.

Original in Von Neumann Papers, Library of Congress, language of original: English

---

March 1, 1950

Dr. I. Kaplansky
The University of Chicago
Department of Mathematics
Chicago 87, Illinois

Dear Dr. Kaplansky,

Very many thanks for your letter of February 11th and your manuscript on "Projections in Banach Algebras". I am very glad that you are submitting it for THE ANNALS, and I will immediately recommend it for publication.

Your results are very interesting. You are, of course, very right: I am and I have been for a long time strongly interested in a "purely algebraical" rather than "vectorial-spatial" foundation for theories of operator-algebras or operator-like-algebras.[108] To be more precise: It always seemed to me that there were three successive levels of abstraction – first, and lowest, the vectorial-spatial, in which the Hilbert space and its elements are actually used; second, the purely algebraical, where only the operators or their abstract equivalents are used; third, the highest, the approach when only linear spaces or their abstract equivalents (i.e. operatorially speaking, the projections) are used. There is a possible intermediate stage between the second and the third one, when only Hermitean operators or their abstract equivalents are used. After Murray and I had reached somewhat rounded results on the first level, I neglected to make a real effort on the second one, because I was tempted to try immediately the third one. This led to the theory of continuous geometries. In studying this, the third level, I realized that one is

[108]See Section 3 of the Introductory Comments on von Neumann's work on operator algebras. The Editor.

led there to the theory of "finite" dimensions only. The discrepancy between what might be considered the "natural" ranges for the first and the third level led me to doubt whether I could guess the correct degree of generality for the second one. In addition I assumed that the best intuitive guide in this field is the analogy with quantum mechanics, and the quantum mechanical interpretation seemed to justify the approach on the third level, or on the intermediate one between the second and third levels (Hermitean operators) rather than on the second level. As you know, I made one more effort in this intermediate direction, about half of this is described in the Mat. Zbornik.[109] The other half of this work is a theory of dimensions within the algebraical framework of the first half. I have never published it. I used an equivalence theory based on the "bisecting" projections which appear in your proof of your lemma 3.1 (cf. your preliminary remarks on page 7). I obtained the usual subdivision into the types *I*, *II*, *III* – finite and infinite – and used this to pass to a Hilbert space in the cases other than *III*.

Your treatment of the purely algebraical case – that is, what I would call the second level – strikes me as very elegant and very satisfactory. In looking at your simple proof of the additivity of finiteness in section 6, the lemma 6.1 seems to me to contain the decisive trick. This is certainly very pretty, and impresses me, too, as being the logical way to handle this problem. Murray and I at this point certainly made our lives harder than necessary. The sequence 5.2 to 5.4 seems to me less radically different from our procedures. In fact, some of the peculiarities of Murray's and my procedure were due to a peculiar idiosyncracy of mine: I was very anxious to derive the concept of abstract, as well as numerical, dimensionality *without* a need for a prior distinction between the discrete and continuous cases. In other words, I felt it was a "better and purer" proof, which permitted one to get all the general properties of dimensionality first, and permitted one to derive its isomorphism to rational integers or the real numbers afterwards, "in the Euclidean manner". You are obviously uninhibited about making the classification into discrete and continuous cases earlier in the deduction.

I hope that these anatomical discussions will not obscure the fact that I am very impressed by your paper, and feel that it is very interesting and beautiful. The conciseness and simplicity of your postulates A,B (page 3) are certainly all one can wish for.

Your questions regarding the cardinal (Cantorian aleph) dimensions in the non-separable case are very interesting. I did also consider the non-separable case, but on the first level only – that is, in the case of operator rings in non-separable Hilbert spaces. I obtained, of course, the same cardinal dimensions which you do. If I remember those things correctly, the difficulties to which you refer do not arise in that case because the alephs of the dimensionality can then be connected with certain topologically defined alephs (minimum alephs of everywhere dense sets and the like). Since this was more than ten years ago I do not remember the details well, but I could try to work them out. Unluckily, they may not be very helpful in that case which concerns you, since your investigations take place on the second, purely algebraical level. I will think more about the subject and see whether I am able to come out with anything that is useful.

---

[109] J. von Neumann: "'On an algebraic generalization of the quantum mechanical formalism", Collected Works Vol. III. No. 9. The Editor.

I hope that you will let me know more about your work on these subjects as it progresses. I can assure you that I am very glad to be reminded about these things.

With best regards,

Sincerely yours,

John von Neumann

Original in Institute for Advanced Study Archives, Historical Studies-Social Science Library, language of original: English

Letter to C.E. Kemble

December 6 [????]

My dear Professor Kemble,

Please excuse my delay in answering your letter. I was out of town when it arrived, and found it only later in a mislaid part of my mail.

If I understand you correctly, your question is this:

Given a differential operator

$$H = -\left(\frac{\partial^2}{\partial x_1^2} + \ldots + \frac{\partial^2}{\partial x_n^2}\right) + V(x_1, \ldots, x_n) \tag{1}$$

when is it possible to apply the general theorems of operator-theory to it, i.e., when is it "hypermaximal"?[110] In particular: What is the situation if

$$n = 3m \tag{2}$$

$$x_{3i-2} = \frac{2\pi\sqrt{2M_i}}{h} X_i \tag{3}$$

$$x_{3i-1} = \frac{2\pi\sqrt{2M_i}}{h} Y_i \tag{4}$$

$$x_{3i} = \frac{2\pi\sqrt{2M_i}}{h} Z_i \tag{5}$$

$$V(x_1, \ldots, x_n) = \sum_{i<j} \frac{\epsilon_i \epsilon_j}{R_{ij}} \tag{6}$$

$$R_{ij} = \sqrt{(X_i - X_j)^2 + (Y_i - Y_j)^2 + (Z_i - Z_j)^2} \tag{7}$$

(This being the wave equation of an m-particle-system, particle No i $(= 1, \ldots, m)$ having mass M_i, charge ϵ_i, and the cartesian coordinates X_i, Y_i, Z_i. The forces are Coulombic.)

The honest answer is: I don't know.

But I am perhaps not too boring, if I add some comments to this shameful confession. The excuse for writing the 12 pages which follow is to be found – if it is any good at all – on page 14.[111]

A.: If $V(x_1, \ldots, x_n)$ had no singularities, and if the range of variability of the $x_1, \ldots, x_n$ was restricted, then the self-adjointness of (1) could be, I think, established with our present methods.

---

[110] For the notion of "hypermaximal" (unbounded) linear operator and other relevant notions related to unbounded operators see Section 4 of the Introductory Comments. The Editor.

[111] The last page of the letter. The Editor.

This would mean in (2)-(7): Restrict the X_i, Y_i, Z_i $(i = 1, \ldots, m)$ to a finite "box", and replace the "singular" Coulomb potential

$$\sum_{i<j} \frac{\epsilon_i \epsilon_j}{R_{ij}} \tag{8}$$

by some "regular"

$$\sum_{i<j} \epsilon_i \epsilon_j \phi(R_{ij}) \tag{9}$$

where $\phi(x)$ is a function of $x > 0$ very similar to $\frac{1}{x}$, but "regular" for $x \to 0$.

B.: The really interesting case, that is (2)-(7) unmitigated, must be quite tricky. Because replacement of (8) by

$$\sum_{i<j} \frac{\epsilon_i \epsilon_j}{R_{ij}^2} \tag{10}$$

in (6) makes the operator H in (1) non-hypermaximal (see below in **C.**) Thus if there is any general theorem which covers (2)-(7) with (8), it must be complicated enough to exclude (2)-(7) with (10).

C.: As to my statement concerning (2) with (10), put $m = 2$, and "separate" the coordinates of the center of gravity as well as the polar coordinates. Then for

$$R = R_{12} = \sqrt{(X_1 - X_2)^2 + (Y_1 - Y_2)^2 + (Z_1 - Z_2)^2} \tag{11}$$

the differential operator

$$H' = -\frac{h^2}{8\pi^2 M}\left(\frac{\partial^2}{\partial R^2} + \frac{2}{R}\frac{\partial}{\partial R} - \frac{n(n+1)}{R^2}\right) + \frac{\epsilon_1\epsilon_2}{R^2} \tag{12}$$

results, where $n = 0, 1, 2, \ldots$ is the rotation-quantum number. Replacing the wave-function by its R-fold, and H' by its $\frac{8\pi^2 M}{h^2}$-fold,

$$H'' = -\frac{\partial^2}{\partial R^2} + \frac{a}{R^2}, \qquad a = n(m+1) + \frac{8\pi^2 M \epsilon_1 \epsilon_2}{h^2} \tag{13}$$

obtains.

Now it is well known that the operator H'' in (13) is a peculiar one, but its properties are usually not correctly interpreted. Specifically: replacing R by θR $(\theta > 0)$ clearly carries H'' over into $\theta^{-2}H''$. Therefore one should believe, that if λ is a (point- or continuous spectrum) proper-value of H, then $\theta^{-2}\lambda$ is one of the same sort. So the $\lambda > 0$ are either all point-spectrum proper values, or all continuous-spectrum proper-values, or none of them is a proper value. Similarly for the $\lambda < 0$.

This is alright for the $\lambda \geq 0$: They are easily seen to be cont.-sp. pr.-v.'s for each a. And if H'' is positive definite, as it is for all $a \geq -\frac{1}{4}$, then this is the entire spectrum. But if $a < -\frac{1}{4}$, then H'' is not definite, pr.v.'s $\lambda < 0$ should exist, by the physical nature of the problem they must be *point-spectrum* p.v.'s, and by the above argument all $\lambda < 0$ must be it. But this is absurd, as a continuum of p.-sp. p.-v.'s cannot exist. What, then is wrong?

As the continuum of p.-sp. p.-v.'s exists, they cannot be all orthogonal, and so H'' cannot be Hermitean without further explanations. Now when is H'' Hermitean?

$$\begin{aligned}\int_0^\infty H''\phi(R)\overline{\psi(R)}\mathrm{d}R \quad &- \quad \int \phi(R)\overline{H''\psi(R)}\mathrm{d}R = \\ = \int_0^\infty \Big[\frac{\mathrm{d}^2}{\mathrm{d}R^2}\phi(R)\overline{\psi(R)} \quad &- \quad \phi(R)\frac{\mathrm{d}^2}{\mathrm{d}R}\overline{\psi(R)}\Big]\mathrm{d}R = \\ = \lim_{R\to 0}\Big\{\frac{\mathrm{d}}{\mathrm{d}R}\phi(R)\overline{\psi(R)} \quad &- \quad \phi(R)\frac{\mathrm{d}}{\mathrm{d}R}\overline{\psi(R)}\Big\}_R^\infty = \\ = \lim_{R\to 0}\Big(-\frac{\mathrm{d}}{\mathrm{d}R}\phi(R)\overline{\psi(R)} \quad &- \quad \phi(R)\frac{\mathrm{d}}{\mathrm{d}R}\overline{\psi(R)}\Big)\end{aligned}$$

So the Hermitean character of H'' requires , that only such solutions $\phi(R), \psi(R)$ should be (simultaneously) permitted, for which

$$\lim_{R\to 0}\Big(-\frac{\mathrm{d}}{\mathrm{d}R}\phi(R)\overline{\psi(R)} + \phi(R)\frac{\mathrm{d}}{\mathrm{d}R}\overline{\psi(R)}\Big) = 0 \tag{14}$$

What is the asymptotic behavior of the solutions $\phi(R)$ of

$$H''\phi(R) = \lambda R, \qquad \lambda < 0 \tag{15}$$

at $R \to 0$? The characteristic exponent ν ($\phi(R) = R^\nu \cdot \mathit{regular\ function}$) must fulfill

$$\nu(\nu-1) + a = 0, \qquad \nu = \frac{1}{2} \pm \sqrt{a + \frac{1}{4}}$$

As $a < \frac{1}{4}$, say $a = -\frac{1}{4} - b^2$, $b > 0$, so $\nu = \frac{1}{2} \pm ib$. So

$$\phi(R) \sim R^{\frac{1}{2}\pm ib} = \sqrt{R}\ \begin{matrix}\cos \\ \sin\end{matrix}\ \begin{matrix}(b\ln R) \\ (b\ln R)\end{matrix}$$

That is, asymptotically at $R \to 0$:

$$\phi(R) \approx \mathcal{A}\sqrt{R}\sin(b\ln R + \alpha) \tag{16}$$

for suitable constants $\mathcal{A}, \alpha$. Similarly

$$\psi(R) \approx \mathcal{B}\sqrt{R}\sin(b\ln R + \beta) \tag{17}$$

and therefore

$$\begin{aligned} -\frac{\mathrm{d}}{\mathrm{d}R}\phi(R)\overline{\psi(R)} + \phi(R)\frac{\mathrm{d}}{\mathrm{d}R}\overline{\psi(R)} \quad &\approx \\ \approx -\mathcal{A}\Big[\sqrt{R}\frac{b}{R}\cos(b\ln R + \alpha) + \frac{1}{2\sqrt{R}}\sin(b\ln R + \alpha)\Big]\overline{\mathcal{B}}\sqrt{R}\sin(b\ln R + \beta) \quad &+ \\ +\mathcal{A}\sqrt{R}\sin(b\ln R + \alpha)\overline{\mathcal{B}}\Big[\sqrt{R}\frac{b}{R}\cos(b\ln R + \beta) + \frac{1}{2\sqrt{R}}\sin(b\ln R + \beta)\Big] \quad &= \\ = \mathcal{A}\overline{\mathcal{B}}b\Big[-\cos(b\ln R + \alpha)\sin(b\ln R + \beta) + \sin(b\ln R + \alpha)\cos(b\ln R + \beta)\Big] \quad &= \\ = \mathcal{A}\overline{\mathcal{B}}b\sin(\beta - \alpha) \end{aligned}$$

Thus (14) means $\sin(\beta - \alpha) = 0$, $\alpha = \beta \ldots \mathit{mod}\ \pi$, or, as α, β are only defined (by (16)-(17)) $\ldots$ *mod* π, so

$$\alpha = \beta \tag{18}$$

In other words: H'' is no Hermitean operator at all, except when further "boundary conditions" are imposed at $R \to 0$:

$$\phi(R) \approx A\sqrt{R}\sin(b\ln R + \alpha) \qquad \text{for a fixed } \alpha \tag{19}$$

Denote this new operator H'' by $H''_{(\alpha)}$. (It would not do to prohibit singularities at $R \to 0$ altogether, because these would make (15) insolvable, and therefore certainly keep H'' from being hypermaximal.)

Now if $\phi(R)$ satisfies (19), then $\phi(\theta R)$ $(\theta > 0)$ satisfies (19) with $b\ln\theta + \alpha$ in place of α. Thus the operator $H''_{(\alpha)}$ will apply to it if and only if $b\ln\theta = k\pi$, $k = 0, \pm 1, \pm 2, \ldots$, that is $\theta = e^{\frac{\pi}{b}}k$. Thus the only $\theta^{-2}\lambda$ $(\lambda < 0)$ which will be point-spectrum proper values along with λ are those with $\theta = e^{\frac{\pi}{b}}k$, that is the $e^{-\frac{2\pi}{b}}k\lambda$. So the negative p.-sp. p.-v.'s of $H''_{(\alpha)}$ form an *enumerable sequence* only.:

$$\ldots, e^{-\frac{4\pi}{b}}\lambda,\ e^{-\frac{2\pi}{b}}\lambda,\ \lambda,\ e^{\frac{2\pi}{b}}\lambda,\ e^{\frac{4\pi}{b}}\lambda, \ldots$$

D.: As all this refers to the fictitious potential

$$V = \frac{\epsilon_1\epsilon_2}{R^2} \tag{20}$$

it is practically not important. But it is worth pointing out, that if we replace Schrödinger's energy-operator (1) (resp. (12), (13)) by Dirac's, the conditions change. In this case, already the Coulombic potential

$$V = \frac{\epsilon_1\epsilon_2}{R} \tag{21}$$

does the same tricks. In this case the energy operator is (for two particles, after the coordinates of the center of gravity have been "separated")

$$H = \frac{hc}{2\pi i}\left(\alpha_1\frac{\partial}{\partial x} + \alpha_2\frac{\partial}{\partial y} + \alpha_3\frac{\partial}{\partial z}\right) + Mc^2\alpha_4 + \frac{\epsilon_1\epsilon_2}{R} \tag{22}$$

(α_1, α_2, α_3, α_4 being the fourth-order Dirac-matrices), and after the "separation" of the polar-coordinates (including replacement of the wave-function by its R-fold and of the energy by its $\frac{2\pi}{hc}$-fold):

$$H'' = \beta_1\frac{1}{i}\frac{\mathrm{d}}{\mathrm{d}R} + u\beta_2 + v\beta_3\frac{1}{R} + w\frac{1}{R} \tag{23}$$

Here β_1, β_2, β_3 are the second order Pauli-matrices, and u, v, w are numerical constants:

$$u = \frac{2\pi Mc}{h}, \qquad v = k, \qquad w = \frac{2\pi\epsilon_1, \epsilon_2}{hc}$$

$k = \pm 1, \pm 2, \ldots$ being Dirac's "auxiliary" rotation-quantum number.

Now a discussion of (23) along the same lines, as the one of (13) in **C.** gives, although not quite so easily, similar results. Write the two components of the wave function as $\phi_1(R), \phi_2(R)$ and choose

$$\beta_1 = \begin{pmatrix} 0 & i \\ -i & 0 \end{pmatrix}, \ \beta_2 = \begin{pmatrix} 0 & 1 \\ -1 & 0 \end{pmatrix}, \ \beta_3 = \begin{pmatrix} 1 & 0 \\ 0 & -1 \end{pmatrix}$$

Then (23) becomes:

$$H''\{\phi_1(R), \phi_2(R)\} = \tag{24}$$

$$= \left\{\frac{\mathrm{d}}{\mathrm{d}R}\phi_2(R) + u\phi_2(R) + \frac{w+v}{R}\phi_1(R), -\frac{\mathrm{d}}{\mathrm{d}R}\phi_1(R) + u\phi_1(R) + \frac{w-v}{R}\phi_2(R)\right\} \tag{25}$$

Thus the analogue of (15),

$$H''\{\phi_1(R), \phi_2(R)\} = \lambda\{\phi_1(R), \phi_2(R)\} \tag{26}$$

becomes

$$\frac{\mathrm{d}}{\mathrm{d}R}\phi_2(R) + u\phi_2(R) = (\lambda - \frac{w+v}{R})\phi_1(R) \tag{27}$$

$$-\frac{\mathrm{d}}{\mathrm{d}R}\phi_1(R) + u\phi_1(R) = (\lambda - \frac{w+v}{R})\phi_2(R) \tag{28}$$

The characteristic exponent ν must therefore, as is well known, fulfill

$$\begin{vmatrix} \nu & w+v \\ w-v & -\nu \end{vmatrix} = 0, \qquad \nu = \pm\sqrt{v^2 - w^2}$$

and the ratio of the leading terms of $\phi_1(R), \phi_2(R)$ becomes $-\frac{\nu}{w+v} = \frac{w-v}{\nu}$. So if $|w| > |v|$, then $\nu = \pm i\sqrt{w^2 - v^2}$ is imaginary, and the above ratio is $\pm i\sqrt{\frac{w-v}{w+v}}$. So we have asymptotically for $R \to 0$:

$$\begin{aligned} \phi_1(R) &\approx A_1\sqrt{|w-v|}\; iR^{i\sqrt{w^2-v^2}} - A_2\sqrt{w-v}\; iR^{-i\sqrt{w^2-v^2}} \\ \phi_2(R) &\approx A_1\sqrt{|w+v|}R^{i\sqrt{w^2-v^2}} + A_2\sqrt{w+v}R^{i\sqrt{w^2-v^2}} \end{aligned}$$

or putting

$$A_1 = \frac{1}{2}Ae^{i\alpha}, \; A_2 = \frac{1}{2}Ae^{-i\alpha}$$

$$\phi_1(R) \approx -A\sqrt{|w-v|}\sin(\sqrt{w^2 - v^2}\ln R + \alpha) \tag{29}$$

$$\phi_2(R) \approx -A\sqrt{|w+v|}\cos(\sqrt{w^2 - v^2}\ln R + \alpha) \tag{30}$$

The Hermitean character of H'' means, as

$$\begin{aligned} &\int_0^\infty \Big[H''\{\phi_1(R), \phi_2(R)\}\Big]_1 \overline{\psi_1(R)}\mathrm{d}R + \\ &+ \int_0^\infty \Big[H''\{\phi_1(R), \phi_2(R)\}\Big]_2 \overline{\psi_2(R)}\mathrm{d}R - \\ &- \int_0^\infty \phi_1(R)\overline{\Big[H''\{\psi_1(R), \psi_2(R)\}\Big]_1}\mathrm{d}R - \\ &- \int_0^\infty \phi_2(R)\overline{\Big[H''\{\psi_1(R), \psi_2(R)\}\Big]_2}\mathrm{d}R = \end{aligned}$$

$$= \int_0^\infty \Big\{ \Big[\frac{\mathrm{d}}{\mathrm{d}R}\phi_2(R) + u\phi_2(R) + \frac{w+v}{R}\phi_1(R) \Big] \overline{\psi_1(R)} +$$
$$+ \Big[-\frac{\mathrm{d}}{\mathrm{d}R}\phi_1(R) + u\phi_1(R) + \frac{w-v}{R}\phi_2(R) \Big] \overline{\psi_2(R)} -$$
$$-\phi_1(R) \Big[\frac{\mathrm{d}}{\mathrm{d}R}\overline{\psi_2(R)} + u\overline{\psi_2(R)} + \frac{w+v}{R}\overline{\psi_1(R)} \Big] -$$
$$-\phi_2(R) \Big[-\frac{\mathrm{d}}{\mathrm{d}R}\overline{\psi_1(R)} + u\overline{\psi_1(R)} + \frac{w-v}{R}\overline{\psi_2(R)} \Big] \Big\} \mathrm{d}R =$$
$$= \int_0^\infty \Big[\frac{\mathrm{d}}{\mathrm{d}R}\phi_2(R)\overline{\psi_1(R)} + \phi_2(R)\frac{\mathrm{d}}{\mathrm{d}R}\overline{\psi_1(R)} -$$
$$- \frac{\mathrm{d}}{\mathrm{d}R}\phi_1(R)\overline{\psi_2(R)} - \phi_1(R)\frac{\mathrm{d}}{\mathrm{d}R}\overline{\psi_2(R)} \Big] \mathrm{d}R =$$
$$= \lim_{R\to 0} \Big\{ \phi_2(R)\overline{\psi_1(R)} - \phi_1(R)\overline{\psi_2(R)} \Big\}_R^\infty =$$
$$= \lim_{R\to 0} \Big(-\phi_2(R)\overline{\psi_1(R)} + \phi_1(R)\overline{\psi_2(R)} \Big)$$

that

$$\lim_{R\to 0}(-\phi_2(R)\overline{\psi_1(R)} + \phi_1(R)\overline{\psi_2(R)}) = 0 \tag{31}$$

Substituting into this a ϕ by (29) and a ψ by

$$\psi_1(R) \approx -\mathcal{B}\sqrt{|w-v|}\sin(\sqrt{w^2-v^2}\ln R + \beta) \tag{32}$$
$$\psi_2(R) \approx \mathcal{B}\sqrt{|w+v|}\cos(\sqrt{w^2-v^2}\ln R + \beta) \tag{33}$$

we obtain

$$-\phi_2(R)\overline{\psi_1(R)} + \phi_1(R)\overline{\psi_2(R)} \approx$$
$$\approx \mathcal{A}\overline{\mathcal{B}}\sqrt{w^2-v^2}\Big[\cos(\sqrt{w^2-v^2}\ln R + \alpha)\sin(\sqrt{w^2-v^2}\ln R + \beta) -$$
$$-\sin(\sqrt{w^2-v^2}\ln R + \alpha)\cos(\sqrt{w^2-v^2}\ln R + \beta)\Big] =$$
$$-\mathcal{A}\overline{\mathcal{B}}\sqrt{w^2-v^2}\sin(\beta-\alpha)$$

and so (31) becomes $\sin(\beta-\alpha) = 0$, $\alpha = \beta$ ... *mod* π, or, as α, β are only defined (by (29)-(30), (29)-(30)) ... *mod* π, so

$$\alpha = \beta \tag{34}$$

Thus the H'' of (23) and (24)-(25) must again be restricted: It is only then a Hermitean operator, when subjected to the further boundary conditions at $R \to 0$

$$\phi_1(R) \approx -\mathcal{A}\sqrt{|w\quad v|}\sin(\sqrt{w^2-v^2}\ln R + \alpha) \tag{35}$$
$$\phi_2(R) \approx \mathcal{A}\sqrt{|w+v|}\cos(\sqrt{w^2-v^2}\ln R + \alpha) \tag{36}$$

for a fixed α. This then is the operator $H''_{(\alpha)}$.

I have *heuristic* ways to see, that while H_α has continuous spectra for $\lambda \geq u$ and $\lambda \leq -u$ (energy $\geq Mc^2$ resp. $\leq -Mc^2$), it has a point spectrum for $-u \leq \lambda < u$ ($-Mc^2 <$ energy $< Mc^2$), which depends on the choice of α.

These critical phenomena occur for $|w| > |v|$, that is for

$$|\epsilon_1\epsilon_2| > \frac{hc}{2\pi}|K| \qquad |K| = 1, 2, \ldots \tag{37}$$

If $\epsilon_1 = \epsilon$, $\epsilon_2 = z\epsilon$, ϵ being the electronic charge, this is

$$z > \frac{hc}{2\pi\epsilon^2}|K| = 137|K| \tag{38}$$

Thus the above phenomena are the quantum mechanically correct and consequent description of the well-known "relativistic falling of the electron into the nucleus" for molecular charges $z\epsilon$ with $z > 137$.

E. If the potential $\frac{\epsilon_1\epsilon_2}{R^2}$ in **C.**, or the potential $\frac{\epsilon_1\epsilon_2}{R^2}$ in **D.**, is replaced by an approximately equal one, which does not increase after R falls below a certain $R_0 > 0$ – for instance if we replace these potentials by

$$\frac{\epsilon_1\epsilon_2}{R^2 + R_0^2} \quad \text{resp.} \quad \frac{\epsilon_1\epsilon_2}{\sqrt{R^2 + R_0^2}} \tag{39}$$

then this happens:

As $R_0 \to 0$ the spectrum of the corresponding energy operator H'' (or H) converges to the spectra of the H'' from **C.** resp. **D.**, but the α introduced there varies all the time, roughly as

$$constant \cdot \ln R$$

Thus the potentials

$$\frac{\epsilon_1\epsilon_2}{R^2} \quad \text{resp.} \quad \frac{\epsilon_1\epsilon_2}{R} \tag{40}$$

are not clearly defined limiting cases any more: When approximated by a potential (39) their behavior will depend *essentially* on the value of R_0, even if R_0 is very near to 0.

Applying this to $\epsilon_1 = \epsilon$, $\epsilon_2 = -z\epsilon$ at the end of **D.**: If an electron rotates around a z-fold charged nucleus, then its behavior will depend *essentially* on intra-nuclear conditions, as soon as (38) holds:

$$z > 137|K| \tag{41}$$

K being Dirac's "auxiliary" rotation-quantum number, $K = \pm 1, \pm 2, \ldots$.

Please excuse the cannibalistic length of my letter, and its contents, which are not too new. I merely wanted to point out, that even the non-hypermaximality of an energy-operator may have a good physical meaning, and that the Coulombic field is not very far from such a state of affairs.

I hope to hear from you again, and to meet you in a not-too-distant future – I will be in Cambridge on Jan. 24.-27.

With the best greetings,

Yours sincerely,

John von Neumann

Original in the Archives for the History of Quantum Physics; copy used for the volume from "Philosophisches Archiv" of the Library of the University of Konstanz, Konstanz, Germany, language of original: English

Letter to J.R. Killian

February 24, 1956

President James R. Killian, Jr.
Massachusetts Institute of Technology
Cambridge 39, Massachusetts
590 Madison Avenue
cc. Dr. J.A. Stratton

Dear President Killian,

When I last had the pleasure of meeting you and Dr. Stratton at the Cosmos Club, and subsequently, with respect to the same subject, Mr. Mervin Kelly in my office, you offered me an "Institute professorship" at the Massachusetts Institute of Technology, with a salary of $20,000 a year and with what I understand to be the privileges of such a position, i.e. no fixed teaching obligations, no fixed departmental affiliation, but general responsibility for the subject that my interests represent. We left our negotiations entirely open at the time. I would now like to tell you that I am considering this offer very seriously. I wish to make a decision about my future states and your offer, of course, figures in my decisions very importantly.

Of course, I am at present a Member of the U.S. Atomic Energy Commission and any plans that I may have to leave the government service must be strictly confidential. Nevertheless, in order to clarify our mutual ideas, I want to tell you – for planning purposes only – that if I went to the Institute, I would think in terms of the date of January 1, 1957, for the change in my status.

There are also some questions regarding detail that I would like to discuss with you. These are very important for the decision that I will have to make. They are as follows:

(1) What are the arrangements for retirement on the Institute faculty, and what part of these would specifically affect me in view of my particular status?

(2) What are the arrangements for the Institute faculty with regard to sickness, hospitalization, and disability insurance? Again, what would the terms be as applied to me? May I add that, as you probably know, at this moment my health is not good; therefore, this particular part of the Institute's arrangements and policies is of quite especial importance to me.

(3) As you know, the scientific interest which has characterized my work the past several years has been high-speed computing machines and, specifically, their application to problems of geophysics and meteorology. I take

it for granted that I can expect the availability of such facilities at M.I.T. in connection with the work of my associates in this field. Could you kindly confirm this point.

(4) In the past I have done a considerable amount of industrial and governmental consulting – of course, always in the areas in which my scientific and intellectual interests lie. I assume that such consulting is in conformity with the policies and desires of the Institute. Would you kindly confirm this. Would you also let me know whether there are any special rules about consulting which apply in particular to a person of my status.

I realize that there would have to be certain upper limits to the time which can be devoted to such activities. I would appreciate your telling me what they are. If you prefer a gentlemen's agreement on this subject, please let me know what its guide lines would be.

Needless to say, I will await your elucidations on these points with the greatest of interest.

I assume that you will consider our correspondence strictly confidential and to be communicated only to those members of the government boards and administrations of M.I.T. who are properly involved.

Sincerely yours

John von Neumann

Original in Von Neumann Papers, Library of Congress, language of original: English

Letters to H.D. Kloosterman

March 25, 1953

Prof. Dr. H.D. Kloosterman
International Congress of Mathematicians
2d Boerhaavestraat 49
Amsterdam-0
The Netherlands

Dear Professor Kloosterman:

I have just received your letter of March 20, and the copy of your letter of November 27 which was attached to it.[112] I am extremely sorry that what appears to have been a piece of exceptionally bad luck has delayed your work and that of the Program Committee, and has caused me to be unpunctual. A very thorough search of my memory and of that of our files shows your letter of November 27 did not reach me. Needless to say, I would otherwise have answered immediately.

I am deeply appreciative of the great distinction that the invitation and the considerations contained in your letter imply. As to which of the three alternatives that you mention would be best, this is a very difficult problem. I must admit that the task implied in alternative (1) is a staggering one. In view of the exceptional confidence that your invitation expresses, I do not see how I can do otherwise than accept your invitation. May I, however, ask you to permit me to consider the matter further for about a week, and then to write you again, as to whether I can undertake the task entirely in the spirit of your alternative (1), or whether some compromise like alternative (2) or some latitude in interpretation might not be worth considering.

I am,

[112]In that letter H.D. Kloosterman, organizer of the International Congress of Mathematicians September 2-9, 1954, invited J. von Neumann to deliver a talk on open problems in mathematics, a talk that was to be similar to Hilbert's famous 1900 lecture in Paris. Kloosterman also mentions that the organizing committee also considered two other alternatives to a talk by a single person: A small team of mathematicians prepares and one mathematician delivers an address; and a small team prepares the address and the members of the team report individually. For a detailed account of von Neumann's talk see: M. Rédei: " 'Unsolved Problems of Mathematics' J. von Neumann's address to the International Congress of Mathematicians, Amsterdam, September 2-9, 1954" The Mathematical Intelligencer **21** (1999) 7-12. The Editor.

yours most sincerely,

John von Neumann

Original in Von Neumann Papers, Library of Congress, language of original: English. Published in full first in M. Rédei, M. Stöltzner (eds.): John von Neumann and the Foundations of Quantum Physics (Kluwer Academic Publishers, Dordrecht, Boston, London, 2001.) p. 228.

April 10, 1953

Prof. Dr. H.D. Kloosterman
International Congress of Mathematicians
2d Boerhaavestraat 49
Amsterdam-0
The Netherlands

Dear Professor Kloosterman:

Since I wrote you on March 25, I have thought a great deal about the possibilities described in your letter of November 27. The conclusion that I have reached is as follows.

If this is the preference of your Committee, and if it is also otherwise acceptable to you, I will give an address on the basis of alternative (1) that you mentioned – that is, an individual address "On Unsolved Problems in Mathematics."

The total subject of mathematics is clearly too broad for any one of us. I do not think that any mathematician since Gauss has covered it uniformly and fully, even Hilbert did not, and all of us are of considerably lesser width (quite apart from the question of depth) than Hilbert. It would, therefore, be quite unrealistic not to admit, that any address I could possibly give would not be biased towards some areas in mathematics in which I have had some experience, to the detriment of others which may be equally or more important. To be specific, I could not avoid a bias towards those parts of analysis, logics, and certain border areas of the applications of mathematics to other sciences, in which I have worked. If your Committee feels that an address which is affected by such imperfections still fits into the program of the Congress, and if the very generous confidence in my ability to deliver continues, I shall be glad to undertake it. The task represents a very interesting and inspiring challenge, and I would certainly try to make the limitations that I have described above as palatable to the audience as I can.

I shall be very much interested in your and the Committee's views and comments concerning these matters.

I am,

yours most sincerely,

John von Neumann

Original in Von Neumann Papers, Library of Congress, language of original: English. Published in full first in M. Rédei, M. Stöltzner (eds.): John von Neumann and the Foundations of Quantum Physics (Kluwer Academic Publishers, Dordrecht, Boston, London, 2001.) p. 229.

Letter to H. Kuhn

April 14, 1953

Professor Harold Kuhn
Department of Mathematics
Bryn Mawr College
Bryn Mawr, Pa.

Dear Kuhn:

Many thanks for your letter of April 13, and your report on the second day of the recent games conference at Princeton University. Since the matter is urgent, I hasten to reply.

The report seems to me excellent and, in particular, the rendition of my remarks is entirely satisfactory to me. I am therefore not suggesting any additions or changes.

I am not aware of any experimental effort towards the determination of actual human behavior in actual n-person games, other than the ones you are familiar with. The problem is certainly a difficult one. I have, of course, thought about it in the past, and the picture which I used was this: I think that nothing smaller than a complete social system will give a reasonable "empirical" picture. Here, over relatively long periods of time, one can meaningfully assert that the "system" has not changed, while the positions of various participants within it may have changed many times. This would seem to me to be the analogue of a single solution and an "exploration" of the imputations that belong to it. After relatively long times, there occur discontinuous changes, "revolutions," which produce a different "system." It would seem to me to be true, that the imputations belonging to a single "system," i.e., solution, have a certain stability relatively to each other (i.e., they do not dominate each other), while no such condition is satisfied in the relationship between imputations belonging to consecutive, different "systems."

With best regards,

Sincerely yours,

John von Neumann

P.S. I am enclosing your typescript.

Original in Von Neumann Papers, Library of Congress, language of original: English

Letter to J. Lederberg

August 15, 1955

Dr. Joshua Lederberg
Professor of Genetics
Department of Genetics
The University of Wisconsin
Madison 6, Wisconsin

Dear Dr. Lederberg:

Thank you for your letter of August 10.

I see that we have some terminology difficulties, and I must confess that they are equally great on my side, since I do not seem to have a clear idea of what the terms "self-sufficient," "indifferent," and "information," that you are using, mean. Doubtless, they are explained in the other parts of your paper, but since I only saw a few pages of it, I am confused.

In view of this, it would seem best if I described precisely what my remarks meant. In other words, I will describe in detail to what sort of construction I am referring.

To be more exact, I have made several constructions which differ in the models of "parts," "organisms," and the spacial relationships of these to each other, that they assume. During the last few years, I have come to believe that one particular procedure is logically more economical than the others, and I will, therefore, use this in what follows.

Under this dispensation, then, I mean by the environment M, a quadratic lattice, as shown in the attached figure, and thought to be infinitely extended in both directions.

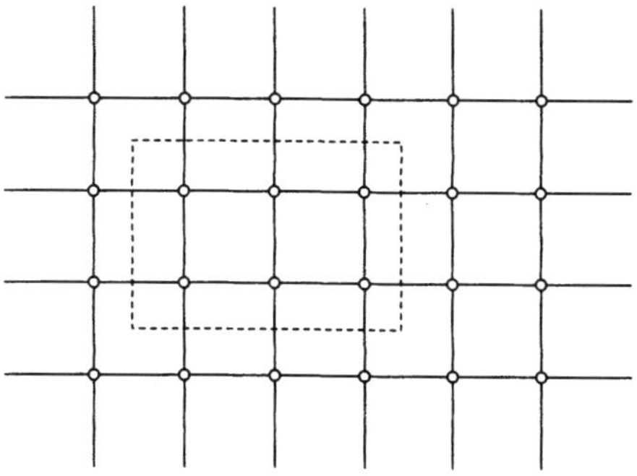

The circles in that figure are basic organs, the lines connecting them are connections over which impulses can travel. (Please ignore the area surrounded by a dashed line, shown in the figure, for this first part of the discussion.) All these basic organs are identical among each other. Every one has a finite number of states, say N – in the construction that I used, the N is less than 30. The basic organ also has a definite scheme of behavior. By this, I mean a complete set of rules which specify that if an organ is in a state i $(= 1, \ldots, N)$, and if its 4 neighbors (enumerated, say, in the order north, east, south and west) are in the states j, k, l, m, respectively, then after the lapse of one time unit, the first mentioned organ will go over into a specified state $i^1 = F(j, k, l, m)$. (Of course, the same happens to its neighbors, with respect to their own neighbors, and generally to all basic organs in the lattice.) Thus, the function F is a complete specification of behavior for the basic organ. I repeat, all basic organs have the same function F, i.e., the same behavior. In addition to this, one of the states, say the state $i = 1$, to be called the "rest state," has the property, that if an organ, as well as its 4 neighbors, are in this state, then the first mentioned organ will be in this state a unit time later, too. (Of course, anyone of its neighbors may itself not be entirely surrounded by neighbors in the rest state. Therefore, this neighbor may not be in the rest state after a unit time, and therefore, the original organ may not be in the rest state after 2 units of time.) Note, that I am treating time as an integer, i.e. I am only considering moments of time $t = 1, 2, 3, \ldots$.

The normal condition of M is one, in which all of its basic organs are in the rest state.

The connecting lines on which, as I said before, impulses are supposed to be traveling, have no further significance. I am only using them to indicate that each basic organ is immediately affected only by the 4 neighbors that I enumerated above, i.e. precisely by those to which it is directly connected by such lines.

An "organism" A is an area in M, like the one shown on the figure, surrounded by a dashed line, in which the states of the basic organs have been prescribed in some definite way – i.e., in which they are not necessarily all in the rest state.

Coming to the theorem that I mentioned in my previous letter, the organisms A and A^1 are disjunct. The theorem is, that given any A, I can construct an A^1 such that if A^1 is left to itself for a sufficient length of time, there will appear in M (of course, at disjunct, and if desired, widely separated locations in M) the original A, plus an additional copy of A^1 (displaced), and a copy A (also possibly displaced).

Of course, the construction of A^1 will depend on what A is. To be more precise, most of A^1 is the same, no matter what A is; however, there will be a part of A^1 which is determined by the structure of A. This part of A^1, by the way, is not a copy A. It is, in a certain peculiar notation, a description of A – i.e. it is related to A in somewhat the same way as the genetic material determining the structure of an organism as related to that organism.

I should add, that for all of this to be true, F has to be chosen in a certain particular way. However, this choice of F is quite independent of the choice of A (and A^1), i.e., it is the same for all possible A (and A^1).

I would appreciate if you would let me know whether this is sufficient clarification. I apologize if I have just re-described things that you had inferred already. I would be much obliged to you if you could tell me how this is related your discussion and ideas.

Sincerely yours,

John von Neumann

Original in Von Neumann Papers, Library of Congress, language of original: English

Letter to W.E. Lingelbach

April 23, 1938

Mr. William E. Lingelbach, Secretary
American Philosophical Society
Independence Square
Philadelphia, Pa.

Dear Mr. Lingelbach:

May I express my appreciation of the great honor done to me by election to the American Philosophical Society, which I accept with the greatest pleasure. I am,

Sincerely yours,

John von Neumann

Original in Von Neumann Papers, Library of Congress, language of original: English

Letter to S. MacLane

May 17, 1948

Dr. Saunders MacLane
13 Eversley Avenue
Norwalk, Connecticut

Dear Saunders:

I have received your letter of May 4th. I answer it only now because I assumed you will not want any correspondence on the first days after your return.

I can well understand your reservations regarding this particular committee and all committees. There is obviously a great deal to be said on the negative side, and I had to face doubts along similar lines before I accepted the responsibilities I have in connection with this committee – and several times before, when I accepted other responsibilities of the same kind. I feel nevertheless, that this is an enterprise which deserves a trial, and since your "no" is not an absolute one, that I should recite to you the positive side of the case.[113]

Before answering the four points in your letter specifically, let me say this in a general way:

All of us felt during the war that the Government – Army – Navy – OSRD[114] did not appreciate mathematics and mathematicians as they should. That it treated some of us as individuals as appendages to groups in other fields, and did not at all recognize the "corporate" existence of the body of mathematicians, and the specific contribution that they could make, par excellence. This situation got somewhat better toward the end of the war with the organization of the Applied Mathematics Panel of OSRD, but I think we all agreed that this, too, was both little and late.

About eight months ago the Army Staff did request the American Mathematical Society to suggest members for an Army Advisory Committee on Mathematics – with competence in all mathematical matters that might come up in the Army's work. This was certainly the result of the assertions, complaints, arguments, etc. of all mathematicians separately and together during and after the war, and of the realization that we did, after all, make important contributions, and might make more important ones. I think that we now have some kind of a collective responsibility to respond to the General Staff's and the American Mathematical Society's request. If we don't, and if the Army – or more generally, the Government – will in the future treat us with less attention and consideration than we think

[113]See Section 1 of the Introductory Comments for von Neumann's advising activity. The Editor.

[114]Office of Scientific Research and Development. The Editor.

we deserve, we will have to admit that we contributed to this ourselves. I may be exaggerating, and it is in any case very difficult to make precise statements about these matters, but I think that what I say is substantially correct, and I am very anxious to learn whether you agree with it.

The things which we can do, in an Army Committee on Mathematics, fall into two main classes:

(1) We can give advice on definite technical matters, on questions of how to distribute emphasis in such research which is predominantly mathematical, or where at any rate mathematics plays a major role.

(2) We can advise the Army on how to use mathematical talent which may become available for its purposes, both under normal conditions and in an emergency. We all felt during the war that the available mathematical talent was not very efficiently used. We now have a chance to suggest correctives, and I think we should do it.

    There is no doubt that the world is imperfect, and all large-scale organizations are *a fortiori* imperfect. This goes in particular for the Army, which is a very large-scale organization, with a number of specific limitations besides, which we know all too well. It is, therefore, unavoidable that anybody who is prepared to help such an organization has to resign himself to the fact that a considerable fraction of his actions will be wasted. I don't think that this is in itself a sufficient reason for those of us who are in the position to give certain types of very necessary help, to withhold our help. Personally, I am quite prepared to participate in such a venture, even if the efficiency is at a very non-negligible distance from 100 per cent.

These are generalities, and I am sure that you are just as well aware of them as I am. I still think they deserve some emphasis, and I hope that you will not consider me unrealistic if I stress them here.

I now come to your specific questions.

(1) I have no doubt that the members of the committee will have access to classified information whenever this becomes necessary in the course of our work – in the case of very high classifications the definitions of "necessary" becomes more rigoristic than for lower or medium classifications. I have no personal doubt that the rules of common sense will prevail in this respect. If you want me to, however, I can get a statement of policy on this question from the Army authorities.

(2) All of us view work on this committee as a "part-time" activity. Obviously, all of us have full-time occupations, and any of us have other obligations toward the Government besides. For my part, I am satisfied that something like three or four meetings of the committee per annum, plus a very moderate amount of correspondence, is the optimum for the organization into which this committee is likely to develop. In any case, I have no doubt that you will be able to make a very useful contribution within such time limitations.

    I don't think that the fact that you have not been connected with other phases of Army work since the war, and in particular, with those phases of mathematics which play the greatest role there, need reduce your usefulness seriously. Your war-time experience is more than most mathematicians possess, and in addition, there is a great deal in educational and connected questions which is not "military" or "applied" mathematics at all, and which should concern us greatly.

(3) I don't think that the point you mention proves more than that you will view the committee's work with a healthy, realistic skepticism. I think that a member with such an attitude is actually necessary in any well-balanced group.

(4) I can only repeat what I said above: the Army is a large and non-homogeneous organization. I think that we have a chance to do some work which is useful, both to the Army and to the mathematical community. We can do this in one particular sector of the Army where the authorities have "seen the light". I don't think that we should be influenced too much by the inadequacies in other sectors.

Please don't view this as a sales-talk. I repeat, each one of us has his hesitations in this field, as you do, and I certainly had mine. After all is said and done, I came to the conclusion that this work is well worth doing, and in fact, that we have, as mathematicians, a certain moral responsibility. Please accept the membership on the committee only if the contents of this letter seem more than 50 per cent convincing to you. I can only hope that they will. Whatever you decide, I shall be very happy to hear from you about this subject.

I am,

Cordially yours,

John von Neumann

Original in Von Neumann Papers, Library of Congress, language of original: English

Letter to J.C.C. McKinsey

February 18, 1949

Dr. J.C.C. McKinsey
The Rand Corporation
1500 Fourth Street
Santa Monica, California

Dear Dr. McKinsey,

The finite problems, connected with the search for solutions of n-person games, which you mentioned in your letter of the 16th of February, 1949, are certainly interesting and significant. I am in complete agreement with you that it would be very desirable to find means – mechanical or otherwise – by which problems of this type can be answered for values of n and r of reasonable size. Regarding the possibility of doing it mechanically, and doing it particular with a machine based on Tarski's decision method for elementary algebra, my comments are these:

I do not see any difference in principle between a machine built expressly for carrying out Tarski's decision method and a "general purpose" mathematical machine of the type which is being now developed in various places. Both of them should be (or at least might be) "universal" in the sense of A.M. Turing (PROCEEDINGS OF THE LONDON MATHEMATICAL SOCIETY, Series 2, Vol. 43, 1937, pages 544-546). The only difference would be differences of convenience: The mathematical "all purpose" machines have special arrangements to accelerate certain particular processes which are exceptionally frequent in normal mathematics (like the four species of digital arithmetic), while a "decision" machine would probably give such a preference to other processes (like the fundamental operations with finite sets, rather than the arithmetical operations mentioned above). In this way these two classes of machines differ only in their characteristics of speed, in dependence on the problems that they are given. That is, they develop their maximum speeds on different classes of problems ("ordinary" mathematics in one case, finite-set-theory-type combinatorics in the other).

In view of this, the mere observation that a particular type of finite problem can be solved with either type of machine is not very relevant. Either one of them, if properly planned, is "universal" in the sense of Turing and will, therefore, solve every problem for which finitistic, constructive method of solution is known. The crucial question is that of the time required for the solution. In order to judge the suitability of any given problem for either kind of machine, and in order to decide whether either kind has serious advantages over the other in connection with a given problem, time-estimates must be made. As a first preliminary to such time-estimates, it is of course necessary to determine the number of elementary logical steps which are involved in a finitistic solution of a given problem.

The preliminary questions that have to be asked in connection with your proposal, therefore, seem to me to be these: Let N be the number of steps required by Tarski's decision method in the case of either one of the problems. N is, of course, a function of n and r. What kind of a function is it? How rapidly does it increase? How large is it for moderately large values of n and r? Or conversely: In what range can n and r be chosen without giving prohibitively large N's?

In this connection, the following remark too is appropriate: Machines will be constructed in order to solve problems with "large" numbers of steps. The concept of "large" must, however, be interpreted somewhat specifically. In the present context it should mean the following: The number of steps in question must be impractically large for human operation, but it must become manageable through the acceleration affected by a fast machine. Since this acceleration is likely to be of the order of 10^4 to 10^5 or so, the above statement can also be formulated as follows: In order to justify the use of a fast machine the number of steps involved must be too large for human operation, but not too large by more than a factor of at most 10^5. The real "figure of merit" for the possibilities of a fast machine with respect to a certain class of problems is therefore this: How wide a family of special cases of that problem involve numbers of steps which lie within the five powers of 10 immediately above the domain of reasonable human operation?

My skepticism regarding the suitability of any variety of machine which is at this moment in sight for combinatorial-logical problems is just due to this: I suspect that while the five powers of 10 referred above include a lot of territory in the sense of problems of a mathematical-analytical character, they cover very little material in the case of combinatorial-logical problems. I am inclined to believe this – until I see a proof of the opposite – because the number of steps that are needed to solve a problem increases with the characteristic parameters of a problem much more quickly for problems of a high logical type than for others of a lower type. This is a rather natural conclusion from the old work of K. Gödel. Now ordinary mathematical analytical problems are never of a high logical type: They are usually of the next type after that one of the arithmetical fundamental variable. Logical-combinatorial problems like those which you mentioned in your letter, on the other hand, are almost always of a higher type. I have not determined the type of either one of your problems, but I suppose that it will be the second or third one above the arithmetical fundamental variable.

I do not want, however, to over-emphasize these general remarks. Since you are considering specific problems, it will be much better to consider them specifically and in detail. I would, therefore, be very much interested in learning from you what is known about the questions of numbers of steps which I have formulated further above. If N lies in a range that is reasonable for machine operation ($N = 10^k$ elementary logical operations, where k is perhaps 10 to 12 of possibly a little more), then machine treatment of these problems may be indicated. In this case, there will then arise a very interesting question as to what type of special machine is best suited to them. I am quite sure, however, that this question of the size of k must be answered first.

Hoping to hear from you again on this subject, which interests me very greatly, I am

Sincerely yours,

John von Neumann

Original in Von Neumann Papers, Library of Congress, language of original: English

Letter to M.M. Mitchell

December 14, 1934

Miss M.M. Mitchell
Editorial Secretary
American Institute of Physics, Inc.
11 East 38th Street
New York City

Dear Miss Mitchell:

Complying with a direction which Mr. Austin H. Clark, Director of Press Service of the A.A.A.S.[115] gave me, I am sending you enclosed one copy of the abstract of the talk which I am going to give on December 29 in the Symposium on "Group theory and quantum mechanics" at the Pittsburgh meetings.

As it is not my intention to publish the material of the talk, I cannot send you a manuscript.

I am,

Sincerely yours,

John von Neumann

REPRESENTATIONS AND RAY-REPRESENTATIONS IN QUANTUM MECHANICS
by John von Neumann
(Delivered at the Symposium on "Group Theory and Quantum Mechanics" at the meeting of the A.A.A.S. at Pittsburgh, December 29, 1934

It is known that any symmetry property of an intuitively described physical system finds its mathematical expression in the existence of a certain group under the operations of which the mechanical determining equations of the system are invariant. This group-theoretical principle of symmetry has been particularly important in quantum mechanics where the mathematical means of description – the wave functions subject to the linear superposition principle – are particularly fit to be treated with group-theoretical methods. The mathematical theory, the application of which has in this connection led to many results in various fields of quantum physics, is the so-called "Theory of Representations". As the states of the quantum mechanical system are really described, not by a uniquely defined wave function, but by one which is only known – and has a physical meaning only – up to a constant factor of absolute value 1, the discussion must be based not on the

[115] American Association of Advancement of Science. The Editor.

theory of representations proper, but on the representations "up to a constant factor of absolute value 1". These have been called Ray-representations. If the group of symmetry operations is a continuous (Lie-) group, the representations may be discussed by considering the so-called infinitesimal representations of the group. These are characterized by certain commutation properties which, in the case of a Ray-representation, will only be valid up to an additive constant multiple of unity. In some cases, these extra terms can be eliminated ("transformed away"). In some other cases they vanish automatically. There exist groups, however, for which neither is the case and the "extra terms" play an essential role. These possibilities are discussed for various main types of symmetry-groups.

Original in Von Neumann Papers, Library of Congress, language of original: English

Letter to T.V. Moore

April 9, 1953

Mr. T.V. Moore
Standard Oil Development Co.
15 West 51st Street
New York 19, N.Y.

Dear T.V.:

This is the letter that I promised you, restating the proposal that I made regarding the 'La Salina Operations Problem', that you described in your letter of February 3, 1953, and that we discussed when we last met in New York at the end of March.

This problem, as described in your letter, deals with the operations of 18 tankers between La Salina and Las Piedras and Aruba. It involves determining the economical value of increasing the number of berths for the loading of these tankers in La Salina from 3 to 4 or to 5. The comings and goings of the tankers are described in statistical terms only, i.e. they are subject to fluctuations which depend on fortuitous events like weather, conditions in the ports of call, etc. The evaluation is, therefore, necessarily a problem involving the calculus of probabilities.

Let me restate the probabilistic elements more specifically and precisely:

(a) Each one of the tankers has a separate and characteristic mean round-trip time for its assigned run, which may be La Salina-Las Piedras-La Salina (for 6 tankers), or La Salina-Aruba-La Salina (for 12 tankers). The times for the former fluctuate around 56 hours, while those for the latter fluctuate around 66 hours, but, as stated, they differ from tanker to tanker. Moreover, the duration of an actual trip of a specific tanker will not be precisely the above mentioned "mean round-trip time". Instead of this, it will fluctuate around this mean by an amount which may vary from -8 hours to +8 hours and depends on chance factors (weather etc.). This "fluctuation of duration" may be presumed to have a purely statistical distribution, to the extent to which it is due to weather and to similar factors. I assume that, to some extent, it is an independent statistical variable for each tanker, since different tankers may run into different weather conditions at different points of their routes. On the other hand, to some extent these time fluctuations will be correlated for all tankers in transit at the same time, since they are all exposed to the same overall weather conditions.

(b) The fluctuations of duration referred to under (a) are also affected by another common factor. Under certain conditions, the captains or the crews of all ships heading for a certain port may desire to make port at a definite moment. (You mentioned the effects of a good picture or amusement in port.) Such events should, I assume, be taken to be more or less chance events with a statistical distribution. Whenever such an event occurs at a certain moment which is within the range of possibility for any particular tanker to make port, the statistical distribution of the tanker's arrival must be presumed to be biased towards making port at that time. The degree of probability bias thus introduced may be taken to be known.

(c) Whenever several tankers enter La Salina at the same hour, under the control of the probabilistic rules described under (a) and (b) above, the number and availability of the berths must be considered. This will determine the time each one of these tankers will have to spend in port.

(d) On the basis of the above criteria, the mean turnover of the entire tanker fleet over some reasonably long period of time should be evaluated. The use of the expression "mean turnover" shows again that the problem is of a probabilistic nature.

My impression is that this problem, as described under (a) to (d), is one of considerable difficulty from the point of view of a strict analytical-mathematical treatment. That is, I think that it will be very difficult to derive complete formulas for the probabilities and means involved, and to penetrate to the ultimate mean that is desired – namely, to the "mean turnover" of the entire fleet referred to under (d) above, as a function of the number of berths (3 or 4 or 5).

I would, therefore, suggest that the problem be treated as a "statistical experiment". This is a technique which was used during the war and thereafter in solving a variety of operations–evaluation problems, and I am convinced that it has wide applicability in industrial operations-evaluation as well.

The procedure would have to be somewhat like this:

Represent each tanker by some suitable form of record, e.g. by a punch card, showing its exit time from La Salina (that is, the date and hour of this exit as the basic datum). Program calculations which will develop the further history of this tanker, always deriving those quantities which depend on chance (see (a) and (b) above), with the use of suitable tables of random numbers. For instance, assume that a tanker operating between La Salina and Las Piedras has a characteristic mean round-trip time of 58 hours, and that its fluctuation of duration around this mean varies from -7 hours to +7 hours. Assume that the statistical distribution between these two limits is characterized by a definite and known histogram. (In order to have something specific, let me assume that this histogram is made up as follows: The first 20 percent of the probability correspond to a fluctuation of duration between -7 and -5 hours, say to an average of -6 hours; the next 20 percent to one between -5 and -1, say to an average of -3 hours; the next 20 percent to one of -1 to +2 hours, say to an average of +.5 hours; the next 20 percent to one of +2 to +4 hours, say to an average of +3 hours; and the last 20 percent to one of +4 to + 7 hours, say to an average of +5.5 hours.) It is now possible to use random decimal digits, say from an appropriate table of random numbers or from any automatic arithmetical device which can produce random numbers (such devices exist and are in use today), to control the behavior of the fictitious tankers on the punch card in accordance with the desired histogram. (In the case of the example given above, this would be done as follows. If the "random decimal digit" which turns up is 0 or 1, let this represent the "first 20 percent of the probability" referred to above, i.e. assign the time fluctuation corresponding to it, which was seen to be -6 hours. If the random digits 2 or 3 turn up, let this similarly correspond to the "second 20 percent of the probability" referred to above, i.e. let the time fluctuation of -3 hours be assigned. The same procedure will assign to the random digits 4,5 the time fluctuation of +.5 hours; to the random digits 6,7, the time fluctuation of +3 hours, and to the random digits 8,9, the time fluctuation of +5.5 hours – all of this is in agreement with the stipulation made for the histogram in the example described above.) Corresponding procedures should and can be applied with the other random processes described in (a)-(b) above. In spite of the relatively complex probabilistic and causal connections which may exist, and which have been indicated in (a)-(b), all of this can be done with relatively little difficulty. As I mentioned before, there are plenty of problems in operations-evaluation (and also in various parts of mathematical physics that involve complicated statistical processes), which have been successfully treated in this way. Some of these were a good deal more complicated than the problem considered here.

In this scheme, then, each punch card represents a tanker and has to be used to develop the tanker's further history in its dependence on various chance events and their interaction with each other. It is thus possible to trace the history of as many days of operation as desired (say, a few years). One can then work out the behavior of sufficiently large samples for any assumed number of berths (3,4 or 5, cf. above) and thereby get an evaluation of the economic significance of any particular arrangement (i.e., of any particular number of berths).

I realize that the above description is somewhat sketchy, and I will be glad to fill in any amount of details that you may desire. I would only like to emphasize that the data handling and computing equipment to process these punch cards at an adequately high speed exists, and that one could in this way obtain good statistical averages in quite short times.

Do you want more details on these matters?

I think that it would be a good idea to discuss this problem more fully in the near future, and along with it, also, some others which have the same structure. (For instance, the problem concerning long-range exploration and oil reserve policies, that we also discussed when we last met in New York.) I expect to be in Tulsa about April 30 - May 6 or so, and it might be profitable to talk about these matters to Alec Bruce there. I think that he is interested in an extension of mathematical operations into such areas, and Carter is progressing quite fast with respect to its automatic equipment and know-how in this field. Does this seem a good idea to you?

Please let me know what you think of all these matters. Needless to say, I would be very interested to learn about your reactions and comments.

Hoping that your travels are proceeding well and do not prove to be too fatiguing, and looking forward to seing you again before too long, I am,

Cordially yours,

John von Neumann

Original in Von Neumann Papers, Library of Congress, language of original: English

Letter to O. Morgenstern

January 2, 1953

Professor O. Morgenstern
Princeton University
Department of Economics
Princeton, New Jersey

Dear Oscar,

I am enclosing copies of three letters (to M. Fréchet, J. Savage, and W.B. Simpson) and of a note, presumably for "Econometrica".[116] They are self explanatory.

I apologize for writing these without consulting you first, but I wanted to get them out of my system. If you think that I should write more, or less, or in any respect differently, please let me know, so that I can write the necessary corrections or modifications. I assume that you may also want to write to Fréchet.

You see that I am abandoning my principle of not arguing – of leaving bad enough alone.

I like Fréchet, but this is a bit too silly. I cannot resist the itch of answering this rather infantile attempt to attribute everything to Borel and to say that everything that can by no stretch of imagination be attributed to him is wrong or previously known or trivial or irrelevant or all of these together. Well, as I have repeatedly observed in 1905, 1911 and 1920: C'est la vie!

All is well here, we have a fine time and some snow. Klari will probably ski. We will be back before January 15.

With best regards, from house to house,

As ever,

John von Neumann

Enclosures: Copy of a letter to M. Fréchet.
Copy of a letter to J. Savage.
Copy of a letter to W.B. Simpson.
Copy of a note.

Original in Von Neumann Papers, Library of Congress, language of original: English

[116]The note was published under the title "Communication on the Borel notes", see Collected Works, Vol. VI. No. 2. See Section 9 of the Introductory Comments for the details of the conflict between Fréchet and von Neumann about who the initiator of game theory was. The Editor.

Letters to M. Morse

April 23, 1952

Professor Marston Morse
Institute for Advanced Study
Princeton, New Jersey

Dear Marston:

I am returning Stone's letter of April 13 which you had sent me. I had a copy made of it, is it all right to keep it? If not, I will have the copy destroyed.

The subject mentioned by Stone is not an easy one. Plans to standardize and publish codes of various groups have been made in the past, and they have not been very successful so far. The difficulty is that most people who have been active in this field seem to believe that it is easier to write a new code than to understand an old one. This is probably exaggerated, but it is certainly true that the process of understanding a code practically involves redoing it de novo. The situation is not very unlike the one which existed in formal logics over a long initial period, where every new author invented a new symbolism. It took several decades until a few of these found wider acceptance, at least within limited groups. In the case of computing machine codes, the situation is even more difficult, since all formal logics refer, at least ideally, to the same substratum, whereas the machine codes frequently refer to physically different machines.

I think, nevertheless, that if a competent mathematician like E.G. Givens is interested in working on this problem, he ought to be encouraged. The task may turn out to be a thankless one, but it is always possible that a competent and energetic man will in the end come up with something useful. My personal doubts are limited to the near future, and even if it will take a non-trivial number of years to produce something, there is no harm in starting early.

I think that the best procedure would be to encourage Givens (or anyone else whom Marshall Stone may have in mind) to make a contract proposal. Instead of having a meeting of the NRC[117] Committee on High Speed Computing, I would circularize this proposal among the committee and get their comments. These comments could then be evaluated and submitted to the NRC Division, which would then be in a good position to decide how to proceed further.

Regarding the related subject of the reactivation of the NRC High-Speed Computing Committee, I would like to add the following. I am sorry that I have not taken any action on this so far, but the Givens-Stone proposal that you have mentioned is the first specific instance that has come up lately where the committee

[117] National Research Council. The Editor.

might do something immediately useful. I would suggest, as I said above, to circularize this proposal among the committee, as well as any other proposal which may come up in the nearer future. If, at any point, the committee develops the impression that the business in hand can be transacted better by having a meeting than by mail, then let us set a date for the meeting. I would, however, suggest that we do not call a meeting until evidence of the type that I mentioned has evolved.

As ever,

John von Neumann

Original in Von Neumann Papers, Library of Congress, language of original: English

---

March 2, 1955

Professor Marston Morse
Institute for Advanced Study

Dear Marston:

In connection with your inquiry regarding possible nominations of Gibbs lecturers for 1956 and 1957, the following occur to me:

1956: R. Feynman or J. Schwinger, on the newer forms of quantum mechanics.

These involve very interesting manipulations of functional operators and rather new ways to achieve relativistic invariance.

From the mathematical point of view, the first mentioned aspect of the work is interesting, since it is not at all rigorous the way it is done, but it must either have a rigorous equivalent or something is totally wrong in the way quantum theory is now proceeding. (I think that "public opinion" among physicists has accepted too easily that a rigorous equivalent must exist.) Either alternatives would be interesting and significant and the sooner a broader mathematical audience familiarizes itself with this challenge, the better are the chances that the matter will be resolved.

1957: S. Ulam, on the uses of the "statistical method" in analysis.

This is a complex of procedures for which the well known approach to the potential problem by way of the "random path" method is the very simplest example. A systematic exposition of what can be done by such methods would seem to me to be useful and interesting at this time.

With best regards,

Cordially yours,

John von Neumann

Original in Von Neumann Papers, Library of Congress, language of original: English

Letter to E. Nagel

December 9, 1953

Professor Ernest Nagel
Department of Philosophy
Columbia University
New York 27, N.Y.

Dear Professor Nagel:

Thank you for your very kind letter of November 28.

Before going into the subject matter as such, please let me tell you how much I appreciate the expression of your warm and friendly feelings regarding the possible contribution that I might make to the Conference on "The Methods of Knowledge" that you are arranging, and your kindness in wanting to make my participation possible. I am also very gratified in what you tell me about the feelings of your colleagues on this matter.

My trouble is, that I know from past experiences that it is completely uncertain whether I can complete the manuscript that appears to be necessary, the more so because the possible delays between June and October do not help me, since I will be abroad during that period. I do not want to disturb the orderly progress of your arrangements by making a promise on which you will count, and which I probably cannot keep. In addition, unfulfilled promises on publications have usually had a disrupting effect on my work, because while I am not writing the thing that I have promised to write, I am also inhibited from doing other work that could otherwise go quite well. I really owe it to you and to all my friends in all sincerity not to make such promises, since I have every reason to doubt whether I could fulfill them in an adequate way.

I hope you will not view the lines which follow as a criticism. I know that you have considered the arrangements for the Conference very carefully, and that you have weighed all pros and cons. It would therefore certainly be quite out of place if I would start making suggestions on a subject that you have studied very carefully and I have not. I would, nevertheless, like to mention that I think it is better for someone who does not feel the spontaneous desire to write on a subject not to be induced by external circumstances into taking on such a project. I have participated in the past in group discussions and symposia and derived much benefit and satisfaction from them, and would be more than happy to participate in the one that you are now arranging, if my part in it could stay on the oral level. If it has to be connected with publication, then I am afraid I must excuse myself. I am sure I could not produce anything satisfactory for the occasion unless I felt a strong subjective desire to write on the topic, i.e. unless I felt that I had ideas on the subject, of sufficient scope, which have not been expressed before – and this is not the case. I think the maxim "verba volant, scripta manent" is not necessarily an argument against "verba".

I hope you will not think too badly of me for my inability to contribute a paper, and my unwillingness, in view of this, to promise one. I also hope you will not consider my last remarks too severely.

I am,

Yours very sincerely,

Original in Von Neumann Papers, Library of Congress, language of original: English

Letter to R. Oppenheimer

Institute for Advanced Study
Princeton, New Jersey

February 19, 1948

Dear Robert,

In response to your suggestion, I am giving in what follows my reactions to the paper containing a proposal for the definition of "full-time employment" by the Institute,[118] which you communicated to the Faculty in your Memorandum of February 11.

Let me state first, that I am in full agreement with the view that employment by a pure research institution like the Institute for Advanced Study should impose on the appointee a general obligation, beyond the hours or days or months actually occupied by explicit official duties, to regulate his life and intellectual interests at all times so that the research interests are the dominant ones. Next, however, it seems to me that it is not a desirable way for implementing this principle, to formalize the appointee's relationship with the Institution as one of strict "full-time employment", i.e., to lay down the maxim that the Institution has a basic claim on every hour of his time, every day of the year – even if it is understood that the Institution will make the most general use of this underlying claim only in a prohibitory or potentially prohibitory way.

If the institution does not postulate any such claim, exercises no control or potential control whatever the total activities of its members, then there is, of course, a risk that a research appointee may divert much or all of his efforts in a manner in which they do not contribute to the research purposes of the institution. This risk varies from one subject to the another, and it has to be dealt with according to its size and nature, and therefore, possibly differently in different fields. In the mathematical and physical sciences, at any rate, this risk is now present, although it is not as serious as in some other subjects. It has certainly greatly increased since the 1930's. In what I am going to say, I have primarily the situation in mathematics and physics in mind.

In dealing with this risk, I feel very strongly that the basic principle should be this: A research appointment is essentially a position of trust, expressing the belief of the appointing group in the appointee's ability and desire to do productive research in a reasonably extended future. Such an "expression of faith" should not be *a priori* combined with strict legal and administrative safeguards of its fulfillment or attempts at such safeguards. The appointing group should, of course,

[118]Institute for Advanced Study, Princeton. The Editor.

keep a reasonable watch on the performance of the appointee "on trust", and if it becomes clear that his performance is not of the quality that could have been reasonably expected, and that this is to any important part due to his voluntary shifting of his interests and efforts, and quite particularly, to outside activities, then an administrative limitation of these is obviously necessary.

To sum up: I think that the primary approach ought to be one of non-intervention and of trust, and administrative intervention and legally formalized delimitations should be secondary. The point where the transition from the former attitude to the latter should take place can, of course, not be defined precisely and reproducibly. It will have to be left in each particular case, as it arises, to the insight and judgement of the administration.

To come from generalities to more specific details:

I think that the way in which the official Institute vacations are spent should not be regulated.

I think that a rigid distinction between Federal and other work will be hard to maintain. In this connection it should be noted, that most of the technical and scientific work of the Federal Government is actually handled by private contractors, many of whom are industrial or semi-industrial.

I think that the proviso that the outside activities of an appointee must not invade his research time and interests, and that their field must be in accord with those interests, is a very wise one. I feel that this should be the prime controlling principle. Any essential deviation from one's primary research interests should be fought. On the other hand, a certain contact with the strivings and problems of the world that surrounds us is desirable and even necessary in certain important parts of the sciences when and to the extent to which it is complementary to and continuing the underlying research interests.

I think that at this moment the realistic approach for the Institute to treat this complex is to deal with its professors according to principles similar to those which are in effect for the major positions at the best Universities: E.g. Harvard and Columbia. I realize that these are in conflict with the very rigid "full-time employment" principles of the University of Chicago, but I do not believe that these will, in the long run, produce the desired results unless very exceptional provisions can be made in other respects.

These thoughts are neither complete nor in a well-balanced literary form. The letter is probably both too long and too short. Please forgive me, especially the former, and use it as you see fit.

I am,

Yours sincerely

John von Neumann

Original in Von Neumann Papers, Library of Congress, language of original: English

Letters to R. Ortvay

Princeton

April 17, 1934

Dear Rudolf,

Yesterday I received a telegram from my mother in which she asks me to inform you as soon as possible about the details of the Cambridge invitation and about the organizational structure of Princeton University and the "Institute".

As for the Cambridge invitation, the simplest is if I send you the letter of R.H. Fowler (a physicist who is now the "chairman of the department of mathematics and physics" in Cambridge); the letter contains the invitation, its history, reason and conditions. I only add that they invited me originally for the whole year 1934/5; however, in view of my obligations here, I could not accept it. Enclosed is the mentioned letter.

As to the nature of organization in Princeton, I am sending you the three catalogues, which so far have appeared, of the "Institute for Advanced Study." The first one contains generalities, the 2nd and 3rd the faculty and the lectures. Weyl and Dirac are featured in the third, since they were not yet here in 1933.

As you know, I was invited by Princeton University for one semester in 1929. In January of 1930 I was offered a permanent professorship, which I did not accept. Upon this they entrusted me to substitute for 5 years in the "Jones professorship for mathematical physics" on a one semester per year basis. In January of 1933 the then beginning "Institute" invited me to occupy a permanent chair. Before me, Einstein and two professors at the University (the topologist J.W. Alexander, and O. Veblen) had been appointed. As visible from the catalogue, the 5 chairs created so far at the Institute are independent of each other and are of the same rank.

I thank you once more for your efforts and I hope these data will be useful to you.

Since your reply can hardly reach me here, I refrain from asking about the war situation. We leave on May 5 with the ship "Roma" via Gibraltar - Naples - Genova. We arrive in Pest on the 16th or 17th and I will again be reachable by mail in Gibraltar around May 11.

We are well, although I am a bit pumped dry, for I held three lectures last week at Yale, Harvard, and Brown Universities.

We have already started getting in the "travel mood".

I do not have any news on physics; in mathematics there is some excitement about topology-group theory because the theory of topological groups has been given a great push. It is tragic that Haar, who initiated this development, did not live to see it.

Many greetings until our upcoming reunion:
Jancsi

Original in Department of Manuscripts and Rare Books of the Library of the Hungarian Academy of Sciences, call number K 785/76. Language of original: Hungarian. Translated by A. Lenard.

The Institute for Advanced Study
School of Mathematics
Fine Hall
Princeton, New Jersey

March 17, [1938 ?]

Dear Rudolf,

Please excuse me for replying only now, after I have received already two letters from you. As an explanation, if not excuse, may serve that I have traveled a tremendous amount: I have delivered lectures at about 7 southern and western universities. *Nota bene*, 4 more are ahead, plus a small astrophysics conference in Washington, where I give a talk. Thus my life has been pretty disorderly.

The Hungarian news was very interesting. The invitation of K. to Budapest was, I think, if one adds all the positive and negative factors, all in all a happy event. Will the minister approve it?

I am very grateful for your efforts at the Academy,[119] but I completely agree with all of you that it's best to let this sleep. I am truly pleased by the step taken by the Association of Mathematics and Physics.

Born's things also interest me very much. It is an old conviction of mine that "reforming" the momentum-space (and of course, with it the Lorentz group), more precisely, rendering it "finite" is what is needed for "quantum electrodynamics". But of course one has to see the detailed publication first.

Nota bene, Born takes the momentum-space to be "Einsteinian" (= cylinder) space. This worries me because the symmetry-group of this space has only 7 parameters. A "De-Sitter" (= hyperboloid) space is more suitable: Here the symmetry group has 10 parameters, and thus is "equivalent" with the Lorentz group.

I tried this once (2-3 years ago):[120] The solutions of the Dirac equation in the De-Sitter space (such an equation exists, its description appeared about one year ago in an article of Dirac in the "Annals of Mathematics") may be re-interpreted so that among the De Sitter spatial 10 infinitesimal "angular momentum" operators 4 (usually interpreted as the energy-momentum vector) are now interpreted as

[119]Ortvay tried to induce the Hungarian Academy of Sciences to nominate von Neumann for membership in the Academy. The Editor.

[120]See von Neumann's letter to Dirac (January 27, 1934) for the detailed computation and Section 5.4 of the Introductory comments. The Editor.

coordinates (and time). This way a 3 + 1 dimensional x, y, z, t space arises in which

(1) The x, y, z coordinates and the time t are *non-commuting* operators.
(2) The order of magnitude of their commutators is $\frac{h}{mc}$. (That is to say, this is the uncertainty associated with a simultaneous measurement of coordinates.)
(3) The whole structure has the Lorentz-symmetry.
(4) Each of the x, y, z coordinates has a discrete spectrum: $\pm 1/2, \pm 3/2, \ldots$
(5) The spectrum of the time t is continuous, from $-\infty$ to $+\infty$.
(6) When 4. and 5. are combined with 3. this comes out:
    Given four real numbers $\alpha, \beta, \gamma, \delta$, the spectrum of the operator

$$\alpha x + \beta y + \gamma z + \delta t$$

is as follows:
    a) If $\alpha^2 + \beta^2 + \gamma^2 - \delta^2 > 0$ then it is discrete:

$$\pm \epsilon/2, \pm 3\epsilon/2, \ldots$$

where $\epsilon = \sqrt{\alpha^2 + \beta^2 + \gamma^2 - \delta^2}$.
    b) If $\alpha^2 + \beta^2 + \gamma^2 - \delta^2 < 0$ (indeed even when it $=0$) then it is continuous from $-\infty$ to $+\infty$.

So this is a discrete "crystalline" space with "continuous" time, which has not only spherical symmetry (even though it is a "crystal"!), but is even invariant with respect to changes of the reference system given by Lorentz transformations, and so shows the proper Lorentz-Fitzgerald contraction phenomena. (This is made possible, of course, by the non-commuting nature of the coordinates.)

I did not examine this model in deeper detail because I considered it very artificial and arbitrary – and so I still think today. But I believe that it points at an interesting possibility.

I was much interested in your philosophical remarks, with which I agree on the whole.

I was also much interested in your political comments. Only I am even more pessimistic than you are.

I don't believe that the catastrophe will be avoidable. The arms race is even more intensive than it was before 1914. The view (anyhow a bit artificial and contradicted by all historical experience) that the (contemporary) dictatorships are by nature more pacific than the monarchies (of yesteryear) has been contradicted by the events of the recent past. Thus, since we don't know the "real" mechanism anyway, in my opinion the strictest empiricism is called for. What did happen in 1914 will happen again now a fortiori. What needs proof is not why this or that thing will happen (although such a proof would not be difficult, using the common devices of dialectic) but rather why this or that thing will not happen: And for that, I see no sufficient reason.

That the U.S.A. will end up again intervening on the side of England (when an English victory is not achievable otherwise) I find indubitable. The main guaranty of the security of the U.S.A. (especially in the Far East) is the existence of the English world empire. This is even more so now than it was in 1917.

I would be much interested in your opinion about Hungarian internal politics; also, about how the disappearance of Austria impinges on it.

I will likely come again to Pest during the summer. N.b. at the end of May there will be a theoretical physics conference in Warsaw with about 30 participants, to which I was invited. It appears that N. Bohr, Dirac, Fermi, Heisenberg, Langevin, and Schrödinger will give talks. (N.b. Did you hear anything about Schrödinger?) Bohr's talk addresses some of my older stuff. As I see it, I shall go there.

You will be interested to learn that Jenő Wigner will come back to Princeton. The university offered him the theoretical physics research chair called "Jones professorship", and he accepted. I am glad this happened, and, quite apart from anything else, we needed this here.

I greet you until our reunion

Jancsi

Original in Department of Manuscripts and Rare Books of the Library of the Hungarian Academy of Sciences, call number K 785/97. Language of original: Hungarian. Translated by A. Lenard.

The Institute for Advanced Study
School of Mathematics
Fine Hall
Princeton, New Jersey

January 26, 1939

Dear Rudolf,

I was very glad to receive your kind letter, forgive me for replying only now. We were in Florida and your letter was forwarded there – with a good deal of delay. We have been back in Princeton for about a week only.

I am replying to your questions concerning the "Prize" as follows: sent me the article in the "Est"[121]. Of course the cliches put in there by the journalist are pure silliness: "American Noble Prize", "5,000 dollars," etc. I don't know where he got all that from. Generally, I am fully astonished that in our – not exactly eventless – world how he got information so quickly: This prize became public at the time when it was awarded on Dec. 29, and the article of the "Est" could have appeared about on the 30th. Right? Truth is this: This prize is given by the American Mathematical Society once every 5 years for the best paper published in America during the last 5 years in Analysis or its border fields. It is called "the Bôcher Prize", the name comes from the American mathematician Maxime Bôcher, who was, about one generation ago, full professor at Harvard and the leading light there before the times of Birkhoff - Osgood. This was the fifth time the prize was awarded, the previous recipients have been: 1923: G.D. Birkhoff – 1924: E.T. Bell and S. Lefschetz (jointly) – 1928: J.W. Alexander – 1933: M. Morse and N. Wiener (jointly) - 1938: myself. The prize is \$100 (and not \$5,000!).

[121] A Hungarian newspaper. The Editor.

What you wrote about the Sommerfeld Jubilee – especially your stylistic observations about it – was very interesting.

It is difficult to write about politics and to convince myself that my diagnosis is not ± "wishful thinking". But it is my belief that it is nevertheless possible to remain approximately objective. I do not believe that universal warfare is avoidable, and I do not believe that the argument that "it is unnecessary," and that "it does not solve the acute problems" gets to the heart of the matter. The whole thing is a pathological process, and within it war is a plausible stage of development indeed. Emotionally it also is "necessary," if one allows oneself the use of this word in this context. It will bring the acute problems closer to a solution to the extent that it will decrease the moral and intellectual weight of the European continent and of its neighbors, something which is deserved, given the current structure of our world. May God grant that I should be wrong.

I would be very happy to again hear from you and of academic matters in Hungary.

Until that hearing, my best greetings

Jancsi

P.S. How is your ex-assistant doing?
I forgot to write about us: we are doing fine and we are looking for a house. Princeton is like always but there is some European excitement here too.

Original in Department of Manuscripts and Rare Books of the Library of the Hungarian Academy of Sciences, call number K 785/98. Language of original: Hungarian. Translated by A. Lenard.

Princeton

February 2, 1939

Dear Rudolf,

Thank you for your kind letter of the 15th, I believe you received my answer to your previous letter in the meantime.

Now I am writing hastily to reply to your questions regarding the Ergodic Theorem[122] in thermodynamics, next time I shall write more extensively.

I am replying to your questions in the order in which you asked them.

To No. 1: We are in complete agreement. Today's formulation of the ergodic theorem is that, apart from a set of measure 0, the "Zeitmittelwert"[123] *does* exist as a limit and = the "Raummittelwert".[124] All "disorder" assumptions are implied by the one hypothesis that we are *not* in that measure-0 set. No separate "H-theorem" is needed.

---

[122]See Section 7 of the Introductory Comments for von Neumann's work on ergodic theory. The Editor.

[123]Time average. German. The Editor.

[124]Phase average. German. The Editor.

To No. 2: Things are roughly this way: The situation in quantum mechanics is better. Nevertheless one has to analyze the situation there further. I wrote about this question in sufficient detail in a paper in "Zeitschr. für Physik" at about 1931: "Über das H-Theorem in der neuen Mechanik."[125] Pauli and Fierz investigated the question anew about 2-3 years ago (based on my Z.f.Ph. paper). (I think in the "Annalen der Physik".)

To No.3: As to the question of what systems are metrically transitive, most interesting is a result achieved a few months ago by S. Ulam and J. Oxtoby (Harvard): Any system can be made transitive by an arbitrarily small "perturbation". Apart from this, we only know that force-free motion on closed parts of certain surfaces of constant negative curvature is metrically transitive.

Many greetings to Bay.

"I'll hear you soon"

Jancsi

P.S. The article by Novobátzky interests me *very much*, I am reading it now.

What do you think about the Uranium → Barium disintegration? Around here this is thought to be of great importance.

Original in Department of Manuscripts and Rare Books of the Library of the Hungarian Academy of Sciences, call number K 785/99. Language of original: Hungarian. Translated by A. Lenard.

Princeton

February 26, 1939

Dear Rudolf,

Thank you very much for your letter of the 14th.

The two topics you write about in it: the political and the biological, especially the brain-anatomical, in a phenomenological point of view – both interest me very much. Unfortunately it is about the le ss productive, politics, that I can contribute considerations of greater length.

I believe that the difference between our views is gradually diminishing. I completely agree with you that a war will not benefit anyone in Europe; indeed I go further: From a certain point of view, neither will it outside of Europe. To be more precise:

(1) To believe that a war, of whatever outcome, could help the Jewry stuck in Europe, I regard as naivite. There are two possibilities, one of which – from this point of view - needs no further analysis. But the other case also is hopeless: In the losing countries such chaotic social conditions are to be expected that, once chaos ends, there will be such distancing from everyone who had anything to do with the

[125] Reference is to J. von Neumann: "Beweis des Ergodensatzes und des H-Theorems in der neuen Mechanik", Collected Works Vol. I. No. 25. The Editor.

other side that I find it inconceivable that an equilibrium concerning the issue will ever become possible.

Of course, here and there, perhaps among the neutrals, there may be some exceptions, but I believe that my picture is right on the whole.

But in the majority of places in Europe something quite different may happen. It is enough to think back about the Turkish-Armenian events during the World War. (In his last speech, H. actually hinted at this.)

(2) That the war would be a catastrophe for all stateforming nations of Europe, is pretty obvious. I find it probable that the Western powers would win, after huge efforts, but their victory would be Pyrrhic: On the one hand the German-Russian complex would become at least as disquieting as it is today, indeed worse, for it would entail sliding closer of these two major "dissatisfied ones" – and on the other hand the position of the Western powers with respect to their extra-European allies and dominions (the distinction keeps diminishing) would be again weakened, at least as much as it did after 1914-18. And that is by very much indeed.

(3) Although according to paper calculation the war could perhaps benefit the U.S.A., but one must take this too with a pinch of salt. Because:

(a) The profit from "military supplies" - in magnitudes of this order - is a pure bookkeeping act. The majority of debts so incurred will never be payed back, but even if they were to be paid back the lenders would never accept them: For in the last analysis they could only be payed back in goods, and it would be impossible to allow the importation of so much goods without heavily upsetting one's own economy. (All this happened just after 1918.) Of course the growth of war industries would cause a temporary boom, but this only means that one part of the country (through the rise in prices) would be paying taxes to the other half (that obtains more employment). And in the end fasting comes necessarily in the form of a post-war depression.

(b) Of course, war loosens the social structure of victorious powers as well.

(c) One can of course imagine that victory would, as in 1918, much increase the world standing of the U.S.A., and would mean the beginning of U.S.A. imperialism on a grandiose scale. (This could only occur if the war would also liquidate Japan – but in these days this is not such a far-fetched prospect.) Against this, I think one can exclude the idea that anyone in the U.S.A. should harbor such ambitions.

Ordinary people certainly do not. And both politicians and "big business," as evidenced by their whole terminology and symbolism, seek to satisfy their ambitions in quite other directions. Maybe after a victorious war such ways of thinking will become fashionable but at the moment these are quite foreign to them.

It is, for instance, a total misunderstanding of the U.S.A. to believe that it intervened in the World War from such (imperialist) motives: Wilson's complete failure after the peace treaty, and such minutiae as the refusal by the U.S.A. to accept any "mandates" (I believe the Entente had wanted to dump Armenia on the graces of the USA) are sufficient proof of this. The direct motive in 1917 from the point of view of the USA was – and if it comes to it, it will likely be again – that a collapse of the British Empire cannot be allowed to happen from the American point of view. It would usher in a very dangerous world condition. And this is a very negative motive, more the avoidance of damage than the acquisition of profit.

I admit that the USA could be imperialist. I would not be surprised if in 20 years it would become so. But today it is not yet. The Roman Empire too became so only quite late. They only "acquiesced" in the second century B.C. that the permanent annexation of at least the Balkans was "inevitable" from the point of view of their "security." I could imagine that somehow the USA will reach that point, but it has not reached it today, and I regard it as unquestionable that no one is led by such an idea, consciously or unconsciously. On the contrary: All instincts here are isolationist, even the "defense" of South America or "saving" the British Empire is based on isolationist desires.

(4) What you write about the anti-European resentment etc. in the U.S.A. has, as I see it, some truth in it. That is: I too believe that there exists such a feeling here. This is most explicit in the lower middle class but it continues upwards in sublimated forms.

On the other hand: I can only partially accept your views on what Europe's decline would mean for the World.

I too think that it would mean a gigantic cultural minus (actually we may speak of this already in the present indicative tense), and the U.S.A. could not completely substitute for it.

But I think that the Greco-Roman analogy of the ancients is valid in this regard. Of course by Rome's "transplantation" of Greek culture much fine detail got lost, indeed important elements too. Yet this way the ancient civilization still could continue to exist essentially intact for at least 300 more years, and in many modified forms even far beyond.

Judging the European Hellenes it must not be forgotten also that most of them are Thracians, Macedonians and Persians... Hence by strict Greek standards barbarians.

Having said all this, I still find the war as probable, and indeed with an early participation of the U.S.A. For it is a pathological process, which comes to be not because someone figured it out, with reason, that it is in his own interest, but because certain abnormal spiritual tensions – certainly existing in today's world – find in it their "resolution." And because, from a rational point of view, England and France cannot allow each U.S.A. cannot allow England to be destroyed. Really, I can only hope that southeastern Europe will be left out of this, partly because the possibilities there are very confused, partly because one always hopes for "miracles."[126]

Original in Department of Manuscripts and Rare Books of the Library of the Hungarian Academy of Sciences, call number Ms 5723/7. Language of original: Hungarian. Translated by A. Lenard.

---

[126]This letter does not have a closing and a signature. The Editor.

Princeton

March 29, [1939 ?]

Dear Rudolf,

I am ashamed that I reply only now to your last letter. The more so because this correspondence means very much to me. It is a rarity in this paranoid world, where everybody – myself included – is partisan, to be able to hear a calm and objective voice and to realize that it is possible to distance oneself from such a world.

I very much hope that you will not follow my bad example and that you will conduct our correspondence with smaller intervals than my last Creativity Pause.[127]

As an extenuating circumstance I would like to bring up one thing: I already wrote you twice (4 weeks ago, and then again 2 weeks ago) approximately complete replies. In both cases, when I re-read the letters I saw that I wrote more about the "war and peace" topic than I liked in retrospect.

Now only a little to this topic; after all, I believe that the events themselves provide a more complete commentary.

The pessimistic diagnosis on which we seem to have agreed finally is obviously closer to reality than the illusions of last October.

As to the U.S.A., I would like to say only this: I think that in Europe the U.S.A. is understood just as little as Europe is understood here. I don't believe that the general public here is motivated in its judgement of world politics by directly egotistic motives: For the average American honestly believes (and in this he may be right) that the World War was a bad deal even for the U.S.A. That public opinion today, in spite of this, is at the stage where it was in 1915 (and that is a fact!) has different causes: It is feared that South America is endangered and that England will get into trouble. The latter will always be a casus belli for the U.S.A., the former even more so, perhaps even a suspicion of it.

What you write about biology and quantum theory interests me a great deal and I very much sympathize with it. Your remarks about the anatomy of the brain are especially interesting. I also believe that this possibility must be taken seriously: That processes essentially connected to life cannot be made geometric, they cannot be described spatially. That the spatial position of our physical bodies, and within it, the geometric localization of all processes, is possible only approximatively. And if we push this localization beyond a certain point (which itself is defined only with some vagueness, as shown by the example below) then difficulties, pseudo-problems, and contradictions appear, just as in special relativity when one looks at the simultaneity of distant events, or in quantum mechanics when one tries to measure p and q at the same time. N.b. I have thought a great deal since last last year about the nature of the "observer" in quantum mechanics. This is a kind of quasi-psychological, auxiliary concept. I think I know how to describe it in an abstract manner divested from its pseudo-psychological complications, and this description gives a few quite worthwhile insights regarding how it might be possible to describe intellectual processes (therefore ones essentially connected to life) in a non-geometrical manner (without locating them spatially).

[127]Von Neumann uses the German term "Kunstpause". The Editor.

I would be much interested in your further thoughts regarding the psychophysical parallelism. By the way, is there any brain-anatomical literature useful from this point of view? Doesn't the workings of muscles, or particularly the tendency of the body to repair injuries or surgical interventions, seem to point to something like this: That it is hard to describe spatially what actually happens but it is perhaps easier on the basis of something else, of some teleological postulate, for instance?. I would think though that a teleological approach also is a bad approximation only. As is shown by the example of quantum mechanical transformation theory, other non-spatial schemes also exist.

I would have the following, somewhat extravagant request to you:

Since the exaggerated reports in the "Est" on the Bôcher Prize appeared – which were then taken over by other Hungarian language publications abroad – I have been inundated by the most varied letters asking for my help. But recently I did get a seemingly more sensible one: A Slovak or Hungarian workingman by the name of János Trenke who, it appears, is some kind of supervisor in a Chattanooga, Tennesee, factory, wrote me. He told me that he had learned of my existence (of course in connection with the B. Prize) from the "Journal of Verhova" (can you imagine that such a thing exists in Tennessee?), and he asks for my advice as to how he could learn some secondary school mathematics in the Hungarian language if possible. I would like to help him.

So I would like to ask you to purchase some high school mathematics textbooks (I assume that only two are in use in grades 1 - 8, I would like to have both) and send them to me. My mother will compensate you for the price when you let her know; I write her now to this effect.

Sorry for bothering you with this and I thank you for it in advance.

What's new (beyond what can be read in newspapers) in academia, particularly about individuals?

Many greetings

Jancsi

P.S. I forgot to reflect on a topic you brought up, though it much preoccupies me too. Namely the future of white – or rather industrial-scientific – civilization and America's and Europe's role in it. Do you exclude that there could be an analogy with Greece – Rome?

The last 50 years of Europe showed a similarity to the history of the Peloponesian War and the period following it. This analogy is of course not always perfect: it seems that the roles of Athens and Sparta got interchanged 20 years ago.

It is true of course that many fine points got lost by the Romans' taking over Greek civilization; but, looking at this with the eye of someone living 150 B.C., that process wasn't all that sad: At any rate it prolonged by some 350 years the existence of the Greek way of life in the world. In many respects it even improved on it.

And when we judge today's Europeans as the "Hellenes" we must not forget that most of them – to continue with the analogy – are Thracians, Macedonians, or Persians.

Original in Department of Manuscripts and rare Books of the Library of the Hungarian Academy of Sciences, call number K 785/100. Language of original: Hungarian. Translated by A. Lenard.

Princeton

July 18, [1939 ?]

Dear Rudolf,

I must again apologize for the delay in replying, especially since in the meantime I got already two of your letters.

I am grateful for having received the mathematics textbooks, I believe they will be very appropriate. I will send you the Gilette blades you mentioned as soon as I find a man acquaintance traveling to Budapest.

There is not much news around here. We have been pretty preoccupied with buying a house and re-modeling it. My wife left on July 8 for Europe to visit her parents and her sister, and she is expected to stay for 2 months. Whether in her journey she will reach Hungary is quite uncertain. She might only go to England and Belgium. I myself will definitely not go to Europe this summer: my mother is likely to come for a visit during the summer; as things stand now she is scheduled to arrive in New York on Aug. 3. Since I certainly want to be here when she arrives, and since she – whom we originally expected much earlier – delayed her departure twice, I did not even leave Princeton. My plans at present are that around August 10 I will visit several acquaintances in Canada and in Maine.

Right now of course Princeton is quite empty, though this year by chance it is not so hot and humid as it is supposed to be.

Your remarks on epistemology were very interesting to me. I too think that Gödel's results mean that there is no "complete" axiomatic system, not even in mathematics, and I believe that there is actually no other consistent interpretation of this complex of questions.[128] From the point of view of the discussion and evaluation of the truly mathematical, resp. mathematical-logical questions that they discuss and evaluate, I regard Carnap's things as naive and feeble. Carnap simply does not possess the knowledge minimally needed to address such matters – let alone to say something new. So for instance he expresses totally naive, simplistic views on the issue of the "completeness" of the axiomatics of mathematics ("categoricity") with an air of terrible self-importance. If things were so simple as he imagines them then "research on the foundations of mathematics" ("Grundlagenforschung") would not be needed – at least from a mathematical point of view! Is it your impression that Carnap has something new or scientifically interesting

[128]See Section 2 of the "Introductory Comments" for Gödel's theorem and von Neumann's evaluation of it. The Editor.

to say about the structure of languages; or generally that what he does may perhaps be evaluated as preparing the ground for later serious work? I am unable to judge whether Carnap deserves credit for his leading "school" philosophers to take seriously philosophical issues in the exact and natural sciences. Obviously, many around here think that the answer is "yes". At any rate, it is a pity that we have to be informed about a very solid topic from such a confused source. N.b. I am specifically annoyed that, while Gödel' s name is constantly on Carnaps' lip, it is obvious that he absolutely did not understand the real meaning of Gödel's results.

Some years ago Menger wrote an axiomatics for the field of "human relations" (ethics?), and I found it to be quite flat and empty. It is my feeling that trying an axiomatic discussion would be most promising (relatively!) in the field of certain particular areas of psychology or perhaps of economics.

I don't know whether I wrote you that Gödel proved the following: If the (customarily axiomatized) set theory is free of contradictions (whether this is the case can, precisely on account of Gödel's first results, not be decided) – free of contradictions by leaving out the "Axiom of Choice" (= "Well Ordering Theorem") – then it is possible to add the following extra axioms, without thereby producing contradictions: 1. The so called "Axiom of Choice"; 2. The so called "Generalized Continuum Hypothesis", which is: The cardinality *immediately* following the cardinality of any set is the cardinality of the set of all subsets of that set; 3. A certain point set – effectively given by G.[ödel] and, featuring as "projective set of 2nd degree" in a known, important classification – is not Lebesgue measurable. All three results are important, and the third opens up a quite unexpected connection.

Did you know that G.[ödel] was suspended in Vienna? ("Aryan") Next year he will be back here again.

In politics I would like to say two things in defense of the Roosevelt and the new Chamberlain (?) policies. One must not misunderstand intentions: For then the most justified regulations can create an impression of naivete, frivolity, or rashness. Probably no one on the side of the Western powers would now want to make a serious sacrifice to save today's transitional state of affairs (perhaps Ch.[amberlain] might, but obviously today he is no longer in the position to enforce his own policies) – for since last winter most people, who were so far indifferent or, even more, neutral with good will (!), were convinced by their experiences that concessions only produce more demands. Today's regulations are already war regulations and, as such, are naturally one-sided and have unforeseen secondary effects, just as such regulations usually do.

Many greetings until our hopefully soon to be reunion

Jancsi

Original in Department of Manuscripts and Rare Books of the Library of the Hungarian Academy of Sciences, call number K 785/101. Language of original: Hungarian. Translated by A. Lenard.

Letter to W. Overbeck

November 28, 1945

Mr. William Overbeck
1211 Marshall Avenue
Richland, Wash.

Dear Mr. Overbeck:

Our common friend, Professor John A. Wheeler, told me about your interest in high-speed computing and in the work which you have done in the past in fields which have connection with this topic. He was also so good as to show me a copy of a letter which you wrote him on November 1 in connection with this matter. I am very happy that I have had this opportunity to make your indirect acquaintance and to get in touch with you.

As Wheeler has probably told you, we were interested in starting an automatic computing development here in Princeton. In the meantime things have developed considerably further. The Institute for Advanced Study, the Radio Corporation of America and Princeton University have undertaken a joint project to build an automatic high-speed computer, and I have been asked to coordinate and direct this project. We are planning to spend the next two to three years on building such a device. We are thinking of a digital, rather than a continuous-variable, device, for scientific purposes primarily. That is, we hope to use it in order to further the theory and practice of partial differential equations, also work in fluid dynamics, turbulence, suitable parts of atomic physics, and a number of other applications which suggest themselves, considering the possibilities of such a machine. We are planning to use the binary system, presumably combined with decimal-to-binary and binary-to-decimal conversion facilities. We are tentatively thinking of a precision of 30 binary digits (equivalent to about 9 decimal digits). We want to carry out the arithmetical operations digitally, essentially by vacuum tube coincidence circuits. We expect to be able to work somewhere in the megacycle range, and count on addition times of the order of 30 milliseconds, and multiplication times of about a millisecond or less. We visualize that the greatest difficulty in building such a machine is to provide it an adequate "inner" memory, – it seems that in order to work efficiently on the type of problem which seems important, it is necessary that the machine should be able to store about 100,000 binary digits. This storage must be such that a digit can be stored, or sensed, or cleared, in times which never exceed a millisecond and which in certain critical situations may have to be up to 30 times shorter. Considering the size of the memory indicated (100,000 digits), ordinary vacuum tubes arrangements would lead to forbidding numbers of vacuum tubes and forbidding switching problems. However, the television and the radar industries comprise devices which will probably meet this problem adequately. This "fast" inner memory must be supplemented by somewhat less fast input and output possibilities, – these must, however, still be fairly fast by conventional standards. We are thinking of digital inputs and outputs on magnetic tape, together with continuous variable inputs of some nature not yet determined, and at least one type of continuous variable output: projection on an oscilloscope screen.

After what Professor Wheeler told me about your work, I wish there were an opportunity for me to tell you in more detail about these things. I think that you would find them of considerable interest. I think that the field contains many interesting problems, from the electronic as well as from the servo-mechanism point of view.

Is there any chance of your coming to the East sometime in the next months? I would anticipate that we should find a great deal that is of mutual interest, and I am quite sure that both the Institute for Advanced Study and the Radio Corporation would want to make an effort to interest you in participating in this project. At any rate I think that it would be well worth our while to get together and to discuss these possibilities.

I expect to be in Los Alamos from December 3 until about December 31. Is there perhaps any possibility that you may visit there, and that we could perhaps talk with each other there? In any case I would be much obliged if you would let me know whether these things appeal to you, and when and where there might be a chance for a more detailed discussion. If you want more detailed information, I will also be very glad to write you more fully about the subject. Perhaps you could let me know which aspect you want to see discussed most. In the course of December I can most easily be reached at Los Alamos, that is P.O. Box 1663, Santa Fe, New Mexico.

Sincerely yours,

John von Neumann

Original in Von Neumann Papers, Library of Congress, language of original: English

Letter to H.H. Rankin

August 26, 1954

Lt. Col. H.H. Rankin, Ord. Corps
Adjutant General, Department of Defense
Armed Forces Special Weapons Project
P.O. Box 2610
Washington 13, D.C.

Dear Colonel Rankin:

I am writing in reply to your letter of August 25, and Standard Form 50 (Notification of Personnel Action) in connection with my appointment as a consultant to the Armed Forces Special Weapons Project for another year beginning 1 July 1954.

In connection with the last paragraph of your letter I would like to inform you that I am a consultant to a number of other Federal Agencies. Among these the following are the important ones. I am a consultant to the U.S. Atomic Energy Commission, and to a number of its laboratories. I am a member of the Scientific Advisory Board of the U.S. Air Force. I am also a member of the Scientific Advisory Committee of a classified Air Force project connected with the Office of the Special Assistant (Research and Development) of the Secretary of the Air Force. I am further a member of the Technical Advisory Board on Atomic Energy connected with the Office of the Assistant Secretary of Defense (Research and Development). I am also a member of the Scientific Advisory Committee of the U.S. Army Ordnance Corps, Ballistic Research Laboratories, Aberdeen Proving Ground. I also have some other connections of lesser importance.

I am,

Sincerely yours,

John von Neumann

Original in Von Neumann Papers, Library of Congress, language of original: English

Letter to H.P. Robertson

Princeton

January 16, 1932

Dear Robertson!

The story of the proof of the quasi-Ergodic hypothesis[129], about which you were asking me, is this:

Koopman informed me of his method, which is now published (Proc. Nat. Ac., May 1931), and well known to you, during my first stay in this country, in May 1930. He expressed his hope to prove the q.E.h by means of it, but none of us succeeded then. Andrè Weil, who lives usually in Aligarh, India, came to Berlin in Summer 1931, and visited me there. He told me about a method, he worked out, to attack dynamical problems, and hoped to solve the q.E. problem, but he did not succeed with it till then either. This was in July 1931 (since then Weil wrote letters to Koopman and to me, acknowledgeing Koopman's priority).

In August-September 1931 I took this problem up again, and solved it, in the way known to you. I did not publish it, because I wanted to know first, if Koopman had not obtained the same result meantime. I would have considered it as dishonest, if I had interfered with his investigations, using his methods, without his authorization.

In October 1931, I met Koopman in New York, and he informed me, that my result was absolutely new to him. He recommended me, to publish it in the Proc. Nat. Ac., where his method had been published, too. I wrote my paper then, and gave it to him, as he kindly accepted to take care of its translation.

At the New York meeting of the American Math. Soc. (October) M. Stone and E. Hopf saw my manuscript, and we discussed it. At the inauguration of Fine Hall in Princeton (October 1931), Koopman and Birkhoff[130] came here, and Koopman and myself informed Birkhoff about the fact, that I had proved the q.E.H. with Koopman's method, and told him the exact statement. He then learned, what I considered my essential results:

(1) that the limit of the "relativ sojourn" (in the time interval $s \leq \tau \leq t$, for $t-s \to +\infty$ *allways* exists, if it is understood as convergence in the mean,

(2) that the q.E.H. is true *if and only if* there are no measurable integrals.

In November 1931 Birkhoff wrote me a letter, congratulating me for this proof; informing me, that he wanted to talk about this new situation in his general talk

[129]See Section 7 of the Introductory Comments for von Neumann's work on the ergodic theory. The Editor.

[130]George David Birkhoff, not to be confused with Garrett Birkhoff. The Editor.

in New Orleans; asking me to write him the exact statement of my result; and informing me, that he had a direct method to attack the problem, and hoped to succeed with it.

On December 4/5, 1931 several Princeton men, between them myself, went to Cambridge, to give and heare conferences on theoretical physics. I gave there my proof, and in the discussions Birkhoff informed us, that he had another proof, which showed even somewhat more than mine: instead of mean-conv. he could prove conv. everywhere excepted on a set of measure 0. (In a paper to appear in the Proc. Nat. Ac.[131] I show, that the physical statement of the q.E.H. requires mean-conv., and not more.) On the dinner at the Harvard Club, which was given by Harvard and Mass. Tech.[132] faculties to us, I asked Birkhoff, when his paper will appeare, as mine would come out in the January number of the Proc. Nat. Ac. He told me, that he tried to get his into the December number. I asked him to withhold it, for obvious reasons, but he refused that, stateing moreover, that it was dubious, whether it will come out already in December. As this was not the place to argue, I did not insist, he said besides, that he will acknowledge my priority in due form. (I did *not* agree to his quick-publishing, I only gave up on this subject of talk).

Some days before I went to Cambridge, E. Hopf informed me, that he had a new proof of the q.E.H., simpler than mine. When I met him in Cambridge, he told me, that his first attempt contained an error. He had another proof-variant, but this essentially followed my line of argument, only avoiding use of the "spectral resolution" for Koopmans operators U_t. He wanted to publish it in the Proc. Nat. Ac. too, and had given his manuscript to Birkhoff, he naturally agreed with me that it should not come out earlier than mine.

(My paper was delayed to some extent by the fact, that Koopman discussed it with M. Stone, who advised some technical changes, which are very interesting, and of one of which I made use. So the manuscript was only sent in on December 9, too late for the December number).

The December number appeared in January 1932, and Birkhoffs article was in it. His quotation of my result is, according to the judgement of Eisenhart, Alexander, Lefschetz, Koopman, Stone, absolutely insufficient. (I mention only these 5 names, because they all spontaneously told me, how dissatisfied they are, without any attempt on my side, to talk about this matter.) The reason they give is, that it does not show to any person, uninformed about the real history of these things, who of Birkhoff and myself got the other started; that which one of us attacked the unsolved q.E.H, and which one found an independent new proof, after he knew that it was solved, and what the necessary and sufficient conditions for its truth are.

Excuse me, for boring you with the lengthy and teadious details of the story, but you wanted me to inform you reliably and exactly.

[131] J. von Neumann: "Physical applications of the ergodic hypothesis", Collected Works Vol. II. No. 13. The Editor.

[132] Massachusetts Institute of Technology (MIT). The Editor.

With the best greetings Your,

J. v. Neumann

Original in the Institute Archives, California Institute of Technology, language of original: English. Published in full first in J.D. Zund: "Georg David Birkhoff and John von Neumann: A question of priority and the ergodic theorems, 1931-1932, Historia Mathematica **29** *(2002) 138-156*

Letter to E. Schrödinger

The Institute for Advanced Study
School of Mathematics
Fine Hall
Princeton, New Jersey

April 11, 1936

Dear Schrödinger,

Einstein has kindly shown me your letter as well as a copy of the Pr. Cambr. Phil. Soc. manuscript.[133] I feel rather more over-quoted than under-quoted and I feel that my merits in the subject are over-emphasized. I am very glad however that you like the "mixing"- method.

As to the content I would like to remark the following:

(1) You assign the "weight" $(U^{-1}\phi, \phi)^{-1}$ to the wave function ϕ in the mixture U. This can be equal to 0 even when $U\phi \neq 0$. This is because U^{-1} *must* be unbounded (disregarding the highly degenerate case where U possesses only finitely many non-zero eigenvalues, since otherwise U's eigenvalues necessarily converge to 0), and hence $(U^{-1}\phi, \phi)$ will be infinite for some ϕ. Example:

$$U = (u_{mn}), \qquad m, n = 1, 2, \ldots, \tag{1}$$

$$u_{mn} = \begin{cases} 2^{-m} & \text{for} \quad m = n \\ 0 & \text{for} \quad m \neq n \end{cases} \tag{2}$$

$$\phi = (a_m), \qquad m = 1, 2, \ldots, \tag{3}$$

$$a_m = 2^{-\frac{m}{2}} \quad \text{for all } m \tag{4}$$

(2) In a certain sense it would perhaps be more appropriate to set the "weight" of ϕ in U to be

$$w_2 = (U\phi, \phi)$$

instead of your

$$w_1 = (U^{-1}\phi, \phi)^{-1}$$

Conceptually, w_1 and w_2 behave as follows:

[1] w_1 is the greatest weight for which one can find a mixture U' such that mixing ϕ and U' with weights w_1 and $1 - w_1$ yields U.

---

[133]Von Neumann refers to E. Schrödinger's paper: "Probability relations between separated systems", Proceedings of the Cambridge Philosophical Society, **32** (1936) 446-452. The Editor.

[2] w_2 is the probability that a physical quantity $\mathcal{A}$ having sharp value a_0 in state ϕ and having only values $\neq a_0$ in states orthogonal to ϕ, possesses value a_0 if measured in mixture U.

Incidentally, it always holds that

$$0 \leq w_1 \leq w_2 \leq 1$$

Only $w_1 \leq w_2$ is not trivial:

(5) $$\frac{w_2}{w_1} = (U\phi, \phi)(U^{-1}\phi, \phi) = (U^{\frac{1}{2}}\phi, U^{\frac{1}{2}}\phi)(U^{-\frac{1}{2}}\phi, U^{-\frac{1}{2}}\phi)$$

(6) $$= (\|U^{\frac{1}{2}}\phi\| \|U^{-\frac{1}{2}}\phi\|)^2 \geq (U^{\frac{1}{2}}\phi, U^{-\frac{1}{2}}\phi)^2 =$$

(7) $$= (U^{-\frac{1}{2}}U^{\frac{1}{2}}\phi, \phi)^2 = (\phi, \phi)^2 = 1$$

(8) $$w_2 \geq w_1$$

(Eq. (6) is Schwarz' inequality.) w_2 is zero only in the case you give: $(U\phi, \phi) = 0$ means $(U^{\frac{1}{2}}\phi, U^{\frac{1}{2}}\phi) = 0$, $\|U^{\frac{1}{2}}\phi\|^2 = 0$, $U^{\frac{1}{2}}\phi = 0$, $U\phi = 0$, the converse is obvious.

(3) I cannot accept your §4. completely. I think that the difficulties you hint at are "pseudo-problems". The "action at distance" in the case under consideration says only that even if there is no dynamical interaction between two systems (e.g. because they are far removed from each other), the systems can display statistical correlations. This is not at all specific for quantum mechanics, it happens classically as well. I mean the following:

Let S_1 and S_2 be two boxes. One knows that 1,000,000 years ago either a white ball had been put into each or a black ball had been placed into each but one does not know which color the balls were. Subsequently one of the boxes (S_1) was buried on Earth, the other (S_2) on Sirius. So one has this probability distribution:

	S_1 white	S_1 black
S_2 white	$\frac{1}{2}$	0
S_2 black	0	$\frac{1}{2}$

Or for S_2 only:

S_2	white	$\frac{1}{2}$
S_2	black	$\frac{1}{2}$

Now one digs S_1 on Earth out, opens it and sees: the ball is white. This action on Earth changes instantaneously the S_2 statistic on Sirius to

S_2	white	1
S_2	black	0

(The situation reminds one of the well-known joke: *"Au moment même*[134] *mon Colonel était cocu á Madascar."*)[135]

To be sure, the peculiar behavior of "incompatible" measurements is characteristic of quantum mechanics – but this too is free of contradiction. One just has to accept that ϕ describes an ensemble that cannot be further analyzed.

And of course quantum electrodynamics proves that quantum mechanics and the special theory of relativity are compatible "philosophically" – quantum electrodynamic fails only because of the concrete form of Maxwell's equations in the vicinity of a charge.

I very much hope that I will have the opportunity to see you again this summer in some part of Europe. I have again worked on quantum theory and believe to have an approach with which one could get around certain energy-divergence miseries.

I am with the best wishes,

Sincerely yours,

Johnny (=John von Neumann)

Original in Archives for the History of Quantum Physics, copy used for the volume from Archives of Scientific Philosophy, University Pittsburgh Libraries Special Collections. Language of original: German. Translated by M. Rédei.

[134]Von Neumann's insertion: "Als mit seiner Frau etwas in Paris geschah."; i.e. "When something happened to his wife in París."

[135]Together with the insertion mentioned in footnote 134 the French sentence reads something like "The moment something happened to his wife in Paris the colonel was cuckolded in Madagascar". The Editor.

Letter to E. Segre

October 24, 1955

Professor E. Segre
University of California
Radiation Laboratory
Berkeley, California

Dear Emilio:

First of all, I would like to congratulate you and your associates most warmly on the discovery of the antiproton. It is truly magnificent and – to an amateur in physics, at least – it seems that for the first time in over 30 years elementary particle physics is getting simple again.

In this connection, I would like to ask you some questions.

So far the pair electron (e^-) - positron (e^+) and the pair proton (p^+) - antiproton (p^-) exhibit symmetry between positive and negative charges in nature. This does not necessarily mean that that symmetry is complete, but it is a strong indicator in that direction. Let me assume that this is indeed so, i.e., that the symmetry in question is complete. What then is the position of the neutron (n)?

The neutron disintegrates with a certain half-life τ ($\approx$ 10 min.) according to the formula:

$$n \to p^+ + e^- + \nu + .78 \; mev \tag{9}$$

(ν is a neutrino.) By virtue of the plus-minus charge symmetry, the disintegration

$$n' \to p^+ + e^+ + \nu' + .78 \; mev \tag{10}$$

must also exist, and with the same half-life τ as (9). Here n' is a particle with the same mass, charge and spin as the neutron n, and the same must be true for the relationship of ν' and ν. The important point is, however, that for all this n' may or may not be identical with n. Hence two essentially different alternatives exist:

(a) $n = n'$: The neutron disintegrates according to both schemes (9) and (10). To determine whether this is so, the experiments that established the decay of the neutron and its half life must be re-examined. Did they count only processes (9), or would they have counted both processes (9) and (10)? If the second is the case, would they have distinguished between (9) and (10)?

(b) $n \neq n'$: In this case n' is best called an antineutron. The odd situation arises that although the neutron has no charge, the plus-minus charge symmetry nevertheless provides it with a distinct anti-form.

It was pointed out to me by Z. Bay (NBS)[136] that (a) can probably be excluded without reference to experiments. Indeed, if (a) were true, ordinary neutrons would disintegrate according to (10), creating antiprotons, with these two consequences:

(II) High density free neutron regions – e.g., high intensity beams of thermal neutrons – which are surrounded by ordinary (p^+) matter – would also be surrounded by large members of $p^+ - p^-$ annihilations, their "stars", their energy release, etc.

(II) Bound neutrons, in the nuclei of the ordinary elements, would probably also disintegrate according to (10). Indeed, their intranuclear binding energy of a few *mev* inhibits (9), but it cannot inhibit (10) followed by the annihilation of the resulting p^- with a p^+ (in the same nucleus) with its resulting energy release of 2 *bev*. Thus, all (ordinary) elements would have to be unstable.

In either case, one would expect that under the same conditions under which you created $p^+ - p^-$ pairs, one would also obtain $n - n'$ pairs, i.e., a neutron (n) and a neutron-like particle that causes annihilation-stars in ordinary matter (n', cf. (a) above). Is this experiment feasible in your present setup? Or has your work already given indications from which the outcome of such an experiment can be predicted?

Let me assume that the argumentation following (II) is correct. In this case, the following questions arise. There is hardly any doubt that n and n' can annihilate each other. If your present experiment were to show an n, n' yield, then this would be certain. Let me, therefore, assume that n and n' do indeed annihilate.

Then any n' in ordinary (p^+, n) matter would annihilate and release 2 *bev* energy. By the way, this energy release would occur in any case after the decay (10), since the p^- would annihilate with a p^+. Hence the stability of ordinary matter requires that the transition:

$$n \rightleftharpoons n' \tag{11}$$

be prohibited. What prohibits (11)? Mass, charge, spin are the same on both sides. Does some other quantum number differ? If the reaction were possible, it would require no energy. I realize that n' may be viewed as a "hole" in a "sea" of negative mass neutrons, but it is nevertheless, contrary to all experience so far that a transition between elementary particles, that does not conflict with any conservation law, be prohibited. What then is the conservation law that enters here?

Because of (9), the number of neutrons n, say $N(n)$, cannot be conserved. There are combinations of $N(n), N(p^+), N(p^-), N(e^+), N(e^-)$, etc., that might be. Is this all? Do other particles not enter? If they do, then (11) with additions of other particles on both sides would become possible. Is this plausible?

I would be much obliged to you for any views and ideas that you might give me on these matters.

With best regards,

As ever,

John von Neumann

Original in Von Neumann Papers, Library of Congress, language of original: English

[136]National Bureau of Standards. The Editor.

Letters to F.B. Silsbee

November 3, 1944

Dr. F.B. Sillsbee
National Bureau of Standards
Washington 25, D.C.

Dear Doctor Sillsbee:

I have received your very kind invitation to deliver the Fourteenth Joseph Henry Lecture before the Philosophical Society of Washington. I appreciate the honor of this invitation, and I accept it with the greatest pleasure. Your suggestion that the subject of the lecture might be on some application of mathematics to quantum mechanics suits me very well. I am going to give the Josiah Willard Gibbs Lecture of the American Mathematical Society at Chicago on the 24th of this month on "The Ergodic Theorem", - that is a mathematical topic in statistical mechanics. I would therefore prefer to choose for Washington the other alternative you mention, - that is, quantum mechanics. I suppose that the audience which you describe would be most interested in the question of interpretation of quantum mechanics, and I would therefore like to submit for your approval the title: "Causality, Statistics and Quantum Mechanics."

I shall be very glad to submit a manuscript of the address for publication. I assume that you will want it after the lecture.

The month of March 1945 suits me very well. I have a little difficulty in choosing now between the 17th and the 31st because I expect to be in the West part of that month and I am not yet quite certain which part. Would it be agreeable to you if I write to you about this again during the first half of January? At that time I expect to have my arrangements settled.

I am very sincerely yours,

John von Neumann

Original in Von Neumann Papers, Library of Congress, language of original: English

February 14, 1945
VIA AIR MAIL

Dr. F.B. Silsbee
The Philosophical Society of Washington
National Bureau of Standards
Washington, D.C.

Dear Doctor Silsbee:

Many thanks for your letter of February 6, which reached me with some delay since I am at this moment not in Princeton.

I will be very glad to give the Joseph Henry Lecture on March 17, the title "Causality, Statistics and Quantum Mechanics" suits me very well. I need no lantern slides. I would phrase an abstract as follows:

> Quantum Mechanics is more closely connected with Probability and Statistics than any other branch of Physics. An analysis shows, that the relation is of a deeper nature than even in classical statistical mechanics. The structure of logics and the relation of strict logics to probability is also essentially affected by the recognition of the new features of Quantum Mechanics. Finally, Quantum Mechanics provides an example for the possibility of building a complete and self consistent system, in which causality does not hold in the conventional sense.

It is very kind of you to mention my possible arrangements for staying in Washington. I do normally stay at the Cosmos Club, and I planned to do this at this occasion, too.

I am very sincerely yours,

John von Neumann

P.S. I did not quite understand what type of a bibliographical note would be useful to you. I think, however, that Dr. R.J. Seeger of the Physics Department of George Washington University and Section Re 2 of the Navy Bureau of Ordnance can give you any data you may want.

Original in Von Neumann Papers, Library of Congress, language of original: English

June 11, 1945

Dr. F.B. Silsbee
Philosophical Society of Washington
National Bureau of Standards
Washington, D.C.

Dear Doctor Silsbee:

Very much to my regret I have to modify the promise I made to you in my letter of June 6, that is that I should be able to send you the complete manuscript of my paper "Logic of quantum mechanics" by June 20[137]. In the meantime it has turned out that I have to leave Princeton for four days this week; also, I have to go to a meeting in Canada next week, from June 20 to 24. On the other hand, I find that I ought to make some rather careful revisions of what I have so far done on this paper, – I am quite sure that it would be regrettable if I didn't and finished it in a hurry. So I am afraid I shall have to ask your indulgence and delay the date of completion. I am sure that I shall have it done within two weeks of my return from Canada, that is by July 7 at the latest.

Sincerely yours,

John von Neumann

Original in Von Neumann Papers, Library of Congress, language of original: English

---

Institute of Advanced Study
School of Mathematics
Princeton, New Jersey

July 2, 1945

Dear Doctor Silsbee,

It is with great regret that I am writing these lines to you, but I simply cannot help myself. In spite of very serious attempts to write the article on the "Logics of quantum mechanics" I find it completely impossible to do it at this time. As you may know, I wrote a paper on this subject with Garrett Birkhoff in 1936 ("Annals of Mathematics", vol. 37, pp. 823-843),[138] and I have thought a good deal on the subject since. My work on continuous geometries, on which I gave the Amer. Math. Soc. Colloqium lectures in 1937, comes to a considerable extent from this source. Also a good deal concerning the relationship between strict and probability logics (upon which I touched briefly in the Henry Joseph Lecture) and the extension of this "Propositional calculus" work to "logics with quantifiers" (which I never so

[137] See Section 6 of the Introductory Comments for von Neumann's work on quantum logic. The Editor.

[138] G. Birkhoff, J. von Neumann: "The Logic of quantum mechanics" Collected Works Vol. IV. No. 7. The Editor.

far discussed in public). All these things should be presented as a connected whole (I mean the propositional and the "quantifier" strict logics, the probability logics, plus a short indication of the ideas of "continuous" projective geometry), and I have been mainly interrupted in this (as well as in writing a book on continuous geometries, which I still owe the Amer. Math. Soc. Colloqium Series) by the war. To do it properly would require a good deal of work, since the subjects that have to be correlated are a very heterogenous collection – although I think that I can show how they belong together.

When I offered to give the Henry Joseph Lecture on this subject, I thought (and I hope that I was not too far wrong in this) that I could give a reasonable general survey of at least part of the subject in a talk, which might have some interest to the audience. I did not realize the importance nor the difficulties of reducing this to writing.

I have now learned – after a considerable number of serious but very unsuccessful efforts – that they are exceedingly great. I must, of course, accept a good part of the responsibility for my method of writing – I write rather freely and fast if a subject is "mature" in my mind, but develop the worst traits of pedantism and inefficiency if I attempt to give a preliminary account of a subject which I do not have yet in what I can believe to be in its final form.

I have tried to live up to my promise and to force myself to write this article, and spent much more time on it than on many comparable ones which I wrote with no difficulty at all – and it just didn't work. Perhaps if I were not continually interrupted by journeys and other obligations arising from still surviving war work, I might have been able to do it – although I am not even sure of this. As things are my work in all respects has suffered, I have not succeeded in producing any paper I would like to offer to you for publication and I was not able to live up to my other duties as I would have liked to. (I owe the NDRC[139] two long reports which I am unable to start while this "struggle" goes on – I realize that this is an irrational inhibition, but I cannot help it.)

Also, I would like to stress it once more, that apart from my unwillingness to publish something I do not like, and my inhibitions to finish "at any rate" a manuscript which does not satisfy me, it is my respect for the WPS[140] and my appreciation of the honor done to me in inviting me to give the Joseph Henry Lecture, which prevent me from turning in a manuscript. In peace time I would try to stick to a decision of doing nothing else until I have thought out this subject until it is mature for publication as a whole and then written a satisfactory account – this might succeed, although it is a rather painful operation which I did certainly not foresee when I suggested the topic. At present, or in the immediate postwar period, however, I would not enforce such a decision upon myself.

I must therefore ask you to excuse me for not being able to fulfill my promise. I will certainly sometime write up this subject, probably in much more detail. If the Washington Philosophical Society or the Washington Academy should still be interested in it, I will be only too glad to offer it to you for publication. Hoping that you do not think too badly of me for having caused all these complications and hoping to see you again before long,

[139]National Defense Research Council. The Editor.

[140]Washington Philosophical Society. The Editor.

I am sincerely yours

John von Neumann

P.S. I would like to tell you once more what an unmixed pleasure it was for me to talk before the Washington Philosophical Society, and how I appreciate this privilege. On the other hand, I feel that I should not have accepted a honorarium, particularly since I am not able to fulfill in a reasonable time the only really serious obligation: To write the article. Would you let me know, in which way I can correct this most conveniently for the Washington Philosophical Society?

Original in Von Neumann Papers, Library of Congress, language of original: English. Published in full first in M. Rédei, M. Stöltzner (eds.): John von Neumann and the Foundations of Quantum Physics (Kluwer Academic Publishers, Dordrecht, Boston, London, 2001.) pp. 225-226

October 22, 1945

Dr. F.B. Silsbee, Chairman
Joseph Henry Lecture Committee
National Bureau of Standards
Washington 25, D.C.

Dear Doctor Silsbee:

I have received your letter of October 18, and I want to express to you how much I appreciate the exceptional patience which you and the General Committee of the Philosophical Society of Washington are showing toward me.

I am very glad that you are now giving me another opportunity to write a paper on "Quantum logics" for publication in the Journal of the Washington Academy of Sciences. The atmosphere in Princeton is indeed quieting down, and while I still have some work to finish in connection with the the termination of various OSRD[141] projects, I expect now to be clear of all such impediments before the end of this year, possibly even by the end of November. I will then immediately begin to write the paper in question. When do the 1946 numbers of the Journal of the Washington Academy of Sciences appear?

Very truly yours,

John von Neumann

Original in Von Neumann Papers, Library of Congress, language of original: English

---

[141] Office of Scientific Research and Development, a government organization. The Editor.

April 20, 1946

Mr. F.B. Silsbee, Chairman
U.S. Department of Commerce
National Bureau of Standards
Washington 25, D.C.

Dear Doctor Silsbee:

Pardon me for the delay in answering your letter of April 12. It had to follow me to Santa Fe where I am now. I have had a very busy time during the last few months, hardly different from what we all experienced during the war, so I did not get much farther in putting into writing my ideas connected with the subject of my Joseph Henry lectures. My program during the summer is still uncertain. It is possible that I will have to go to the Navy "Crossroads" test in the South Pacific. The things will be clear in a few weeks. May I therefore ask you to forgive me for one more delay. As soon as I know what I am at I will be able to suggest to you a valid dead-line for my manuscript.

Very truly yours,

John von Neumann

CC H.A. Rehder

Original in Von Neumann Papers, Library of Congress, language of original: English

---

December 23, 1946

Dr. F.B. Silsbee, Chairman
Joseph Henry Lecture Committee
National Bureau of Standards
Washington, D.C.

Dear Doctor Silsbee:

My apologies for having strained your forbearance for so long and for still straining it. I will do my best to produce, in the new year, as soon as feasible, a paper on "The Logic of Quantum Mechanics" which is worthy of publication in the "Journal of the Washington Academy of Sciences".

Very truly yours,

John von Neumann

Original in Von Neumann Papers, Library of Congress, language of original: English

Letter to L. Spitzer

June 9, 1949

Professsor Lyman Spitzer, Jr.
Chairman, Scientist's Committee on Loyalty Problems
14 Battle Road
Princeton, New Jersey

Dear Professor Spitzer,

Please excuse my delay in replying to your circular letter of May 24th. I was away from Princeton and only returned recently.

I intend to write to Senator McMahon on the subject of extending clearance requirements to unclassified fields. I need not tell you that I consider this trend to be a pernicious one, and I hope that it will be possible to prevent its progress.

You were so good to ask me my views in somewhat more detail on some definite phases of this subject. They are as follows:

My basic view on the training of talented persons in science is that it should be done with absolutely no consideration of anything other than scientific ability. I realize that this may produce situations which seem unreasonable to the non-scientist, but I am convinced that it will in the long run prove to be the best policy and most profitable to science as well as to the nation. In view of this, I think that it is most deplorable that the entire complex had to come up at all, but since it did come up and is causing serious repercussions, I find it undeniable that some compromise is now necessary. In consideration of this, the requirement of non-communist affidavits may represent a reasonable compromise between what is practical and what is desirable. Requiring loyalty clearance from all scientists receiving financial aid in the same way that it is required from government employees would, I think, be a grave mistake.

I am,

Very truly yours,

John von Neumann

Original in Von Neumann Papers, Library of Congress, language of original: English

Letters to M. Stone

Budapest

the 8.10.1930 [October 8, 1930]

My dear Mr. Stone,

Very many thanks for your kind letter. Having your approval, I will use in my paper the phrase I wrote to you, adding to it, as you suggested, some indications about the difficulty existing in your way of the proof.

Answering the question, you wrote me, about the way I found my proof, I can tell you this:[142]

Any proof of this theorem had to construct with the aid of P, Q or

$$U(\alpha) = e^{i\alpha P}, \qquad V(\beta) = e^{i\beta Q}$$

some operator, which has easily identifyable properties, determining him in a unique way – and which operator on the other hand can be used to determine some vectors in Hilbert space. So your method had the objective, to construct the operator $P^2 + Q^2$ of the "oscillator".

Now a bounded operator would be more desirable than a non bounded one, as his discussion is simpler. The simplest would be a one-dimensional projection-operator (P_ϕ in my terminology). And such an operator E is characterised by the property

$$EAE = cE$$

(for every A and suitably chosen c's.) Such an operator has an integral kernel. Now operators with integral kernels can easily be constructed with help of the $U(\alpha), V(\beta)$, in the usual Schrödinger-realisation

$$U(\alpha)f(q) = f(q+\alpha), \qquad V(\beta)f(q) = e^{i\beta q}f(q)$$

In fact, there is

$$\left(\int\int a(\alpha,\beta)U(\alpha)V(\beta)\mathrm{d}\alpha\mathrm{d}\beta\right)f(q) =$$

[142]Von Neumann refers here to the proof of uniqueness of the Schrödinger representation of the canonical commutation relations in his paper: J. von Neumann: Die Eindeutigkeit des Schrödingerschen Operatoren, Collected Works Vol. II. No. 7. See Section 5.2 of the Introductory Comments for notes on von Neumann's proof. The Editor.

(this is the most general linear aggregate of $U(\alpha), V(\beta)$ – therefore the most general expression, obtainable by the 4 species from $U(\alpha), V(\beta)$)

$$
\begin{aligned}
&= \int\int a(\alpha,\beta)e^{i\beta(q+\alpha)}f(q+\alpha)\mathrm{d}\alpha\mathrm{d}\beta \\
&= \int\int a(\gamma-q,\beta)e^{i\beta\gamma}f(\gamma)\mathrm{d}\gamma\mathrm{d}\beta \\
&= \int\Big[\int a(q'-q,\beta)e^{i\beta q'}d\beta\Big]f(q')\mathrm{d}q'
\end{aligned}
$$

That is an integral operation, with the kernel

$$K(q,q') = \int a(q'-q,\beta)e^{i\beta q'}\mathrm{d}\beta$$

Now this is obviously the most general kernel, because the inversion-formula of Laplace's-Transformation gives:

$$
\begin{aligned}
K(q,q+a) &= \int a(\alpha,\beta)e^{i\alpha\beta}e^{i\beta q}\mathrm{d}\beta \\
a(\alpha,\beta)e^{i\alpha\beta} &= \int K(q,q+\alpha)e^{-i\beta q}\mathrm{d}q \\
a(\alpha,\beta) &= e^{-i\alpha\beta}\int K(q,q+\alpha)e^{-i\beta q}\mathrm{d}q
\end{aligned}
$$

So the relation between $a(\alpha,\beta)$ and $K(q,q')$ is one-to-one. Chosing

$$K(q,q') = e^{-\frac{1}{2}(q^2+q'^2)}$$

th.i. the kernel of P_ϕ, $\phi = e^{-\frac{1}{2}q^2}$ one obtains the operator A of my paper. $ABA = cA$ (for every B and suitable c's) holding in the Schrödinger realisation, must hold too, when computed by means of Weyls Commutation-Relations.

The remainder of my proof is natural enough, I think: it is simply the identification of the properties of $e^{-\frac{1}{2}q^2}$.

I am very sorry, but I will only be back to America in September 1931. I would be very pleased, if you could let me hear about you and your work as much as possible in the meantime.

Begging to give my best greetings to Mrs. Stone,

I am truly yours

John von Neumann

After the 1.11.1930
Berlin-Tiergarten,
Hohenzollerstrasse 23/24

Original in John Hay Library, Brown University, Rhode Island, U.S.A., language of original: English

New York

Jan. 22, 1932

Dear Mr. Stone,

Koopman informs me, that you took an interest in my affair with B.[143], I thank you heartily for this. He asks me to inform you about my talks and correspondence with B.[irkhoff] in this matter. They where[144] as follows:

(1) B.[irkhoff] was told in Princeton, at the inauguration of Fine Hall (Oct. 22) by Koopman and myself that I proved with K.[oopman]'s method this: the q.[uasi] E.[rgodic] Theorem is true if and only if there are no measurable integrals, and that the relative-sojourn limit allways exists.[145]

(2) B.[irkhoff] wrote me a letter sometime in November, expressing his satisfaction about these results, asking me for an exact statement of them, for two reasons: because he wanted to refer to them in New Orleans, and because he *hoped* to prove them directly.
I answered him immediately, giving the exact statement of conv.[ergence] in the mean for all motions, and the nec.[essary] and suff.[icient] conditions for the q.[uasi] E.[rgodic] hypothesis.

(3) In Cambridge (Dec. 5) I gave a talk on my results. In the discussion B.[irkhoff] stated, that he had proved somewhat more: conv.[ergence] excepted on a 0-measure set.

(4) On the same evening, at a dinner given by the faculties, I asked him, when his paper would appear. He said: Proc.[eedings of the] Nat.[ional] Ac.[ademy of Sciences], December of January. I asked him to wait, till mine would appear. He refused this, saying: Nobody likes to delay his own papers, and I will state your priority in due form. I gave up the conversation at this point, as I could not force him to act as I would have expected – naturally without a word which could have been interpreted as my accepting his view.

Hoping to hear from you soon, I am truly yours

John von Neumann.

Budapest, (Hungary),
V. Arany János utca 16

Original in John Hay Library, Brown University, Rhode Island, U.S.A., language of original: English

---

[143]George David Birkhoff. See Section 7 of the Introductory Comments for von Neumann's work on the ergodic problem. The Editor.

[144]Spelling error: should be "were". The Editor.

[145]Von Neumann's insertion: "We told him, that it was 'in the mean', but he may not have paid attention to it".

The Institute for Advanced Study
School of Mathematics
Fine Hall
Princeton, New Jersey

March 31, 1935

Dear Marshall,

There has been one more complication since I last wrote you: Mariette had gall-colic, which was very painful and lasted almost three days. But now it is all over, and she had sufficiently recovered by the end of last week, to make our exodus from New York possible. We are back to Princeton since yesterday, and in a few days Mariette will – presumably – be completely back to "normalcy". The baby is alright and has successfully crossed the 8-pound-line.

In a few days I will return to you a copy of the typewritten manuscript of our paper, with the minor modifications, about which we agreed.

Our programm is now to stay in Princeton until April 18 or 19, then move to New York, from where we sail with the "Paris" on April 20. I will leave the boat in Plymouth, in order to begin my lectures in Cambridge with as little delay as possible (the "May term" begins there on the 23rd), and Mariette will take the baby to Budapest, and then return alone and join me in Cambridge.

Encouraged by your last letter, I am giving you an account of the results Murray and I obtained on rings of operators lately.[146] Here it is:

As you will probably remember, I always suspected, that in order to be able to apply representation-theory and various other things to Hilbert space, the conditio sine qua non would be to know more about rings of operators $\mathcal{M}$, for which

$$(1) \qquad \mathcal{M} \cdot \mathcal{M}' = 0$$

($\mathcal{M}'$ is the ring of those operators A, which commute with every $B \in \mathcal{M}$. 0 is the smallest possible ring of operators: the set of all αI, α a complex number, I the unity operator. Analogously I is the maximum ring of /bounded/ operators: the set of all bounded operators. Denoting the minimum ring containing the rings $\mathcal{M}, \mathcal{N}$ by $\mathcal{M} \cap \mathcal{N}$, and the maximum ring contained in $\mathcal{M}, \mathcal{N}$ /which, by the way, simply is their set-theoretical intersection/ by $\mathcal{M} \cdot \mathcal{N}$[147], I proved in 1929 /Math. Ann. 102/[148] that $\mathcal{M} = \mathcal{M}''$, from which $(\mathcal{M} \cap \mathcal{N})' = \mathcal{M}' \cdot \mathcal{N}'$, $(\mathcal{M} \cdot \mathcal{N})' = \mathcal{M}' \cap \mathcal{N}'$ follow at once. This symbolism differs from the one I used in 1929 on these points: 1) I then wrote $\mathcal{B}$ for I, 2) I did not require that a ring $\mathcal{M}$ must contain the operator I, which I now find preferable to do. 3) I then wrote $\mathcal{R}(\mathcal{M}, \mathcal{N})$ for $\mathcal{M} \cap \mathcal{N}$. Thus if one looks at the entire business sub specie of Boolean Algebras or of non-distributive lattices, then the operation of $\mathcal{M}'$ acts very much like a "negation".

[146] See Section 3 of the Introductory Comments for von Neumann and Murray's work on operator algebra theory. The Editor.

[147] Von Neumann's insertion on the margin: "As you see, this symbolism is directly inspired by the theory of Boolean algebras, which I learned since from your discussion of the subject. These $\cap$ and $\cdot$ do not obey the distributive law, and thus form a *lattice* in the sense of G. Birkhoff."

[148] J. von Neumann: Allgemeine Eigenwerttheorie Hermitescher Funktionaloperatoren, Collected Works Vol. II. No. 1. The Editor.

But a "negation" should have these further properties:

$$(2) \qquad \mathcal{M}\cdot\mathcal{M}' = 0$$

$$(3) \qquad \mathcal{M}\cap\mathcal{M}' = I$$

which $\mathcal{M}'$ does not have in general. Thus there is a good formal reason, to inquire about those *special* $\mathcal{M}$'s which fulfill (2) and (3). As (2) and (3) are dual, and thus equivalent, one is lead[149] back to (1). So you see, that their[150] is a quite nice formalistic reason to inquire about (1).

Of course they[151] are several material reasons, too: (1) corresponds to the "Fragestellung"[152] of Burnside-Frobenius's theorems on representations, and as I since found to the question of the existence of a "trace" in an algebra /I will discuss this later/. Another problem which leads immediately to (1) is this: Which rings $\mathcal{M},\mathcal{N}$ which commute with each other /but may – and will – not be Abelian themselves/ generate I, that is, fulfill $\mathcal{M}\cap\mathcal{N} = I$. This means in the quantum mechanical slang: How can a mechanical system of observables be split up into two independent parts?

I originally surmised, that whenever $\mathcal{M}\cdot\mathcal{M}' = 0$ holds, the underlying Hilbert space $\mathcal{H}$ can always be described as (isomorphic to) the space of all functions $f(x,y)$ of two independent variables x, y (the "finiteness condition" being

$$\int\int |f(x,y)|\mathrm{d}x\mathrm{d}y < +\infty$$

where $\mathrm{d}x$ and $\mathrm{d}y$ are to be understood in terms of some totally additive measure function of the x- resp. y-space, which may even be discrete), in such a manner, that $\mathcal{M}$ consists of all operators which operate on x alone, and $\mathcal{M}'$ of those which operate on y alone.[153] This surmise was shaken by some results of Murray, whom I interested in the problem, and now we are able not only to disprove it, but to give an almost complete enumeration of all possibilities too. They turn out to be much more interesting and attractive, then[154] what the "separation into two independent variables"-case could have been.

The general method to deal with these rings $\mathcal{M}$, which we called *factors*, is this:

Two projections $E, F \in \mathcal{M}$ for which an operator $U \in \mathcal{M}$ exists, which maps the linear closed set of E in a one-to-one and isometric way on that one of F are called equivalent, $E \sim F$. (Another way to formulate this: $U^*U = E$, $UU^* = F$.) Like in set theory, write $E \preceq F$ if $E \sim E' \leq F$. One proves the Cantor-Bernstein and the general-comparability-theorem, using that $\mathcal{M}$ is a factor. And finally one can prove this: There exists one, and up to a constant > 0 factor only one, function $D(E)$ with these properties:

(1) $D(0) = 0$; if $E \in \mathcal{M}$, $E \neq 0$, then $0 < D(E) \leq +\infty$.
(2) If $E, F \in \mathcal{M}$, $EF = 0$, then $D(E+F) = D(E) + D(F)$.
(3) $E \sim F$ if and only if $D(E) = D(F)$.

---

149 Spelling error: it should be "led". The Editor.

150 Spelling error,: it should be "there". The Editor.

151 Spelling error: it should be "there". The Editor.

152 German: "the question asked". The Editor.

153 Von Neumann's insertion: "This answer is, of course, strongly suggested by quantum mechanics!"

154 Spelling error: it should be "than". The Editor.

This is the *relative dimension function* in $\mathcal{M}$.

Now the range Δ of $D(\mathcal{M})$ must be one of the following sets:

(I_n) $[n = 1, 2, \ldots]$	Δ consists of $0, \epsilon, 2\epsilon, \ldots, n\epsilon$; where ϵ is a fixed number $\epsilon > 0, < +\infty$. Of course we can normalize $\epsilon = 1$.
(I_∞)	Δ consists of $0, \epsilon, 2\epsilon, \ldots, \infty$; ϵ as above.
(II_1)	Δ consists of all real numbers $\geq 0, \leq \eta$; where η is a fixed number $\eta > 0, < +\infty$. Of course we can normalize $\eta = 1$.
(II_∞)	Δ consists of all real numbers $\geq 0, \leq +\infty$.
(III_∞)	Δ consists of $0, \infty$ only.

We have constructed lately quite simple examples of cases (I_n), (I_∞), (II_1), (II_∞), but we do not yet know, whether (III_∞) really exists.

The "separation into two independent variables" is possible if and only if the cases I_n or I_∞ hold: Then the range of x is finite (consisting of n distinct values, so that the $f(x)$-space is n-dimensional Eucledean space), resp. infinite (so that the $f(x)$-space is Hilbert space).

Now while the customary generalization of the "elementary" operator rings (I_n) consisted always in passing to (I_∞) (the object of Hilbert's theory of bounded matrices), I am inclined to think that the natural and more promising "infinite limiting case" of (I_n) is (II_1). Their analogy becomes more perfect, if we use in (I_n) the normalization $\epsilon = \frac{1}{n}$.

I am giving a list of properties which I could deduce for both (I_n) and (II_1), and show how similar (II_1) is to (I_n) (contrary to (I_∞)!). It runs as follows:

(1) There exists one and only one function $T(A)$, defined for *all* Hermitean $A \in \mathcal{M}$ and assuming real values only, with the properties of the *trace*, that is:

(α)	$T(1) = 1$
(β)	$T(\alpha A) = \alpha T(A)$
(γ)	$T(A + B) = T(A) + T(B)$ if A, B commute !
(δ)	$T(A) \geq 0$ if A is definite
(ϵ)	$T(U^{-1}AU) = T(A)$ if U is unitary

(2) $\mathcal{M}$ consists prima facie of bounded operators only, but we may say that an arbitrary (possibly unbounded) operator X *belongs* to $\mathcal{M}$, if it is invariant under all unitary transformations $U \in \mathcal{M}'$. Then denote the set of all linear, closed operators X with an everywhere dense domain (in $\mathcal{H}$), which belong to $\mathcal{M}$, by $\mathcal{U}[\mathcal{M}]$.[155]

(3) Every Hermitean operator in $\mathcal{U}[\mathcal{M}]$ is self adjoint and has a unique spectral resolution of unity in $\mathcal{M}$.

(4) If $X, Y \in \mathcal{U}[\mathcal{M}]$ and Y is a continuation of X, then $X = Y$.

(5) If $X, Y \in \mathcal{U}[\mathcal{M}]$, then $\alpha X, X^*, \widetilde{X + Y}, \widetilde{XY} \in \mathcal{U}[\mathcal{M}]$ too. (Observe that this means: The f for which Xf, Yf are both defined, are everywhere

[155]Von Neumann's insertion on the margin: "While this construction is void in (I_n), where $\mathcal{U}[\mathcal{M}] = \mathcal{M}$, it is not so in (II_1): There unbounded operators in $\mathcal{U}[\mathcal{M}]$ do really exist, and in essentially the same way as in the familiar case (I_∞)."

dense; similarly those for which Xf, $Y(Xf)$ are both defined; and the closures $\widetilde{X+Y}, \widetilde{XY}$ can be formed.) All formal rules of computation of matrix calculus hold for the polynomials of any finite number of $X_1, \ldots, X_m \in \mathcal{U}[\mathcal{M}]$, which are built up with the help of these operations.

(6) The "minimax" principle holds: If A is a Hermitean operator in $\mathcal{M}$ (or even in $\mathcal{U}[\mathcal{M}]$), we can form for any $\alpha > 0, \leq 1$ the number

$$\epsilon(\alpha) = gr.l.b._{\left\{ \begin{array}{ll} E \in \mathcal{M}, & E \text{ projection} \\ D(E) \geq \alpha & \end{array} \right\}} \left[l.u.b_{\left\{ \begin{array}{l} f \in \mathcal{H}, \ Ef = f \\ \|f\| = 1 \end{array} \right\}} (Af, f) \right]$$

Then $\epsilon(\alpha)$ has all properties of the "proper-value No. α of A". Thus $T(A) = \int_0^1 \epsilon(\alpha) d\alpha$. In the case (I_n) $\epsilon(\alpha)$ is, what is usually called the "proper value No. v of A (from below)" if $\frac{v-1}{n} < \alpha \leq \frac{v}{n}$, $v = 1, \ldots, n$; but in case (II_1) $\epsilon(\alpha)$ will really vary continuously with α – so that it makes a sense to speak about the "proper value No. α of A" even for an irrational $\alpha > 0, \leq 1$, and a quite natural weight-ing of the continuous spectra results.

The list could be continued, but I think that these are the most characteristic features.

Recently I could prove, that every abstract algebra in which a trace can be defined (cf. 1. on the previous page, replacing however (γ) by this: $T(A+B) = T(A) + T(B)$ for all A, B, and (δ) by this $T(A^2) > 0$ if $A \neq 0$) is isomorphic to a suitable ring $\mathcal{M}$ of bounded operators in a Hilbert space $\mathcal{H}$; and that this trace is unique if and only if $\mathcal{M}$ is a factor of class (I_n) or (II_1) (with a certain restriction concerning (γ)). The notion still haunts me, that this may be applicable to quantum theory: After all this result means that one can enumerate all abstract algebra which contain a uniquely determined probability-theory – and that while the complete matrix-rings of finite-dimensional Euclidean spaces are naturally such (cases (I_n)!), the limiting case for $n \to \infty$ is *not* (I_∞), *but* (II_1)!

What do you think about these things? I am very agreeably surprised, because the pathology of unbounded operators did really not look so, that one could expect such a *dénonement* – if this is a *dénonement*!

I forgot to answer your inquiry about B. Lengyel. I know him quite well. He is not without talent, but I d not think that he is category I. – neither in the international nor in the particular Hungarian market. If Princeton would offer fellowships for foreign students of his degree of maturity, I would have supported him, but this is not the case at present. As I understand from your letter – and as I surmised – the situation in Harvard is similar. Studying with you would, I think, do him a lot of good, first of all as you intend to lecture on Hilbert space next year, second, because he could learn other things than Hilbert space too, third, because he can have all I can give him as well in the summer months from me in Budapest. But these are probably considerations of a remote theoretical value only !

With the best greetings, presently from Mariette and Marina too.

Yours as ever

John.

Original in John Hay Library, Brown University, Rhode Island, U.S.A., language of original: English

---

The Institute for Advanced Study
School of Mathematics
Fine Hall
Princeton, New Jersey

April 19, 1935

Dear Marshall,

there was some delay here with the filling in for formulae in our Found. Math. manuscript,[156] but now it is being done, and the paper will go off next week. I will have the second typewritten copy, so I think it is reasonable, if you keep the original manuscript.

Some of the questions you formulated concerning factors can be answered on the basis of our present results.[157] This leads to the following picture:

(1) The notion of "direct sum" of a (finite or countably infinite) number of unitary spaces can be generalized to a notion of a "direct sum" of a continuum of such spaces. This process is analogous to the passage from a numerical sum to a Stieltjes integral, and even more to that one from a point spectrum to a continuous spectrum. I think, that it would be lengthy and dull to detail this, but it can be done. Every "resolution of unity" $E(\lambda)$ in the unitary space $\mathcal{H}$ generates such a decomposition.

(2) In this sense every ring $\mathcal{M}$ of operators is the direct sum of factors $\mathcal{M}^\lambda$, situated in resp. spaces $\mathcal{H}^\lambda$, of which $\mathcal{H}$ is this generalized "direct sum". This decomposition is obtained by choosing the $E(\lambda)$ so as to generate the (Abelian) ring $\mathcal{M} \cdot \mathcal{M}'$.

This process is the equivalent of the following step in unitary representation theory in Euclidean spaces: To break up the representation-ring $\mathcal{M}$ into aggregates of equivalent irreducible representations. (So that two different aggregates always contain inequivalent ones!)

A factor then corresponds to an aggregate of equivalent irreducible representations. In the cases which we called (I), it is such, too; but in the cases (II), (III) it behaves differently.

(3) So the process of "reducing completely" a unitary representation in Hilbert space consists naturally of two parts: Part one: Represent $\mathcal{M}$ as a direct sum of "factors". (Analogon: proceed as far as the aggregates of equivalent

---

[156]Reference is to J. von Neumann, M. Stone: "The determination of representative elements in the residual class of a Boolean algebra", Collected Works Vol IV. No. 5. The Editor.

[157]Von Neumann refers to results on "rings of operators" obtained in cooperation with F.J. Murray; see Section 3 of the Introductory Comments. The Editor.

irreducible representations.) Part two: Analyze the "factors". (Analogon: resolve these aggregates into the irreducible representations they contain.)

Part one is almost the same in Hilbert space as in Euclidean spaces, while Part two leads to essentially new possibilities (the cases $\neq (I)$ for "factors").

(4) Now it makes a definite sense to inquire what factors are contained in $\mathcal{M}$? The simplest case is if $\mathcal{M}$ is Abelian. This means that all its factors are (I_1), and is a reformulation of the usual spectral theory.

The case where all factors are (I_1), (I_2), ... or (II_1) is probably the most important one after this.[158] I think that groups with such representations can be handled with the methods of "infinitesimal elements", etc., owing to the good behavior of unbounded operators in case (II_1).

(5) Every Abelian ring can be immersed in a factor of class (I_n) or (II_1). So every *single* operator A can be immersed. *Two* given operators or more can not be in general: Thus the P, Q of quantum mechanics, as you mentioned cannot. However certain pairs with very similar properties, somewhat lengthy to discuss, can be.

I am writing up these things now in detail (I hope to finish it on the boat), and I will try to make to you a more detailed report afterwards.

We are sailing tomorrow morning. Mariette and daughter are well, but scared of the sea. I hope, that it will not be half as bad as they expect it to be.

Many thanks for your good wishes. Best greetings from all of us to Emmy and to you, please give some news from yourself from time to time. (The best way to write to me, will be c/o the department secretary here, marked "to be forwarded".) I will write about all "exciting" things that may happen to us., I hope they will not be very "exciting".

John.

P.S. Please excuse the confused style of my letter. The atmosphere of packing which surrounds me does not seem to be good for my epistolar style. Considering the near departure, the house looks very much like an insane asylum!

Original in John Hay Library, Brown University, Rhode Island, U.S.A., language of original: English

---

Budapest

Oct. 4, 1938

Dear Marshall,

I suppose that the place from which I am writing at this time of the year will strike you as somewhat odd. I had to ask for a leave of absence for part of the first

[158]Von Neumann's insertion on the margin: "(I_1), (I_2), ... alone gives the groups which I called 'maximally almost periodic'."

term, in order to be able to finish everything here and to get married. The way things look now, I expect to be married before Nov. 1, and back to Princeton late in November.

I enjoyed the best parts of the September-war-scare in Budapest, and I cannot say, that it was pleasant: It looked "business" more than is good for anybodys nerves. I had been traveling in the Scandinavian countries, and was staying with the Nils Bohr's in Copenhague, when "things began to happen". Since both my mother and my fiancée were in Budapest at that moment, I flew down to Budapest from Copenhague on Sept. 20. (Originally I had intended to go to England for 1 or 2 weeks.) I have been here since, and now I want to stay here until all my affairs are settled – and then take the first boat and the fastest one.

How does the present European settlement impress you? I think, that there is some good in it, since it gives Tchekoslovakia the frontiers which it should have had in the first place, after 1918 – or at least approximately so. But I don't think, that the next general war is more than postponed by it, and not by much, either. England's weakness and indecision has only become more obvious, and this, after all, is the main motive for any European war. In any case, it seems that the armament race is going on at top speed, and that is a more reliable criterium, than anything the "statesmen" say.

I have also done some mathematics, in a rather indirect way. I wanted to construct examples of certain occurrences for matrices, and have failed so far, but from one of my failures I could derive something like a "unitary spectral theory for not-necessarily-Hermitean (or even normal) matrices". I will give a brief discussion of it, since it might amuse you.

I proved first the following theorem, which, I think, is new even in 2 dimensions.

> **Theorem:** Let A be a matrix with $|||A||| \leq 1$, ($c = |||A|||$ is the smallest $c \geq 0$, such that for every vector γ, $\|A\gamma\| \leq c\|\gamma\|$) and $\rho(z)$ a (complex, numerical) rational function, such that $|z| \leq 1$ implies $|\rho(z)| \leq 1$. Then the matrix $\rho(A)$ can be formed and $|||\rho(A)||| \leq 1$.

This theorem is obvious, if A is diagonal, hence if it is normal ($A^*A = AA^*$), but I proved it for *all* matrices A. The proof is the same for a finite dimensional space, as for Hilbert- or any hyper-Hilbert-space. It is closely connected with I. Schur's method of determining all those analytical functions $\sigma(z)$, for which $|z| \leq 1$ implies $|\sigma(z)| \leq 1$.

With the help of this theorem, and of considerations of the familiar type in the abstract theory of bounded Hermitean operators, I can now develop a theory as follows:

(1) Let A be a matrix and $\mathcal{S}$ a closed set (of complex numbers). $\mathcal{S}$ is a *spectral set* of A if and only if this is true:
rational function $\rho(z)$, for which $|\rho(z)| \leq 1$ for all elements z of $\mathcal{S}$, the matrix $\rho(A)$ can be formed, and $|||\rho(A)||| \leq 1$. In Hilbert- or hyper-Hilbert-spaces A is assumed to be bounded. It may be, however, that this restriction could be avoided.

(2) The circle $|z| \leq r$ is a spectral set of A, if and only if $|||A||| \leq r$.
(This follows directly from the theorem.)

(3) The half-plane $\Re z \geq 0$ is a spectral set of A, if and only if $\Re A = \frac{1}{2}(A+A^*)$ is definite.
(This follows by a "Cayley transformation" from 2.)

(4) The real axis ($\Im z = 0$) resp. the unit circle ($|z| = 1$) is a spectral set of A, if and only if A is Hermitean resp. unitary.
(This follows from 3. resp. 2.)

(5) Given a closed set $\mathcal{S}$, a function $\sigma(z)$ on $\mathcal{S}$ is *quasi-analytic*, if it is the uniform limit of a sequence of rational functions on $\mathcal{S}$.

Examples:

α) If $\mathcal{S}$ is the closure of a connected open set, then quasi-analytic means: Analytic in the interior of $\mathcal{S}$, and continuous on its boundary.

β) If $\mathcal{S}$ is the sum of several disjoint sets as described in α), then quasi-analytic means the same thing as in α), but the various pieces of $\sigma(z)$ need not be analytical continuations of each other.

γ) If $\mathcal{S}$ is a Jordan-curve (or arc, or the sum of a finite number of them), then quasi-analytic means: Continuous.

Now if $\mathcal{S}$ is a spectral set of the matrix A, then $\sigma(A)$ can be defined for all those functions $\sigma(z)$, which are quasi-analytic on $\mathcal{S}$.

This notion of $\sigma(A)$ has all heuristically plausible algebraical and analytical properties – I suppose, that I need not enumerate them.

(6) A set $\mathcal{S}$ is *minor* if $\bar{z}$ is quasi-analytic on $\mathcal{S}$. This is equivalent to this: That quasi-analyticity on $\mathcal{S}$ is equivalent to mere continuity. (By 5.γ) every set which is the sum of a finite number of Jordan-curves or -arcs, is minor.)

If A has a spectral set $\mathcal{S}$ which is minor, then A is normal.
(Proof: Put $\sigma(z) \equiv \bar{z}$, $\sigma(A) = B$. Since $\sigma_1(z) = \frac{1}{2}(z + \sigma(z))$ and $\sigma_2(z) = \frac{1}{2i}(z - \sigma(z))$ are everywhere real, so 5. and 4. imply, that $B_1 = \sigma_1(A) = \frac{1}{2}(A + B)$ and $B_2 = \sigma_2(A) = \frac{1}{2i}(A - B)$ are both Hermitean. They commute, being functions of A, by 5. Now $A = B_1 + iB_2$, hence $A^* = B_1 - iB_2$, and so A, A^* commute along with B_1, B_2. Hence A is normal.)

(7) The intersection of all spectral sets of A is precisely the set of all proper values (in the usual sense) of A. (To prove this, 2. is needed.)

(8) A possesses a unique minimum spectral set (which then, owing to 7., is necessarily the set of its proper values), if and only if this is true: The intersection of any two spectral sets of A is again a spectral set of A.

(9) A possesses the property described in 8., if and only if it is normal.

1.-8. hold for all, finite or infinite, numbers of dimension, and the proof is literally the same in all cases. The same is true for the sufficiency (of normalcy) in 9., but I could only prove the necessity in 9. in the finite-dimensional case. I believe, however, that 9. also holds quite generally.

I hope you will forgive me this long story.

Please remember me to Emmy. I am looking forward to see you again – I wish we were so far along already!

Yours as ever

John

Original in John Hay Library, Brown University, Rhode Island, U.S.A., language of original: English

Letters to L.L. Strauss

October 20, 1945

Commodore Louis L. Strauss
Office of the Secretary of the Navy
Room 2032A
Navy Department
Constitution Avenue
Washington, D.C.

Dear Commodore Strauss:

At our conversation on Tuesday, October 16, at the Navy Department, and on Thursday, October 18, at Dr. Aydelotte's house, you expressed the desire that I should give you a somewhat more detailed account in writing of the subject we discussed: the desirability of carrying out a high-speed computer project at the Institute,[159] as a purely scientific undertaking, and the interest which the Government might take in such a project.

As you know, it is possible at present to accelerate the automatic performance of the elementary arithmetical operations (addition, multiplication, etc.) by a factor of about 10,000 against what is ordinary practice today, and a factor of 1000 against the most advanced device in actual use. This can be done by electronic counting methods and by using various television and radar technics which have been developed in the course of the last five or ten years – and it can be done without inventing anything very new from the engineering point of view, just by combining and properly "organizing" existing components. (I am thinking of the very realistic electronic multiplication speed of 1 millisecond for 10-digit numbers, while the present performances to which I referred above are actually between 1 and 10 seconds per 10-digit multiplication.

Electronic machines of such speed are now being built by various Government agencies, in particular by the Acoustic Division of the Naval Ordnance Laboratory for the Navy Bureau of Ordnance, and under a contract at the Moore School of Engineering of the University of Pennsylvania for the Army Ordnance Department. I fully anticipate that these projects will lead to satisfactory results in about two years, and there is no doubt that the machines which they aim to produce will be exceedingly useful in a wide variety of problems, – ballistics, aerodynamics, hydrodynamics, and many other fields of engineering.

[159]Institue for Advanced Study. See Section 8 of the Introductory Comments for von Neumann's work on computer design. The Editor

Nevertheless it seems to me that it would be an essentially incomplete policy to develop such devices only for industrial or government laboratories, which have definite, and necessarily relatively narrowly defined, applied problems to which they must devote all or most of the time of their equipment. In saying "narrow", I may be unfair in the conventional sense; both the Bureau of Ordnance and the Ordnance Department have problems which range over wide areas of mathematical physics and engineering, and possession of such devices will probably broaden further the interest of the sections which have access to them. Nevertheless I think that their interests are unavoidably narrow considering the extremely great scientific importance and absolute novelty of such devices. My point is that all existing methods of computing, or more broadly speaking of "approximation mathematics", as they were developed in the course of the last 150 years, are essentially conditioned by what was practically feasible during this period, that is by the speeds of computation which were possible. These speeds changed considerably during this period, but even at its close, that is in the immediate past, the fastest procedure that was at all assimilated by large groups of mathematicians and computers, was the electrical "desk" computing machine, or the standard I.B.M multiplier, both of which require about 10 seconds for 10-digit multiplication. The electronic devices I am talking about multiply 10,000 times faster, and if one considers the other advantages of fully automatic operation over entirely or partly human operation, the acceleration is even considerably higher. Approximating and computing methods developed under conditions which were 10,000 times or more exigent in time than what we shall now be able to achieve, are certainly not efficient or even reasonable under these new circumstances. It is certain that all our computing and approximating methods will have to be redeveloped from this point of view. I have studied these questions in considerable technical detail, but I do not want to impose on you at this time by going into great length in this respect; I shall be very glad to do it at some other occasion. At any rate it seems to me to be technically quite clear that very considerable mathematical and logical changes will have to be effected in this field. Furthermore, while an effort is being made to estimate and discover these changes "theoretically", that is without the actual existence of such a machine, the main work can only be undertaken "experimentally", that is in using and studying a machine of the type in question. In other words, I feel sure that an electronic machine of the most advanced conceivable type should be constructed, not for use on specific applied mathematical or physical or engineering problems, but with the purpose of experimentation with the machine itself in order to develop new approximation and computing methods, and generally to acquire the mathematical and logical forms of thinking which are necessary for the really efficient operation of such a device, with the methods it will have brought into existence. I have no doubt whatever that we are here on the threshold of very important developments both in pure mathematics and in its applications, and that a pure research institution should spend several years in building a machine and experimenting with it. If we devote in this manner several years to experimentation with such a machine, without a need for immediate applications, we shall be much better off at the end of that period in every respect, including the applications.

The Institute for Advanced Study has already a contract with the Navy Bureau of Ordnance for a theoretical study of computing methods suited for high-speed computing machines. This contract was let by the Bureau and accepted by the

Institute because we all felt that such a theoretical investigation was a necessary preliminary to the building of advanced high-speed computers. It is, however, certainly only a first step, and a well integrated program should include the planning and building of a high-speed device for the experimental purposes which I outlined. I do not think that it would be adequate to ask an agency which is already building such a machine to build two, and to turn over one to a research institution for the experimental purpose which I mentioned. Planning and developing of the device itself is an essential complement of the experimentation, and a group which builds such a device will be vastly better qualified to explore its possibilities experimentally than one which obtains it readymade. There is plenty of evidence in the past experience of "computing laboratories" to justify this assertion.

Obviously there should be the closest liaison between several projects of this nature, and in particular between the project I am proposing and those in the Naval Ordnance Laboratory and at the Moore School of Engineering. For the reasons which we have discussed I have no doubt that we could easily arrange these things satisfactorily. It is also a matter of the greatest importance to have good liaison with those organizations which did most to advance the television and the radar arts and the development of various servo mechanisms in the immediate past.

To conclude, I would like to mention that the importance of accelerating approximating and computing mathematics by factors like 10,000 or more, lies not only in that one might thereby do in 10,000 times less time problems which one is now doing, or say 100 times more of them in 100 times less time, – but rather in that one will be able to handle problems which are considered completely unassailable at present.

I need not tell you that I shall be very gladly at your disposal at any time when you want more specific details about this subject. Thanking you for the attention you are giving the matter, I am

Sincerely yours,

John von Neumann

Original in Von Neumann Papers, Library of Congress, language of original: English

---

October 24, 1945

Commodore Lewis L. Strauss
Office of the Secretary of the Navy
Room 2032A
Navy Department
Constitution Avenue
Washington, D.C.

Dear Commodore Strauss:

You asked me to write a supplement to my letter of October 20, containing a general discussion of those problems which a high-speed computer could handle, while they are practically inaccessible to our present computing methods and equipment. This letter represents an attempt to answer your question. I would like to point out at the start, however, that the discussion which follows is subject to two important limitations:

(1) I am discussing only applications which are already clearly discernible at this moment. This is necessary in order to produce conservative and matter-of-fact analysis. Nevertheless, I should emphasize that this is a very severe limitation, which will probably cause me to err considerably in the direction of conservatism. I am sure that the projected device, or rather the species of which it is to be the first representative, is so radically new that many of its uses will become clear only after it has been put into operation, and after we have adjusted our mathematical habits and ways of thinking to its existence and its possibilities. Furthermore, these uses which are not, or not easily, predictable now, are likely to be the most important ones. Indeed they are by definition those which we do not recognize at present because they are farthest removed from what is now feasible, and they will therefore constitute the most surprising and farthest-going extension of our present sphere of action in mathematics and in applied mathematics.

In other words: I will attempt, in what follows, to enumerate only such problems as are now sufficiently familiar and clear to serve as a motivation in initiating the project under discussion – assuming very conservative standards of acceptance in such a motivation. I do anticipate, however, that after the device in question has been completed and its performances and possibilities studied for a few years, the problems which appear in this motivation will only constitute a part of its uses. I expect that a considerable, possibly the major, part of its uses will consist of problems which we would exclude now as unrealistic or which do not even occur to us. I think that this is typical in the theory of innovations which are as radical as this one. It would be easy to give examples from other fields.

To sum up briefly: I feel sure that what I am stating in the discussion which follows is only a very conservative minimum claim.

(2) I will restrict myself to such uses as are or may be of interest to the Navy. This is not a very severe restriction, since the range of problems referred to includes most of aerodynamics, hydrodynamics, elasticity, plasticity, optics, electrodynamics, dynamic meteorology, and in connection with meteorology and cryptography numerous sorting problems, usually in a particularly critical combination with computations. Nevertheless it does exclude some categories of problems of mathematical or applied mathematical interest.

I now proceed to enumerate applications which would be rendered practicable with a high-speed device of the type under consideration, and which are completely denied to us with our present methods and equipment:

a. Aerodynamics. At present analytical or numerical calculations in this field are possible only in problems which have a maximum of two physically significant dimensions. This means that a stationary phenomenon in which time plays no role must be spatially two-dimensional, and a transient phenomenon which changes with time, must be spatially one-dimensional. "One-dimensional" phenomena may of course occur in three-dimensional space. One-dimensionality merely means that because of some inherent symmetry only one spatial dimension is quantitatively significant; e.g. a phenomenon which has spherical symmetry or which occurs along identical plane slices of space rates as a one-dimensional. Similarly "two-dimensional" covers any phenomenon which has cylindrical symmetry or which occurs in identical linear slices of space. Thus the decay of a blast wave in space (spherical symmetry) rates as one-dimensional, while the flow of air around a projectile (cylindrical symmetry) or around an aerofoil (identical linear slices parallel to the aerofoil's axis) or the flow of a gas through a nozzle (either case according to the nature of the nozzle), rates as two-dimensional. Actual computing of these phenomena is already attended by very considerable difficulties when curvatures are involved (as in the case of spherical or cylindrical symmetries), and also when the phenomena under consideration are not isentropic (e.g. decay of a blast wave or any gas flow in which energy is being generated). Already in these latter cases (curvatures, anisentropy), the difficulties of calculation are sufficient so that while a single problem might be solved occasionally, general surveys can hardly have been carried out with anything like the desirable completeness. Yet such surveys are frequently of great practical importance; in determining optimum dimensions or other details of some projected mechanism (projectile, nozzle, etc.) it may be necessary to vary its disposable parameters and to determine their separate and joint influence on the performance of the mechanism in question.

As soon as more than two physically significant dimensions enter the problem, i.e. as soon as a two-dimensional problem becomes unstationary, or when any truly three-dimensional situation arises, present analytical and computing methods become completely inadequate. Yet even in the one- or two-dimensional problems referred to above, this situation will arise when any perturbation of symmetry or of stationarity occurs. Examples: Non-stationary phenomena around a projectile or an aerofoil or in a nozzle (vibrations or instabilities), introduction of not fully symmetric components into any one of these devices, detailed analysis of organs like rudders or propellers, etc. Also most questions connected with the transmission or reflection of pressure waves by obstacles, and consequently most problems of damage, belong to this category.

b. Hydrodynamics. In all questions of underwater blast the problems of hydrodynamics have very much the aspect of aerodynamics under **a.**, since under the high pressure that occur water must be regarded as compressible. Therefore most of what was said under **a.** applies here too. Otherwise, when water may be treated as incompressible, the appearance of viscosity causes great difficulties. In particular, all cases where turbulence plays an essential role are practically beyond our present abilities in computing. At present all information of turbulence must be obtained from experiments which are expensive, very delicate, and not too precise. They could probably not be replaced entirely by computation, but certainly computing of the type under consideration here would complement them very essentially and contribute in a most important manner to our knowledge. There were also many semi-empirical and statistical procedures which would be rendered possible by computing of this type: discovery of correlations in an extensive but not too precise experimental material could be furthered greatly in this way.

c. Elasticity and Plasticity. The mathematical treatment of these fields is even more difficult than that of aerodynamics, although technically of a similar type. All except the most trivial problems require very massive computing, and only very little is computable with present methods and equipment. By and large the state of the mathematical art in this field is considerably behind the corresponding one in aerodynamics. Fast computing of the type suggested would bring these subjects at least to the level at which we are now in aerodynamics.

d. Optics. Very many questions in planning optical equipment have to be treated today with hit-or-miss methods because systematic investigations of all possible arrangements of optical devices to fit a certain situation would require too lengthy computing. Surveys of the arrangement of lens systems in view of determining optimum arrangements, which now consume several months and are then rather incomplete, could be carried out with great completeness in fractions of an hour or so.

e. Electrodynamics. In this field present computing methods are probably much better than in any of those mentioned before, presumably mainly because the equations of electrodynamics are in general linear. Nevertheless, properly three-dimensional problems are even here occasionally of very considerable difficulty. In designing resonant cavities, antennas, various transmission elements, etc., a great deal could be gained by accelerated computing.

f. Dynamical Meteorology. This is intrinsically a chapter of three-dimensional aerodynamics complicated by the presence of turbulence. A properly theoretical treatment, and thereby mathematical weather prediction, would be possible if a computing equipment meeting the requirements of **a.** and **b.** were available. It is relatively easy to see that the speeds which we were contemplating (multiplication of a millisecond or so) would be about adequate – but only just about adequate – in order to compute the weather over a few thousand square miles in, say, an hour, for a day ahead. This would certainly necessitate more work on turbulence, and here the remarks of **b.** above apply.

g. Sorting Problems. Dynamic meteorology could also be furthered if the present technique, which consists mainly of discovering in the past decades some days with the same weather chart as the present day, could be expanded and accelerated. This is essentially a "sorting approach": very extensive numerical material is given and it is desired to locate in it a small portion which possesses certain analytical characteristics within the total material. The same problem also arises in a different, but analogous, manner in cryptography. Present sorting methods are considerably restricted by the physical limitations in handling punch cards, and by considerable difficulties in meshing the sorting proper with computations. Yet such computations are necessary in the course of sorting, as required in meteorology and in many forms of deciphering; also the numerical aggregate handled may be of greater size than the capacity of a punch card. The electronic devices to which I am referring could handle such problems considerably more quickly than present sorting machines, and combine them with numerical investigations without interruption of sorting or a need for human intervention.

I don't know whether this letter is too long or too short, or both in appropriate places. If you want any point amplified, please let me know; and in using it please omit from it anything you consider superfluous. Hoping that it is some reasonable approximation to what you wanted, I am

Sincerely yours,

John von Neumann

Original in Von Neumann Papers, Library of Congress, language of original: English

December 9, 1946

Mr. Lewis L. Strauss
U.S. Atomic Energy Commission
Washington 25, D.C.

Dear Lewis:

Many thanks for your very kind note concerning the award of the Medal of Merit that I have received.

I need not tell you that I am very highly gratified by this recognition, although I cannot help feeling, over and above the normal knowledge of personal inadequacy that all I did during the war were per se very interesting and stimulating intellectual pursuits. Actually, the war introduced me to great parts of mathematical physics and applied mathematics which I had neglected before, and I feel that I received intellectually a good deal more than I gave.

It is a particularly great gratification to receive, at this moment, a renewed expression of your friendship.

Faithfully yours,

John von Neumann

Original in Von Neumann Papers, Library of Congress, language of original: English

Letter to J. Stroux

September 19, 1950

Dr. Johannes Stroux,
President
Deutsche Akademie der Wissenschaften zu Berlin
Berlin, Germany

Dear Professor Stroux:

I have received your letter of July 9 of this year, in which you inform me that I was elected a corresponding member of the German Academy of Sciences in Berlin.

I would like to assure you that I am taking with the most sincere feelings of joy and appreciation this honor and this expression of trust and appreciation by members of the Academy, and especially by colleagues in the University of Berlin, where I spent the first and formative years of my academic career.

Unfortunately, however, the current state of the world is such that my accepting this honor would be a political act that would prompt interpretations other than the mutual appreciation of colleagues. It would suggest, in particular, that I take a position concerning things that I reject. I do not want to detail these here; I would like to assure you, however, that I have decided a very serious question of conscience in the only way that seems correct to me and I am asking you not to consider me for a corresponding member of the Academy at this time.

I remain most sincerely yours,

J. von Neumann

Original in Von Neumann Papers, Library of Congress, language of original: German. Translated by M. Redei.

Letter to T. Tannaka

August 8, 1939

Dr. T. Tannaka
Mathematical Seminary
Tohoku Imperial University
Sendai, Japan

My dear Dr. Tannaka:

I want to thank you for sending me a reprint of your extremely interesting paper, "Über den Dualitätssatz der nicht-kommutativen topologischen Gruppen" (Tohoku Mathematical Journal, September 1938). I have wanted for a long time to write you concerning it, but I have been away a number of times and kept postponing it – indeed, I am very sorry that it is only now that I am writing.

In the first place may I express again my admiration for your result? I always felt that Pontrjagin's duality theorem about the characters of commutative groups ought to be generalized to non-commutative ones. But it was also clear that while the dual of a commutative group is a commutative group, yet the composition laws of primitive characters (I mean those of Frobenius), and the possibility of combining representations both by direct addition and by direct multiplication, indicate that the dual of a non-commutative group is commutative, but not a group. It is much rather like a ring, but not exactly that either. In any case there is some duality between the notions of "group" on the one hand, and "commutativity" on the other, but the wider self-dual domain in which all this takes place ought to embrace considerably more than groups – something in the nature of, although not exactly the same as, rings.

Your paper is, I am convinced, the first progress in this direction, and may well be the decisive one. I was also very much interested by your surmise on page 5 (the "Vorbemerkung" referring to Fejér's theorem).

Permit me to congratulate you once more on your important and very interesting result.

I should like to ask, however, some more explanation about your Theorem 3, page 10, – that is, in which sense it is to be understood. What do you mean by a "stetig isomorphe Abbildung"[160]? Do you mean a mapping which maps closed sets on closed sets and vice versa; that is, that the inverse images of closed sets are closed? It is only in the latter sense that your Theorem 3 is true.

[160] Continuously isomorphic map. German. The Editor.

This is because the imbedding of the group G into the group $\tilde{G}$ is "stetig"[161] in that sense only. Indeed: your "weak" topology of $\tilde{G}$, as far as it refers to $\tilde{G}$'s subgroup G, is not identical with the original topology of G: Let G be the addition group of all real numbers in its usual topology; then the general neighborhood of x is defined by

$$|x - x'| < \epsilon \qquad (\epsilon > 0)$$

In the topology of G, however, the general neighborhood of x is defined by

$$|e^{2\pi i \alpha_n x'} - e^{2\pi i \alpha_n x}| < \epsilon \quad \text{for } n = 1, \ldots m, \quad \begin{pmatrix} m = 1, 2, \ldots \\ \alpha_1, \ldots, \alpha_m \text{ real}, \\ \epsilon > 0 \end{pmatrix}$$

All closed sets in the latter sense are obviously closed in the former, but not conversely. This does, of course, not affect the validity of your Theorem 2, that is your statement that Theorem 3 implies Theorem 2 is still true. But this is only so because G is then assumed to be bicompact: since for a bicompact group any mapping, which is "stetig" in the second (weaker) sense, is also such in the first (stronger) sense. (Similarly for the topologies of G and $\tilde{G}$: Since G is everywhere dense in $\tilde{G}$ and has a less stringent topology, the fact that G is bicompact implies the identity of the sets and topologies G and $\tilde{G}$.)

I may be misunderstanding you altogether, and you may have meant the second type of "Stetigkeit"[162] all the time, but in this case I cannot help feeling that your formulations are a little misleading. In any case these purely technical remarks affect in no way the importance and beauty of your main results.

I should be very much obliged to you if you could find time occasionally to let me know some more about your ideas on this subject, and also send me any other publications of yours.

Sincerely yours,

John von Neumann

Original in Von Neumann Papers, Library of Congress, language of original: English

[161]Continuous. German. The Editor.

[162]Continuity. German. The editor.

Letter to E. Teller

April 8, 1947

Dear Edward:

Many thanks for your letter. I hope that your trip to the East will still allow our seeing each other. I will be in Washington from April 29th (probably arriving the evening of the 28th) to May 2nd, inclusive, staying either at the Cosmos Club or at the Shoreham Hotel. I expect to get back to Princeton on the evening of May 2nd. Please let me know at which of these places, and when, our paths can be made to cross.

I am very glad to learn that Earl Long will look into the matter of the "reaction time of superconductivity." I would like to add that the following seems to me to be a reasonable "integral experiment:"

Given a superconducting ring R, which was cooled below the "transition point" in a magnetic field $H = 0$, turn on a magnetic field H in the direction of the axis A. This will set up a current in R, which compensates it in the area between A and R, where no lines of force can enter. As H grows larger than the "critical field" H_c, R ceases to be superconducting, the current stops, and the lines of force penetrate the area between A and R. If, now, H is decreased below H_c, R becomes supeconducting again, and the lines of force of H_c are "trapped" between A and R. As H decreases from H_c to 0, a new current must develop in (the now superconducting) R, which maintains the field H_c in the area between A and R (hence this current is opposite to the one mentioned earlier), since now no lines of force can leave that area. Thus, the breakdown and restoration of the superconductivity of R by varying H from 0 to $> H_c$ to 0 leaves (by hysteresis) a resulting current in R. If R had remained superconducting throughout, then there would have been no hysteresis: The current would have adjusted itself at every moment to compensate H between A and R, and at the end (when H is again 0) it would have been again 0.

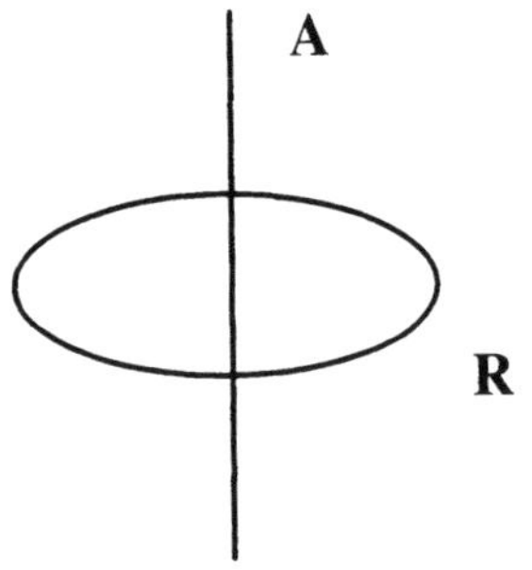

Hence, going with H from 0 to $> H_c$ to 0 in a time t, and seeing whether a permanent current is found afterwards in R, would test whether superconductivity "reacts" in this time t. Besides, the experiment is "integral," since it tests directly the act of "writing" into a superconducting memory. Of course, only one cycle of H (from 0 to $> H_c$ to 0) should be delivered.

I have thought more about the possible forms of superconducting counting-switching, etc., mechanism. Assuming that the "reaction time of superconductivity" is short ($\ll 10^{-6}$ sec.), it seems to me that there are still serious difficulties left. The main point is that if R is the "residual resistance" of the substance used (i.e. its resistance at liquid helium temperatures when it is not superconducting [for magnetic reasons]) and L the inductance, the switching time t of the system must fulfill a condition of the form $t > L/R$, hence R must be $> L/t_0$, if t_0 is the desired switching time. R for Hg or Pb seems to be too low (it looks acceptable for certain alloys [$Pb - Sn$, $Sn - T$]); here, the specific residual resistance may be as high as 10^{-5} ohm cm. Even so, $t_0 10^{-6}$ may require using wires with diameters 10^{-3} cm. Of course, this may well be feasible and, if so, it would lead to arrangements involving various micro techniques which are very tempting.

On the other hand, the W. Thompson effect of Bi at low temperatures looks even better than superconductivity. The zero magnetic field specific resistance of Bi at $10-20K^\circ$ is 2.10^{-5} ohm cm, and a magnetic field of the order which would be critical and reliably usable for the appropriate superconductors (say 1000 Gauss) might easily increase this by factors like 5 or 10. From the point of view of the circuit, it is mainly the difference (and not the ration, provided that the ratio is $\gg 1$) of the specific resistances (with and without a magnetic field) which matters. For a superconductor, this may be at best 10^{-5} ohm cm vs. 0 ohm cm; for Bi it will be 10^{-4} or $2 \cdot 20^{-4}$ or more ohm cm vs. $2 \cdot 20^{-5}$ ohm cm, i.e. better by a factor of 10 or more. Also: Bi will work at liquid H_2 temperatures – but of course it may be worthwile to go to $2K^\circ$ anyway, just to have $HeII$ as a coolant. There, however, the Bi effects will be still larger.

What do you think about the "reaction time" of the Bi - W. Thompson effect? What is known about it? What do you think about the whole matter?

I hope that we will meet before long. I have some definite ideas concerning automata, memory, "clearing" and irreversibility and entropy, somewhat connected with Szilard's treatment of "Maxwell's Demon" (Zschr. f. Phys.,[163] 1929) and my old treatment of entropy and observability in quantum mechanics. I would very much like to tell you about these matters and learn your views concerning them.

With best regards, and hoping to see you soon, I am

Yours,

John von Neumann

Original in Von Neumann Papers, Library of Congress, language of original: English

---

[163]Zeitschrift für Physik. The Editor.

Letters to L.B. Tuckerman

August 25, 1950

Dr. L.B. Tuckerman
Cosmos Club
Washington 5, D.C.

Dear Doctor Tuckerman,

I have just returned to Princeton after a trip to South America, and I find your very kind letter of August 18. The statement that Mr. Evans appears to have made in the AMERICAN MERCURY is incorrect. I have never made any of the assertions quoted. The only results that I have obtained on the subject are those contained in Morgenstern's and my book on THE THEORY OF GAMES AND ECONOMIC BEHAVIOR.[164] I am in complete agreement with your interpretation of those conclusions, possibly with this amendment:

It is conceivable that one of the three outcomes in chess ("white" or "tie" or "black") might be established without a complete enumeration of all sequences, but by some implicit analysis and abstract and general demonstrations, which cover all possible sequences without actually enumerating them. I mean this in the sense in which one can prove that all maps on the sphere can be colored with five different colors or that all maps on the torus can be colored with seven different colors without ever having actually enumerated all such maps (of which there are infinitely many). By the way, if one really tried to enumerate all possible sequences of moves (forming valid games) in chess, the durations would be much longer than a few hundred years, even with the present fastest conceivable machine. In addition, it would be necessary to remember large portions of the classification into positions of the types mentioned above ("white" or "tie" or "black") during the process of attributing all positions consecutively to these classes. The following is an estimate of the size of the numbers that are involved here:

I think that the normal chess position almost always permits more than thirty valid moves, and usually about ten are not obviously wrong. Assuming only ten valid moves, and considering that the length of the average game is about forty moves for each player, one is led to estimate the order of magnitude of the number of possible games 10^{80}. In any conceivable electronic arrangement it would take a good deal longer than a microsecond to test a single game, 10^{80} microseconds are about 3×10^{66} years. This is 10^{57} times the age of the universe now in vogue.

[164] O. Morgenstern, J. von Neumann: *Theory of Games and Economic Behavior*, Princeton University Press, 1944. The Editor.

There are some other games, one of them a variant of chess, in which one can prove that white can force a "win", although no method to achieve this is known. In one case the finding of the method by which white can win is just as difficult as obtaining of the decision in the trilemma of chess. I need not tell you that if this last mentioned matter interests you, I shall be very happy to write to you in more detail.

I am,

Yours very truly,

John von Neumann

Original in Von Neumann Papers, Library of Congress, language of original: English

---

September 25, 1950

Dr. L.B. Tuckerman
Cosmos Club
Washington 5, D.C.

Dear Doctor Tuckerman,

Thank you for your letter of September 8th. I have to apologize for replying to you only now. I have been away for a considerable portion of the intervening period. I agree with the interpretations contained in your letter. It seems to me too, in particular, that the steps (1) and (2) which you enumerate represent the normal mode for knowing how to win in a specific game.

The example of a game where one can prove that "white" can win without knowing how to do it – where in particular the latter problem is precisely the same as "finding out all about the winning possibilities in chess", which, of course, is unknown in the rigorous sense is as follows:

It is somewhat preferable to designate the two players not by "white" and "black" but by two other designations – say, "green" and "red".

The game begins with a move by "green". This move consists of "green" making one of two choices: "white" and "black".

If "green" chooses "white", then the subsequent moves follow precisely the rules of chess, "green" playing for " white" and "red" playing for "black".

If "green" chooses "black", then the subsequent moves again follow exactly the rules of chess, but now "red" is playing for " white" and "green" is playing for "black".

The play ends when the subordinate chess game of either of the two alternatives has ended. In the first case, in which "green" chose "white", "green" will have won the game if and only if "white" (whose side "green" is representing) wins the chess game according to the ordinary rules of chess. In the second case, in which "green" chose "black", "green" wins the game if and only if "black" (whose side "green" is representing now) succeeds in either winning or tying according to the ordinary rules of chess.

Note that "green" can certainly win in this game: It is certain that in chess either "white" can win under all conditions, or that "black" can win, or tie under all conditions. If, therefore, "green" makes his original choice accordingly, and then plays the subordinate chess game correctly, he will certainly win. Note next that the way for "green" to win implies a complete knowledge regarding the situation in chess – already his first move requires that.

I would like to add that Dr. John Nash, now at Princeton University, has discovered a board game which is related to some extent to "Go" and to "Checkers", in which one can also prove that the player who moves first will win, but at this moment no way to force a win is known. Indeed, so far as I know, the statistics of extant plays between strong players have not disclosed any marked bias towards winning by the first player.

Dr. Nash's example is less good than mine, inasmuch as it offers only an existing, practical situation (which may change in time), and not a rigorous proof. It is better than mine, however, in the sense that it describes a genuine, "naturally" invented game, and not at ad hoc artefact, synthesized for the sake of proving a theorem.

Note that both the game that I described, as well as Nash's game, do not allow ties.

I am, with best personal regards,

Very truly yours,

John von Neumann

Original in Von Neumann Papers, Library of Congress, language of original: English

Letters to S. Ulam

The Institute for Advanced Study
School of Mathematics
Fine Hall
Princeton, New Jersey

Oct. 3, 1936

Dear Ulam,

many thanks for your kind letter – both Mariette and I are delighted to see you soon in Princeton again. Oct. 10 – or any other date thereabout, which suits you – will be very convenient to us – and we expect, that you *will* stay with us, and stay as long as possible.

I agree wholeheartedly with your plans to write an up-to-date presentation of measure theory. Carathéodorys exposition, which is perhaps the *relatively* best one existing, is hopelessly obsolete. A thoroughly modern one, as much combinatorial and as little topological as possible, making extensive use of finite and infinite direct products, and – above all – interpreting measure much more as probability and much less as volume, would really be a very good thing. At least I often felt how badly such a thing is lacking in the present literature. What would be the style of your treatise, and its length? I will be very glad, if you can let me see any part of your mscr. In the lectures I gave here on "linear operators" in 1933/34 and 34/35, I tried to deal with measure somewhat in the above spirit, but I was badly handicapped by the fact, that measure was not my primary topic there.

I looking forward, too, with great interest for your mscr. on the general product-operation.

I am expecting to discuss several mathematical questions, when you come here, those you mentioned, and a few others. By then I will have unearthed my two last year mscr's, too, which you mentioned – we are unpacking now, so the excavations do not proceed very quickly.

Expecting to see you soon again, and with the very best greetings from Mariette, too,

I am yours as ever

John von Neumann

Original in S. Ulam papers, Archives of the American Philosophical Society, Philadelphia, language of original: English

The Institute for Advanced Study
School of Mathematics
Fine Hall
Princeton, New Jersey

Febr. 3, 1937

Dear Stan,

many thanks for your valuable suggestions for the "Philokaketic Society". I still find some difficulties in raising an endowment for the society.

It occurred to me, that the example of an "unseparabilisable" group of power $\aleph$" can be made effective – and the necessity of discussing the "relations" avoided – as follows:

Let $\mathcal{S}$ be the set of all functions $\phi(x)$, x a real number, the values of $\phi(x)$ rational integers, but $\phi(x) \neq 0$ for a finite number if x's only. Let F be the set of all rational integer valued functions $\mathcal{A}(\phi)$, ϕ running over $\mathcal{S}$. Define

(1)

$$\begin{aligned} a_\xi \phi &= \psi, \text{ where} \\ \psi(\xi) &= \phi(\xi) + 1 \\ \psi(x) &= \phi(x) \text{ for all } x \neq \xi \\ A_\xi \mathcal{A}(\phi) &= \mathcal{A}(a_\xi \phi) \text{ for every real number } \xi, \end{aligned}$$

(2)

$$B_{\xi_1 \xi_2 \dots} \mathcal{A}(\phi) = \mathcal{A}(\phi) + \phi(\xi_1) + \phi(\xi_2) + \dots$$

for every sequence of real numbers $\xi_1, \xi_2, \dots$.

Let $\mathcal{G}$ be the group of all one-to-one mappings of F on itself. Then the A_ξ, $B_{\xi_1, \xi_2 \dots}$ are elements of $\mathcal{G}$. Let $\mathcal{G}_0$ be the subgroup of $\mathcal{G}$ generated by all $A_\xi, B_{\xi_1, \xi_2 \dots}$. The power of $\mathcal{G}_0$ is clearly $\aleph$.

If a group-topology τ of $\mathcal{G}_0$ existed, in which $\mathcal{G}_0$ is separable, then a sequence $\xi_1, \xi_2 \dots$ (all mutually $\neq$) with

$$\tau - \lim_{i \to \infty} A_{\xi_i} \qquad \text{existing}$$

would exist. Thus

$$\tau - \lim_{i \to \infty} A_{\xi_{2i}} A^{-1}_{\xi_{2i+1}} = I$$

and hence

(1)
$$\tau - \lim_{i \to \infty} A_{\xi_{2i}} A^{-1}_{\xi_{2i+1}} B_{\xi_2 \xi_4 \dots} A^{-1}_{\xi_{2i}} A_{\xi_{2i+1}} B^{-1}_{\xi_2 \xi_4} \dots = I$$

Now identically

$$A_{\xi_{2i}} A^{-1}_{\xi_{2i+1}} B_{\xi_2 \xi_4 \dots} A^{-1}_{\xi_{2i}} A_{\xi_{2i+1}} B^{-1}_{\xi_2 \xi_4} \dots = U$$

where

(2)
$$U \mathcal{A}(\phi) = \mathcal{A}(\phi) - 1$$

(1) gives $U = I$, which contradicts (2).

If you can use my mscr. on measurable mappings, please keep it.

Is there any hope to see you in Princeton in the near future? I need not tell you how glad we will be. Best greetings from all of us and particularly from

John

P.S. Many thanks for the interesting Szilprajn-reprint.

Original in S. Ulam papers, Archives of the American Philosophical Society, Philadelphia, language of original: English

---

The Institute for Advanced Study
School of Mathematics
Fine Hall
Princeton, New Jersey

Oct. 4, [1937 ?]

Dear Stan,

many thanks for your letter of Sept. 30., and particularly for what it contained about my "domestic" complications. I am sorry that things went this way – but at least I am not particularly responsible for it. I hope that your optimism is well founded – but since happiness is an eminently empyrical proposition, the only thing I can do, is to wait and see[165] ... I hope very much to see you soon again. When will you be in these parts again? There is a meeting of the Nat. Acad. in Rochester, N.Y., on Oct. 24-27, which is so near that I may go there, and also the New York meeting of the A.M.S. on Oct. 29-30, apart from these days I will be in Princeton, with the possibility of frequent excursions to New York. Would you come before or after the N.Y. meeting?

Your remark about the extension of finite-set mappings with or without increasing the Lipschitz-Constants is simple, but very much to the point. How do you want to carry on the idea of gradually extending a finite mapping, after this observation? What you wrote about "recurrent" sequences puzzled me, I am not quite sure, what the definition of such a sequence $a_1, a_2, \ldots$ is. If it is what I assumed it to be, namely

$$a_{n+k} = F(a_n, a_{n+1}, \ldots, a_{n+1-k})$$

for fixed $k(= 1, 2, \ldots)$ and $F(x_0, \ldots, x_{k-1})$, then your statement cannot be true, hence you certainly mean something else by a "recurrent" sequence.

I am preparing myself to cause a scandal by making much noise about the "non-distributivity" of logics.[166] I think I know now, how to handle the "quantifiers" $((x), (Ex))$ in such a system. Of course, I should rather do "honest" work on the algebra and arithmetics of "continuous rings" – but after all before God one pastime is as good (or bad) as any other.

---

[165] See Section 1 of the Introductory Comments for von Neumann's marriage and second marriage. The Editor.

[166] See Section 6 of the Introductory Comments for von Neumann's work on quantum logic. The Editor.

In diesem Sinne[167], and with the best greetings

as ever

John

P.S. I would really like to see you again to have a conversation *de rebus omnibus et quibus aliis* ... I hope, that you will be able to spend some time here after the N.Y. meeting!

Original in S. Ulam papers, Archives of the American Philosophical Society, Philadelphia, language of original: English

---

The Institute for Advanced Study
School of Mathematics
Fine Hall
Princeton, New Jersey

March 15, [1939 ?]

Dear Stan,

I hope that you'll soon have the news about your parents for which you are writing. I will suspend making easter-plans, until I will have heard from you.

As to your question about measure-preserving transformations, I know this much: First, a slightly more general problem: Let $\mathcal{S}_1, \mathcal{S}_2$ be two spaces, each with a Lebesgue-measure in it: $\mu_1(\mathcal{M}_1)$ for $\mathcal{M}_1 \subset \mathcal{S}_1$ and $\mu_2(\mathcal{M}_2)$ for $\mathcal{M}_2 \subset \mathcal{S}_2$. I assume that $\mu_1(\mathcal{S}_1) = \mu_2(\mathcal{S}_2) = 1$. Let $\mathcal{T}_1, \mathcal{T}_2$ be measure-preserving transformations (for μ_1, resp. μ_2) of $\mathcal{S}_1$ resp. $\mathcal{S}_2$ on itself. Question: Does a mapping $\mathcal{U}$ of $\mathcal{S}_1$ on $\mathcal{S}_2$ exist, which carries μ_1 into μ_2, such that

$$\mathcal{T}_2 = \mathcal{U}\mathcal{T}_1\mathcal{U}^{-1}$$

Answer (partial): Form for $i = 1, 2$ the $\mathcal{L}_2$ over $\mathcal{S}_i, \mu_i$: $\mathcal{L}_2^{(i)}$. Define in $\mathcal{L}_2^{(i)}$ a linear operator $\mathcal{U}^{(i)}$ by

$$\mathcal{U}^{(i)}\phi(x) = \phi(\mathcal{T}_i x) \qquad (x \text{ in } \mathcal{S}_i)$$

$\mathcal{U}^{(i)}$ is clearly unitary. It has a spectrum on the unit circle, which may by mapped by

$$\lambda = \frac{1}{2\pi i} \ln \alpha$$

on the real axis *mod* 1. Call this the log-spectrum of $\mathcal{U}^{(i)}$.

Now it is clearly a *necessary condition* for an affirmative answer to the above question, that $\mathcal{U}^{(i)}$ and $\mathcal{U}^{(2)}$ have the same log-spectra, incl. multiplicities and (in their continuous parts) their Hellinger types.

If either $\mathcal{U}^{(1)}$ or $\mathcal{U}^{(2)}$ has a pure point log-spectrum, and is ergodic, then I proved, that this condition is also sufficient. In this case the common log-spectrum

---

[167]German: "In this sense". The Editor.

of the $\mathcal{U}^{(i)}$ must be simple, and *an enumerably infinite set of real numbers mod 1, which contains $\lambda - \mu$ whenever it contains λ, μ.* Conversely: Whenever such a set of real numbers mod 1 is given, then a $\mathcal{U}^{(1)}$ exists (given $\mathcal{S}_1, \mu_1$, provided that every point x of $\mathcal{S}_1$ has $\mu_1(x) = 0$), the log-spectrum of which is the prescribed set. (By what I said above, these $\mathcal{S}_1, \mu_1, \mathcal{U}^{(1)}$ are unique up to a transformation $\mathcal{U}$, cf. the bottom of page 1.)

Now your infinite-dimensional torus examples have all them pure point log-spectra, more precisely those, which possess enumerably infinite bases (with integral coefficients). Hence the *necessary and sufficient condition* in your problem is this: In order that the mapping you desire should be possible, it is necessary and sufficient, that the log-spectrum be a pure point spectrum, and possess an enumerably infinite basis (with integral coefficients).

Therefore this is not always possible: An enumerable set of real numbers mod 1, containing $\lambda - \mu$ if it contains λ, μ need not have such a basis (and every such set is the log-spectrum for a suitable $\mathcal{T}_1$): For instance the set of all rational numbers. Furthermore examples of $\mathcal{U}^{(1)}$'s with continuous log-spectra are known.

These things are proved in my paper "On the operator methods in classical mechanics", Annals of Mathematics, 1932.[168] (There I consider not one $\mathcal{T}_i$, but a one-parameter group $\mathcal{T}_i(t)$. $((-\infty < t < +\infty,\ \mathcal{T}_i(t)\mathcal{T}_i(s) = \mathcal{T}_i(s+t)))$. But those results apply even more easily to the $\mathcal{T}_i$, there being only one difference: There the spectra are completely defined (($\mathcal{T}_i(t)$ corresponds to $\mathcal{U}_i(t)$, and by Stone's theorem $\mathcal{U}_i(t) = \exp(2\pi i t H_i)$, H_i being self-adjoint, and thus having a spectrum)), and here only mod 1.)

I have been asked to give a "... Putnam ..." lecture in Cambridge. So I'll probably be there around April 1.

Believe it or not, the "Comité pour la Coopération Intellectuelle Internationale" of the League of Nations (Nebbich! – if you still understand Middle-High-German) organizes a meeting on "modern theoretical physics" in Warsaw, on May 29-June 4, 1938. I have been invited to attend, costs paid, and one of 5 or 6 "reports" there, to be made by N. Bohr, will deal (among other things) with a youthful sin of mine in quantum mechanics. So I'm going to attend. Isn't this a nutty world? Will you be in Poland at that time? I may even cross on a Polish boat, if I should stay over here throughout May – which, of course, is dependent on an entirely different set of considerations!

I think that the idea of the Neugebaerinian book is very good.

Hoping to hear from you (by the way: in what stage is your+Oxtoby's paper?), and with the best greetings

John

Original in S. Ulam papers, Archives of the American Philosophical Society, Philadelphia, language of original: English

---

[168] J. von Neumann: Collected Works, Vol. II. No. 17.

April 2, 1942

Dear Stan,

Excuse please[169] the paper, but this is going to be a sea serpent.

Many thanks for your letter; which I enjoyed a great deal. I was particularly flabbergasted by one of your problems, since I, too, had considered it lately.

This is the question on groups. You consider a permutation group $\mathcal{G}$ of elements of a (finite or infinite) set a. You postulate this:

(*) If $b, c \subseteq a$, $\overline{b} = \overline{c}$, $\overline{a - b} = \overline{a - c}$ ($-$ is the "Mächtigkeit"[170]) then an $x \in \mathcal{G}$ with $xb = c$ exists.

I considered these groups, too, and called them *set transitive on* a or, in abstracto, on $\overline{a}$. To be precise: I considered them only when a is finite, say $a = (1, \ldots, n)$, $\overline{a} = n$.

They arose, oddly enough, in the theory of games. In an n-person game, played by players who form the set $(1, \ldots, n)$, it is important to express, how much a coalition $b \subseteq (1, \ldots, n)$ of players can win from the others, $-b$, good playing on both sides assumed. I denoted this by $v(G)$, the *characteristic function* of the game. (This quantity *can* be defined in a rigorous way, I won't discuss here how.)

There are two ways to define symmetry of the game:

(1) Let $\mathcal{G}$ be the group of all permutations of $a = (1, \ldots, n)$ which leave the game unchanged – the *symmetry group* of the game.
Then the game is *symmetric 1)* if $\mathcal{G}$ is the group of *all* permutation of $a = (1, \ldots, n)$, the symmetry group Σ_n.

(2) The game is *symmetric 2)* if $v(G)$ depends only on $\overline{b}$ – the number of elements of b.

My theory shows, that 2) is the important concept, not 1). Of course 1) implies 2). But much less will imply 2): Clearly 2) is true if the symmetry group $\mathcal{G}$ is set transitive on n, in the sense of (*). (Of course, 2) can be true without any group theoretical background at all.)

The question arises therefore: Which are the set-transitive groups $\mathcal{G}$ for $n = 1, 2, 3, \ldots$? For $n = 1, 2$ the answer is trivial: $\mathcal{G} = \Sigma_n$ only. So the real problem arises for $n = 3, 4, 5, \ldots$. For these the alternating group A_n, too, is set-transitive, as you pointed out. So the final question is this:

(Q) For which $n = 3, 4, 5, \ldots$ do there exist no other set-transitive groups $\mathcal{G}$ than Σ_n, A_n?

I call such an n *regular*.

Now I have the following results:

α) $n = 3, 4$ are easily seen to be regular.

β) $n = 5, 6$ are *not* regular. Proof:
For any n, $\frac{n}{2}$ or $\frac{n-1}{2}$ -fold transitivity in the ordinary sense is sufficient for set-transitivity. (If $\overline{b} = \overline{c}$ then $\overline{(1, \ldots, n) - b} = \overline{(1, \ldots, n) - c}$, and one of these two numbers is $\leq \frac{n}{2}$ or $\frac{n-1}{2}$, so $x \in \mathcal{G}$ with $xb = c$ or $x(-b) = x(-c)$, i.e. again $xb = xc$, exists.) So for $n = 5$ double, and for $n = 6$ triple

[169] Von Neumann's insertion: "This is to demonstrate the difference between a sinicism and cynicism."

[170] German: Cardinality. The Editor.

transitivity suffices.

$n = 5$: Consider the group of all linear transformations

$$L(\alpha, \beta)\colon \xi \to \xi' = \alpha\xi + \beta, \qquad \alpha \neq 0$$
$$\alpha, \beta, \xi, \xi' \text{ restclasses } \pmod 5$$

It is doubly transitive, hence set-transitive. It has $4 \cdot 5 = 20$ elements, while Σ_5, A_5 have 120, 60.

$n = 6$: Consider the group of all rational transformations:

$$R(\alpha, \beta)\colon \xi \to \xi' = \frac{\alpha\xi + \beta}{\gamma\xi + \delta}, \quad \alpha\delta - \beta\gamma \neq 0$$
$$\alpha\beta, \gamma, \delta \qquad \text{restclasses } \pmod 5,$$
$$\xi, \xi' \qquad \text{restclasses } \pmod 5 \text{ or the symbol } \infty.$$

It is clearly triply transitive, hence set-transitive. It has $5^2 \cdot 4 + 5 \cdot 4 = 120$ elements, while Σ_6, A_6 have 720,360.

γ) $n = 7, 8, \ldots$ are *probably* all regular. This is what I can prove about them: Put $k = \frac{n}{2}$ or $\frac{n-1}{2}$. $\mathcal{G}$ is transitive for the sets of k elements, hence its order is divisible by the number of such sets, $\binom{n}{k}$. If p is a prime $> k$ and $n-1$, i.e. if $p > \frac{n}{2}$ or $\frac{n+1}{2}$, then $\binom{n}{k}$ is divisible by p. So we see: The order of $\mathcal{G}$ is divisible by any prime $p > \frac{n+1}{2}$.

Any set-transitive group $\mathcal{G}$ is clearly transitive (simply, in the ordinary sense) and primitive. Hence if $\mathcal{G} \neq \Sigma_n, A_n$, then by Burnside ("Theory of groups of finite order", Cambridge 1897), p. 199, § 141, the order of $\mathcal{G}$ is not divisible by the same power of p as $n!$, if p is a prime $< \frac{2}{3}n$.

If $p > \frac{n+1}{2}$, then $n!$ is divisible by the first power of p only, and the order of $\mathcal{G}$ is divisible by it, too. So n is certainly regular, if a prime p with $\frac{n+1}{2} < p < \frac{2}{3}n$ exists, i.e. if $\frac{3}{2}p < n < 2p - 1$. For $p = 2, 3$ these conditions conflict, so we must have $p = 5, 7, 11, \ldots$. Then our inequalities mean this $\frac{3p+1}{2} \leq n \leq 2p - 2$. So we see:

(R) $\quad n$ is certainly regular, if $\dfrac{3p+1}{2} \leq n \leq 2p - 2$ for a prime $p = 5, 7, 11, \ldots$

Now $p = 5, 7, 11, 13$ establish by **(R)** the regularity of

$$n = 8, 11, 12, 17, 18, 19, 20, 21, 22, 23, 24$$

To these $n = 7$ can be added for the following reason: If $\mathcal{G}$ is set-transitive for 7, then its order is divisible by 5, since $p = 5$ is a prime $> \frac{n+1}{2}$ for $n = 7$. Now we don't have $p < \frac{2}{3}n$, but nevertheless the order of every transitive, primitive $\mathcal{G} \neq \Sigma_n, A_n$ (for $n = 7$) is divisible by 5 by Burnside, pp. 206-207, § 146, case (2), $n = 7$.

So we see:

γ_1) $n = 7, 8, \ldots, 25$ are all regular, with the possible exception of $n = 9, 10, 13, 14, 15, 16, 25$.

γ') Consider now $n = 26, 27, 28, \ldots$. For $p = 17$ the lower limit in **(R)** is $\frac{3p+1}{2} = 26$. Hence from here on all n are regular, as far as for all $p \geq 17$

the next prime p' fulfills the inequality

$$(2p-2)+1 \geq \frac{3p'+1}{2}, \quad \text{i.e.} \quad 3p' \leq 4p-3 \quad , \text{ i.e.}$$

$$\textbf{(P)} \qquad p' \leq \tfrac{4}{3}p - 1$$

(That is: As far as there are no lacunae between any two successive intervals **(R)**.)
Now a chain of primes fulfilling **(P)**, and beginning with $p = 17$, seems to exist:

$$\textbf{(C)} \qquad 17, 19, 23, 29, 37, 47, 61, 79, 103, 131, 173, 229, \ldots$$

and it seems probable that **(C)** can be carried on in infinitum. Anyhow, as far as it goes, **(C)** takes care of all n up to $2p-2$ for $p = 229$, i.e. up to $n = 456$.
So we see:

γ_2) $n = 26, 27, 28, \ldots, 456$ are all regular, and this goes on in infinitum, if **(C)** can be carried on infinitum.

Do you know anything about **(C)**, **(P)**? It is plausible, but worse than Tchebycheff's $p' \leq 2p-2$.

The set-transitivity on $\aleph_0$ is easier to investigate, and it seems to be amenable to treatment without the axiom of choice.

This is an "effective" example of a set-transitive group $\mathcal{G}$ on $(1, 2, 3, \ldots)$, which is not the symmetric group $\Sigma_{\aleph_0}$:

Consider a permutation

$$\pi : n \to n^{\pi} \qquad (n = 1, 2, 3, \ldots)$$

for which $(n = 1, 2, 3, \ldots)$ can be decomposed into a *finite* number of parts $\Omega_1, \ldots, \Omega_k$, on each of which (separately!) n^{π} is a monotone increasing function of n. The smallest k of this π is its *height* $h(\pi)$. If no such finite decomposition exists, put $h(\pi) = \aleph_0$.

One verifies with ease:

a) The π with a finite $h(\pi)$ form a group $\mathcal{G}$.
b) $\mathcal{G}$ is set-transitive.
c) If for a π n^{π} is a monotone decreasing function of n on a set B of s elements, then necessarily $h(\pi) \geq 3$. (Proof: No two elements of B can belong to the same one of the hypothetical sets $\Omega_1, \ldots, \Omega_k$. So $k \geq s$, i.e. $h(\pi) \geq s$.)
d) If for a π there exist numbers $a_1, a_2, a_3, \ldots$ (all in (1,2,...)) with

$$a_1^{\pi} > a_2^{\pi}, \quad a_3^{\pi} > a_4^{\pi} > a_5^{\pi}, \quad a_6^{\pi} > a_7^{\pi} > a_8^{\pi} > a_9^{\pi}, \ldots$$

then π is not in $\mathcal{G}$. (Proof: Put

$$B = (a_1, a_2); (a_3, a_4, a_5); (a_6, a_7, a_8, a_9); \ldots$$

and apply **c)**. This gives $h(\pi) \geq 2, 3, 4, \ldots$ – i.e. $h(\pi) = \aleph_0$.)

By replacing $>$ by an other ordering, any given permutation π, which moves infinitely many elements, can be excluded from $\mathcal{G}$. My method does not seem to work for those π, which move only a finite number of elements.

Well, that's that.

I don't quite understand, what you mean by a "recursive function" $f(x)$. Just one defined by a single induction

I) $f(1) = a$, $f(x+1) = \phi(x, f(x))$, $\phi(x, y)$ a given polynomial, or do you permit

II) superposed inductions (in finite number) as well? Possibly even "transfinite types" like

III)

$$\begin{aligned} f(x,1) &= \psi(x), \qquad f(1, y+1) = \chi(y) \\ f(x+1, y+1) &= \omega(x, y, f(f(x, y+1), y)) \\ \psi(x), \chi(y), \omega(x, y, z) & \qquad \text{given polynomials etc.?} \end{aligned}$$

If you mean III) and similar things included, then there is no difficulty in getting Borel sets of any class? Is there any for II) in the finite Borel classes? I) is unreasonably narrow for any purpose, isn't it?

Your results on near isometrisms are very interesting. Are they difficult to derive?

As to the war: I think, as before, *qu'on les aura*. May take honorable 2-3-4 years. But it's going to be a peculiar little world. Not that I mind it. I'd like to talk about it with you some time ...

I'd like to come to Wisconsin some time, but I don't know yet whether I can get away from here during the summer. Many thanks for the invitation, Mammon's lure on me is unabated, and besides I would really like to come.

Life is getting complicated. I am indeed writing a book on "economics" with O. Morgenstern (at Princeton University, he was formerly the director of the Austrian "Institut für Konjunkturforschung"). It is, as you guessed, mainly on games, with a tendency to apply it to oligopoly etc. and with a distant hope of some application to social phenomena. *Qui vivra verra.*

But I'm getting more and more snowed under by war work.

Best greetings from both of us, and $(\frac{1}{2})^2$ unknown "from house to house", and au revoir,

as ever

John

Original in S. Ulam papers, Archives of the American Philosophical Society, Philadelphia, language of original: English

Cosmos Club
Washington 5, D.C.

November 9, 1943

Dear Stan,

I am very glad that Mr. Hughes "and all he stands for" have come through ... I told them about you, because you wrote me several times in the past that you definitely wanted a war job, and because this is a very real possibility, where you could do very effective and useful work.

The project in question is exceedingly important, probably beyond all adjectives I could affix to it. It is very interesting, too, and the theoretical (and other) physicists connected with it are probably the best group existing anywhere at this moment. It does require some computational work, but there is no doubt, that everybody will be most glad and give you all the encouragement you can wish in doing original research on the subject, for which there is ample opportunity. I can also assure you of my own cooperation in this respect.

The secrecy requirements of the project are rather extreme, and it will probably necessitate your and your families essentially staying on the premises (except for vacations) as long as you choose to be associated with it.

To repeat: If you want war work, this is probably a quite exceptional opportunity.

I may be able to give you a better idea orally, and I would be glad to do so, if we can meet somewhere before you answer – but I suppose that there is no time (I will be in Princeton on 13-15 and in Washington on 16-17 of this month.)

Do you really count on a quite short war? I don't see that from a purely technical standpoint Germany need be broken before next fall. Of course a collapse may come any day from now on for moral and political reasons, but I can't see how to judge that, without knowing much more about the present state and efficiency of the Nazi political machine.

And there is still a year's worth of Asiatic war after that. Anyhow, *qui vivra verra...*

It seems that Morgenstern's and my book on "games"[171] will be out in 3 months or so.

Best regards, and looking forward to seeing you soon – here or "there".

as ever

John

Original in S. Ulam papers, Archives of the American Philosophical Society, Philadelphia, language of original: English

---

[171]O. Morgenstern, J. von Neumann: *Theory of Games and Economic Behavior* (Princeton University Press, Princeton, 1944).

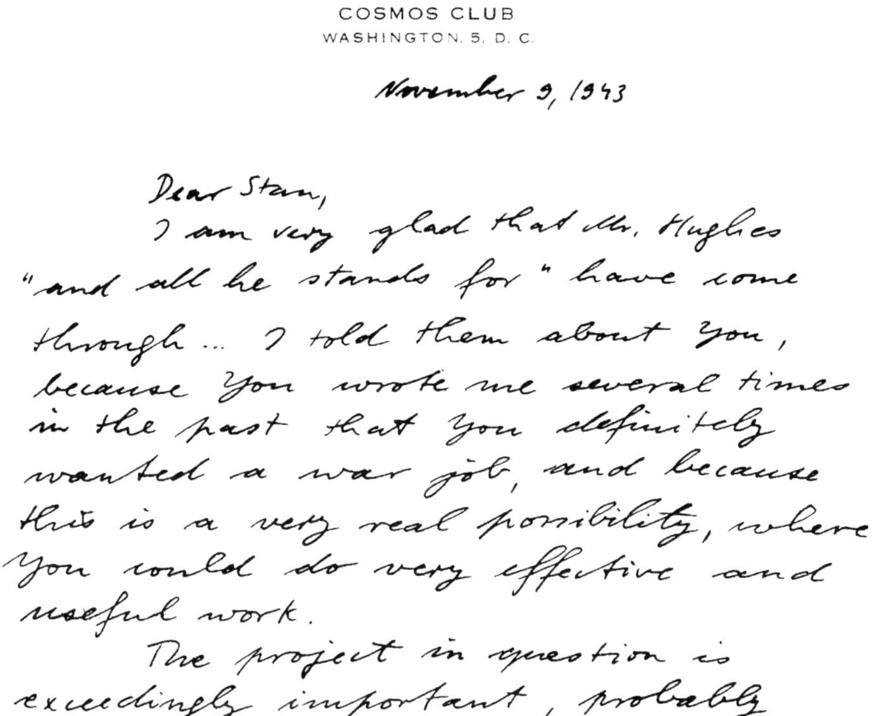
COSMOS CLUB
WASHINGTON, 5, D. C.

November 9, 1943

Dear Stan,

I am very glad that Mr. Hughes "and all he stands for" have come through... I told them about you, because you wrote me several times in the past that you definitely wanted a war job, and because this is a very real possibility, where you could do very effective and useful work.

The project in question is exceedingly important, probably beyond all adjectives I could affix to it. It is very interesting, too, and the theoretical (and other) physicists connected with it are probably the best group existing anywhere at this moment. It does require some computational work, but there is no doubt, that everybody will be most glad and give you all the encouragement you can wish in

The first page of John von Neumann's original three-page handwritten letter to Stanislaw Ulam, November 9, 1943, written while staying at the Cosmos Club in Washington, D.C., a social establishment where intellectuals in the arts, literature, and sciences gathered.

Letter to E.R. van Kampen

November 21, 1934

Professor E.R. van Kampen
Johns Hopkins University
Baltimore, Maryland

My dear van Kampen:

Many thanks for your letter and your note, which interested me much. I think this a very elegant and rational way to get the right topological picture of the seemingly non-topological behavior of almost-periodic functions. In particular, your introduction of an invariant subgroup, the factor of which is automatically separable and "almost" compact, is very suggestive. Does your statement F (according to which a continuous numerical function defined on a bi-compact group is "really" defined on a compact group) not have a broader, purely-topological background somewhat of this sort: If the domain of a continuous function is bi-compact while its range is separable, then suitable identifications in the domain (which do not affect the function's one-valuedness) make the domain separable and compact?

The topologization of the group by means of a given almost-periodic function, and the possibility of reducing the almost periodic case by this means to the Weyl-Peter theory for compact groups + Haar integral, has been noted by Weyl (Annals of Math., vol. 35, No. 3 (July 1934), p. 488-9). The use of the Haar integral, however, causes some technical difficulties, the chief one being that it is necessary, for use in the Weyl-Peter theory as well as in the almost-periodic theory, to prove the interchangeability of x- and y- means:

$$\int_x \Big(\int_y f(x,y)\mathrm{d}y \Big)\mathrm{d}x = \int_y \Big(\int_x f(x,y)\mathrm{d}x \Big)\mathrm{d}y$$

This interchangeability was always a technically weak point of almost-periodic theories: H. Bohr needed certain "uniformity" considerations to establish it for his "integral mean". I proved it in my Transactions paper more simply by using the unicity of my mean (Transactions vol. 36, No. 3 (July 1934), pp. pp. 455-6, theorem 10[172]). As far as I can make out, there is no very easy way to establish it for the Haar measure; probably the best thing is to use the unicity again – the unicity, however, has so far only been proved in the compact case, and there only

[172] J. von Neumann: "Almost periodic functions in a group", Collected Works Vol. II. No. 25. The Editor.

by defining the Haar integral with the method of my Compositio paper (Compositio, vol. 1 (1934), 106-114, in particular footnote 13 on p. 112)[173], – that is, as a "mean" instead of the "integral" – at least if one wants not to loose simplicity. (By the way, do you know any other means of establishing the interchangeability of x - and y - integration for the Haar integral?)

Hoping to hear from you soon again, I am

Yours very truly,

John von Neumann

Original in Von Neumann Papers, Library of Congress, language of original: English

[173] J. von Neumann: "Zum Haarschen Mass in topologischen Gruppen", Collected Works Vol. II. No. 22. The Editor.

Letters to O. Veblen

Budapest
V. Arany János utca 16

23.9.1930 [September 23, 1930]

Dear Professor Veblen!

Very many thanks for your kind letter from Maine.

In the meantime Wigner decided to come for the second term 1930/1931 as well, as in the first one of 1931/1932. So at the latter date we will be together in Princeton. I learned with great pleasure, that Dirac too will be in Princeton at that time.

I am sorry, not to be able to get away this winter already, as you proposed, but it would be hard to get a leave of absence now in Berlin, only 4 month after the end of the last one.

Many thanks for your Quarterly Journal separatum, I am greatly interested in the further results of your work about projective relativity. Are you able to clear up the nature and properties of your field equations? And did you follow further the connection between them and the wave equation? Could you let me know, what your knewer[174] results are, mainly on this last point?

I travelled much in this last month, I was in Salzburg and the north of Italy by car, and will only be the first part of November in Berlin. Meantime I wrote a work about the unicity of Schrödinger operators, i.e. I proved, that two operators having the "commutation property"

$$PQ - QP = i1 \tag{1}$$

(and being "irreducible") must be identical with Schrödingers operators $i\frac{d}{dq}\cdot\ldots$, $q\cdot\ldots$, if the system of coordinates is conveniently chosen. ((1) is not just the most favorable mathematical form of the "commutation property", I use an equivalent form given by Weyl

$$e^{i\alpha P}e^{i\beta Q}e^{-i\alpha P}e^{-i\beta Q} = e^{i\alpha\beta}1)$$

Besides this I am meanly[175] occupied with hydrodynamical questions, with the so called problem of turbulence. It is very complicated, but I hope to get along with it in a not too long time – and write then to you about it, if you find some interest in it. I am very fascinated by this question, which is, as I believe, the central problem of a rational-mathematical dealing with the flow of gases and fluids with little friction, and has numerous very interesting mathematical and physical aspects.

[174]Misspelled word; correctly: newer. The Editor.

[175]Misspelled word. Correctly: mainly. The Editor.

Hoping to hear of you again, and repeating the hope, that I will have a very interesting time next Winter with you in Princeton.
Begging to give my best regards to Mrs. Veblen.

I am yours truly

John von Neumann

Original in O. Veblen Papers, Library of Congress, language of original: English

---

Berlin

11.1.1931 [January 11, 1931]

Dear Professor Veblen!

Many thanks for your kind Christmas card, please express Mrs. Neumann's and my thanks to Mrs. Veblen too.

I have no objection against lecturing about hydrodynamics. As summary of a lecture about this subject, I would propose this:

> "Principles of hydrodynamics and the main forms of its equations. Two-dimensional problems, connexion with conform representation. Vortex theorems, Paradoxon of D'Alambert. Fluids with friction, outlines of their problems and applications."

We have few news here, there was some excitement in quantum-mechanics, whether a statement of Bohr, that length never can be measured with an accuracy exceeding $\frac{h}{mc}$, is correct. It is, as I think, settled now, that this is not the case. Heisenberg will be in Berlin next week, and we are expecting an interesting discussion of this matter.

The economical crisis in Germany is very acute now, and as people do not like to be alone in their miseries, there is much talk about the bad state of things in America. Is anything of this kind true? I mean is there any visible change in American life? From a distance of 3,500 miles I had rather the impression that there must be 90% exageration in all these stories.
With the best greetings

truly yours

John von Neumann

Original in O. Veblen Papers, Library of Congress, language of original: English

---

Budapest

April 3, 1933

My dear Oswald,

It seems, that this Summer will be an endless series of sensations – and not always of the agreeable kind. Now it is the American inflation, or the declaration to inflate, which looks very puzzling, mainly because it is very difficult to understand from here what really happened. Was the new administration not anti-inflationist? How and why did Roosevelt change his mind? Or is it in some way not a change of his policy?

What do people in general believe now about the further developments? Are they optimistic, for the country in general, and the universities in particular?

Please excuse me that I am asking such a lot of questions. But you know, how these things interest me, and how little the newspapers a[re] worth, if you want to find out anything.

There is not much happening here excepted that people begin to be extremely proud in Hungary, about the ability of this country, to run its revolutions and counter-revolutions in a much smoother and more civilized way, than Germany. The news from Germany are bad: heaven knows what the summer term 1933 will look like. The next programm-number of Hitler will probably be annihilation of the conservative-monarchistic-("Deutsch National" $\rightleftharpoons$ Hügenberg)-party.

You have probably read, that Courant, Born, Bernstein have lost their chairs, and J. Frank gave it up voluntarily. From a letter from Courant I learned 6 weeks ago (which is a very long time-interval now in Germany) that Weyl had a nervous break-down in January, went to Berlin to a sanatorium, but that he will lecture in Summer. I did not hear anything about changes or expulsions in Berlin, but it seems that the "purification" of universities has only reached till now Frankfurt, Göttingen, Marburg, Jena, Halle, Kiel, Königsberg – and the other 20 will certainly follow.

I am glad to learn from your letter that these things receive the full attention and appreciation in America which they deserve. It is really a shame, that something like that could happen in the 20-th century.

What is the matter with Einstein? How will his Collége de France – affair be coordinated with his full-time presence in Princeton?

We are going to Italy for 3-4 weeks around May 10, and then for 2 weeks, or less, to Germany. I feel that I have to see Berlin and Göttingen once more – although an expedition to the North Pole would be a much nicer thing under the present conditions.

Mariette[176] is very well, but we are far too social now. So the trip to Italy will do us much good.

Hoping to hear from you soon (please write to me to the Budapest address),

I am sincerely yours

John von Neumann

[176] M. Kövesi, von Neumann's wife. The Editor.

P.S. Please excuse once more the aggregate of silly questions I am asking. Now when I re-read my letter, I am quite afraid of their number.

Original in O. Veblen Papers, Library of Congress, language of original: English

---

Budapest

July 6, 1935

Dear Oswald,

Please excuse the size of my letter, which will probably be quite considerable before I am through. But there is quite a number of things I should write about.

Von Neumann's remark on the margin:
I see now with horror, what I did: My letter developed into a monster of 21 pages! But pp. 7-20 are mathematical, so please do not read them seriously unless you are very bored.

Proceeding from the subjective to the objective, I have to report first, that we are all well, although Mariette[177] is again engaged in her usual battle with her elders, wether to put on weight or not. Marina[178] is well, and even gained 1 lb. between New York and Budapest, so that there is every reason to assume that she crossed the ocean without noticeing it.

Cambridge was very beautiful and interesting. The main non-architectoric sensation is of course still Hardy. He was somewhat disturbed by Milne's theological amplifications of his (not so hot) cosmology, and in particular about the fun Milne seems to find in connecting Creation with a singularity at $t = 0$. Hardy wishes to avoid this by introducing a parameter $\tau = \ln t$, which begins with $\tau = -\infty$, and thus satisfies his philosophical needs ...

I met Newman several times, and I am very glad of having made his acquaintance, he is very attractive both from the topological and from the human side. He has postponed his sabbatical leave for 1936/37, and seems to be quite anxious to come to Princeton then.

By the way, there seems to be quite a traffic-jam on the road to Princeton. There are 4 or more advanced students or PhD.'s who will come next year to Cambridge: A Commonwealth fellow M. Price, who is very able, he worked in quantum theory until now, but he wants to change to group theory; an NRC[179] fellow, whose name is, I think Lewisohn, who is a pupil of N. Wiener (!) and spent his first year in Cambridge (Hardy thinks that he is very good), Touring[180], whom you mentioned, and who seems to be strongly supported by the Cambridge mathematicians, for the Proctor fellowship (I think that he is quite promising); and one or two more, whose names I forgot.

---

[177] Mariette Kövesi, von Neumann's wife. The Editor.

[178] Von Neumann's daughter. The Editor.

[179] National Research Council. The Editor.

[180] Von Neumann means A. Turing. The Editor.

There is quite a considerable number of good men in the young group in C.[ambridge]. I was particularly interested in 3 of them: Two fellows of Trinity: L.C. Young, who works in real functions, Stieltjes integrals, etc.; and S. Chandrasekhar (a Hindu) who is primarily an astrophysicist, but who also has a considerable knowledge of quantum theory and connected mathematics, too; and a fellow of Caius: a theoretical physicist with the name F.C. Powell.

I had an audience of $\geq 15, \leq 20$ people, which showed no "radioactive decay", until the examination period (around June 3.) came along. Then the number reduced by several orders of magnitude, until I closed on June 14.

Fowler is looking forward with much expectation to his expedition to Princeton. We made the acquaintance of the Cooks, who live with him, and who are very charming. They will come to America in April, to join him for a "transcontinental tour".

The views about Eddington are essentially homogeneous in C[ambridge].: The astrophysicists are scared, the physicists (like Fowler) very afflicted; and advanced students highly amused. Nb. he is really going from bad to worse, he wrote lately a paper on "relativistic degeneracy of gases", which is almost worse than his quantumtheoretical papers – a scarcely credible feat! He seems to have lost contact with theoretical physics, and the methods which are in use for the last 50 years, completely.

I met E. once or twice, but succeeded avoiding discussions on anything more serious than graphology.

Summing up, the entire experience was a very agreeable one, and the new people we met were interesting and nice throughout.

We went to Oxford, too (for 3 days), where we saw Schrödinger, Whitehead and family. Mrs. Whitehead is very charming, but Whitehead's classical definition of her ("I married a pianist") although literally true, bears false implications: She has a definitely "bourgeois" background. We stayed at Whitehead's house, and had a wonderful time. Whitehead's chief occupation is presently to knot tori, he constructed some very funny examples.

Weyl stayed 3 days with us in Cambridge.

I preached a sermon in Paris. There seems to be there a considerable interest in operator theory, boosted by G. Julia, and the younger group. Nb. the latter includes several absolutely first class people (A. Weil, Leray), and some who are at least interesting (C. Chevalley, perhaps D. Possel). A. Weil's wish to have a chance to come to Princeton in 1936/37 is unchanged.

Mathematically I was laboring very hard in writing down a long-long paper on the theory of reduction of operator-rings (or which is equivalent: of unitary-orthogonal group representations) in Hilbert space. The results of Murray and myself are really the "second step" in such a theory, and as this is completed, I had to fill in the first. (This is, of course, not the reasonable order in which to do things, but the "second step" was more interesting, and more different from the finite-dimensional case, and therefore we took it first. It corresponds roughly to the reduction of representations which consist of equivalent irreducible ones, to irreducible parts. As compared to the finite-dimensional case, it gave essentially new phenomena. My present work achieves the reduction of an arbitrary representation to such blocks of equivalent irreducible ones (in the terminology Murray and I used: "factors")[181]. This is full of epsilontico-technical catches. It is really a generalisation of the theory of spectra of Hermitean operators, and therefore it leads to all sorts of messes, which are very similar to the "continuous spectra". But one finally meddles through, and proves, that the perfect analogon of the finite dimensional situation holds.)

One of the peculiarities of the subject is that one gets *effectively constructed* functions, of which it is difficult (and in some of the cases as yet impossible) to prove the Lebesgue-measurability. (This connects with the so-called "higher-projective sets" of Lusin-Sierpinski.)

Now that I have finished it, I feel definitely relieved. I think that this theory will permit several applications to groups and to infinite algebras.

Garrett Birkhoff and I worked together on the following subject, which seems to be quite amusing to me. It connects logics, quantum theory, and projective geometry in a somewhat surprising way. The idea is this:[182]

In classical mechanics the state of a mechanical system γ is represented by a point P in its "phase space" Φ. A statement a concerning the system γ amounts always to this, that the "representative point" P belongs to a certain subset A of Φ. Thus there is a one-to-one correspondence between the statements about γ, $a, b, c, \ldots$ on one hand, and the subsets $A, B, C \ldots$ of Φ on the other. The fundamental operations of logics are:

$$\begin{array}{rcl} a \cup b & : & a \textit{ or } b \\ a \cap b & : & a \textit{ and } b \\ a' & : & \textit{not } a \end{array}$$

They correspond to the set-theoretical operations:

$$\begin{array}{rcl} A \cup B & : & \text{set-theoretical sum of } A \text{ and } B \\ A \cap B & : & \text{set-theoretical intersection of } A \text{ and } B \\ A' & : & \text{set-theoretical complement of } A \end{array}$$

(All this follows from the interpretation of the correspondence: a corresponds to the set of all points P of Φ which fulfills a.)

Now $\cup, \cap, '$ fulfill for the sets $A, B, C, \ldots$ the rules of "Boolean algebras", and therefore they do it for $a, b, c, \ldots$ too. These rules are:

[181] See Section 3 of the Introductory Comments for von Neumann's work on operator algebra theory. The Editor.

[182] See Section 6 of the Introductory Comments for von Neumann's work on quantum logic. The Editor.

Rules **A**

(Zero law)	$0 \cap a = 0$	$0 \cup a = a$
(Commutative law)	$a \cap b = b \cap a$	$a \cup b = b \cup a$
(Associative law)	$(a \cap b) \cap c = a \cap (b \cap c)$	$(a \cup b) \cup c = a \cup (b \cup c)$
(Distributive law)	$(a \cap b) \cup c = (a \cup c) \cap (b \cup c)$	$(a \cup b) \cap c = (a \cap c) \cup (b \cap c)$
(Law of inference)	$a \cap b = a$ equivalent to	$a \cup b = b$
(Involutory character of negation)	$a'' = a$	
(Duality of $\cap$ and $\cup$)	$(a \cap b)' = a' \cup b'$	
(Exclusion)	$a \cap a' = 0$	

(The first and second columns are $\cap, \cup$-dual, thus they imply each other by the "Duality law".)

Conversely: Starting with a set of entities $a, b, c, \ldots$, for which three operations $a \cup b$, $a \cap b$, a' are defined fulfilling the rules **A**, one can always find a set Φ, and put the $a, b, c, \ldots$ in a one-to-one correspondence with subsets of Φ, so that

$a \cup b$	corresponds to the set theoretical sum of a and b
$a \cap b$	corr.[esponds] to the s.[et] th.[eoretical] intersection of a and b
a'	corr.[esponds] to the s.[et] th.[eoretical] complement of a

Thus the rules **A** are those rules of logics which characterise the system of all statements concerning a classical-mechanical γ by "inner properties".

Passing to quantum mechanics, the following changes occur: Here the states of a mechanical system γ are represented by the directions P^* in a unitary-orthogonal (Hilbert) space G^*, that is by the points of the projective space G arising from G^* by identification of all points on every (complex) straight line through the origin (direction).

A statement concerning the system γ is again equivalent to the statement, that the "representative point" P^* belongs to a certain subset A^* of G^* (or, using the projective picture: P to a certain subset A of G). But here the most characteristic feature of quantum-mechanics comes in: This set A^* (resp. A) cannot be any subset of G^* (resp. G), it must be a linear set (containing the origin if we use the G^*-picture, unrestricted if we use the G-picture). This is the famous "superposition principle": If the wave functions ϕ and ψ have a (physical) property in common, then they must have it with every wave function $c_1\phi + c_2\psi$ (c_1, c_2 complex constants), too. The customary interpretation of quantum mechanics leaves no doubt, that we must now continue as follows:

The fundamental operations of logics are:

$a \cup b$	:	*a or b*
$a \cap b$	:	*a and*
a'	:	*not a*

They correspond to the set-theoretical operations:

$$a \cup b : \text{Linear sum of } A^* \text{ and } B^*.$$

(That is: set of all $\phi + \psi$, ϕ in A^*, ψ in B^*. The set theoretical sum would be: Set of all ϕ in A^* and of all ψ in B^*.)

$a \cap b$: Set theoretical intersection of A^* and B^*.

a' : Unitary-orthogonal complement of a.

(That is: Set of all ϕ for which $\phi \perp \psi$ if ψ is in A^*. The set theoretical complement would be: Set of all ϕ not in A^*.)

This is the G^*-picture, in the G-picture the two first operations are the same, but the third one becomes a polarity with a purely imaginary quadric. (By the latter statement I mean merely that a and its polar a' never have common points, $aa' = 0$.[183]

One may now try wether the rules **A** still hold. It is easy to see, that they all do, except the distributive law. And one can understand, why this law

(1) $(a \cup b) \cap c = (a \cap c) \cup (b \cap c)$ cannot be true for physical reasons. Indeed its special case $(c = a')$

(2) $b = (a \cap b) \cup (a' \cap b)$ can be seen to be equivalent to what is called in quantum mechanics the "simultaneous observability of a and b". On the other hand one verifies that a part of (1) still holds with the present interpretation. this is the case where a is contained in c $(a \cap c = a)$, that is

(3) $a \cap c = a$ implies $(a \cup b) \cap c = a \cup (b \cap c)$. (Dedekind's law.)

Now call *rules* **B** this set of rules: The rules **A**, without the distributive law, but with Dedekind's law (3) instead.

Now it is again possible to start at the other end: Consider a set of entities $a, b, c, \ldots$, for which three operations $a \cup b$, $a \cap b$, a' are defined, fulfilling the *rules* **B**. One may do this with the purpose, to consider the $a, b, c, \ldots$, as the set of all statements concerning a quantum mechanical system. There are good phenomenological reasons to postulate the rules **B** in this case: The rules **A** without the distributive law are true in the abstract picture which one usually makes of quantum mechanics (bearing the peculiarities of "simultaneous observability" in mind). And the law (3) can be derived – by using some previous results of G. Birkhoff – from the existence of so called "a priori thermodynamical weights of states", that is, from a physical assumption, too.

Now one can prove this:

A system which fulfills the rules **B** is the direct sum of systems of these two kinds:

(1) *First Kind*:

Two elements, 0,1, rules:

$$0 \cup 0 = 0 \quad , \quad 0 \cup 1 = 1 \cup 0 = 1 \cup 1 = 1$$
$$0 \cap 0 = 0 \cap 1 = 1 \cap 0 = 0 \quad , \quad 1 \cap 1 = 1$$
$$0' = 1 \quad , \quad 1' = 0$$

(Mnemotechnic rule: 0=false, 1=true.)

(2) *Second Kind*:

The $a, b, c, \ldots$ are all linear subsets of a projective space G,

$\cup$ is the linear sum,

$\cap$ is the set-theoretical intersection,

$'$ is a polarity with a "purely imaginary quadric" (meaning as above: polars have no points in common).

[183] One should write $a \cap a' = 0$. The Editor.

(I omitted certain simplifying assumption of finite-dimensionality, which we hope to eliminate later.) The classical-mechanical systems are built up by direct-summing of solutions of the first kind. But in quantum-mechanics those of the second kind are essential. For these one can proceed further as in projective geometry:

Let N be the dimensionality of G. If $N \geq 3$, then there exists a not-necessarily-commutative field $\mathcal{A}$, so that G consists of all left-ratios, $x_0 : x_1 : \ldots : x_N$ ($x_0, x_1, \ldots, x_N \in \mathcal{A}$; not $x_0 = x_1, = \ldots = x_N = 0$; the left ratio $x_0 : x_1 : \ldots : x_N$ must be identified with all $\rho x_0 : \rho x_1 : \ldots : \rho x_N$, if $\rho \in \mathcal{A}$, $\rho \neq 0$). The existence of a polarity in G means, that an antiisomorphism exists in G (carrying $x + y$ into itself, but xy into yx), with certain "definitivity" properties.

It is quite amusing that a non-commutative algebra may come in, and that one must consider polarities nevertheless!

In actual quantum mechanics $\mathcal{A}$ happens to be commutative: the system of all complex numbers; but it is well-known, that the non-commutative system of all quaternions would lead to a very similar theory. I am presently trying to restrict the possibilities for $\mathcal{A}$ by making use of the existence of numerical "transition probabilities" between the states.

I think, too, that the best policy concerning the vacancy left by Wheeler, is to wait till September. Besides it is probable, that we will have some more cases like this until then, and it is preferable to deal with possibilities we have together.

Do you know, why Fermi cannot come, and if there is any chance of having him in a subsequent year?

How far did your résume of the spin-geometries develop? Do you plan to include something about the possibility of > 4-component theories? I am afraid that you give me too much credit for my 1928 paper, which was after all hopelessly special relativistic, but I am very glad that you find that the discussion of the tensor properties of the various α-products can still be used.

We met at the house of friends in London Mr. Richardson Jr. and his wife., and had a very good time.

I hope, that your extension of the home is completed and that you enjoy it. With many greetings, from Mariette and my mother, too.

Yours truly,

John

P.S. Mariette sends her love to Elizabeth and swears, that she will write soon. She is still somewhat upset by the home-coming.

Original in O. Veblen Papers, Library of Congress, language of original: English

Budapest

October 3, 1938

Dear Osvald,

Please excuse the delay with which this letter follows up my two letters from Copenhague. Since a period of political "tension" intervened, during which it made very little sense to talk about events farther in their future than, say, 24 hours, I hope that you will forgive this delay.

As to "great" politics I can only say, that Mr. Chamberlain obviously wanted to do me a personal favor, since I needed a postponement of the next world war very badly. Otherwise: The liquidation of the Tchecoslovak[184] affair may not be unreasonable in itself, possibly Tchecoslovakia will get now those frontiers which she ought to have had in the first place, after 1918. But I can't see any reason, why European politics should not be in the same mess 6 month's hence, as they were 6 month's ago. At any rate I feel closer to your views about the British conservatives, than before.

In my matrimonial affairs there is at last some palpable progress: The 6 month's "waiting period" ended during the last week of September, the formal-legal proceedings were started on September 27, and they are going now at such a pace, that I expect that the divorce will be completed around October 20.[185] Getting married involves some more red tape, but I expect, that we will be married before Nov. 1, and able to sail around Nov. 10-15. I suppose, that I need not tell you, how glad I am that things are now in this shape.

At last I know what it will look like from the inside, when the next European war gets started – which is very good for me, since I hope and trust to look at it from the outside, when it really starts.

It was very lovely: Air-raid drills, black-outs, a dearth of gas-masks, pleasant meditations on the topic, which part of the town would be safest during an air-raid (the Schwabenberg seems to be just the last word!), and for me personally many warnings – and even a letter – from the US consulate to get out while the going is good.

I got some additional information from N. The Editor. on the Ztrbl.[186] It seems that printing in Denmark is not essentially cheaper than the Waverly Press. The Ztrbl. publishes about $2\frac{1}{2}$ volumes yearly (*not* 2, as we originally assumed) each = 10 numbers, each of 3·16 pages. Hence 1 Volume = 480 pages, 1 year = 1200 pages. (The "Annals"[187] and "Transactions"[188] are only 900-1000 pages per year each.) Printing this would cost 25,000 danish crowns, which is the equivalent of about \$5,500. N. thinks, that the Ztrbl. is a rather expensive kind of printing,

---

[184] Von Neumann mixes here German and English spelling of the country "Czechoslovakia". The Editor.

[185] See Section 1 of the Introductory Comments for von Neumann's private life. The Editor.

[186] "Zentralblatt für Mathematik und seine Grenzgebiete", a mathematical journal. The Editor.

[187] Annals of Mathematics. The Editor.

[188] American Mathematical Society Transactions. The Editor.

because it contains many short articles, which must be set at different times, and then the set-ups kept for several weeks before they can be printed.

My impression is, that producing the Ztrbl. (including salaries) should cost about \$10-15,000, of which sum about $\frac{1}{2}$ has to be raised, or at least guaranteed in US or England. Since the Ztrbl. sells 500 copies, i.e. 1250 Volumes per year, it might perhaps even become an almost self-supporting affair – but a guaranty would be needed in any event.

H. Bohr and N. agree now, that it would be very good to move N. to US if this can be done. His historical work is developing very beautifully.

I have been in Budapest since Sept. 20. (I flew here from Copenhague) when the mess began to thicken.) Now I expect to stay here, until everything is settled, and then take the first boat!

Please let me have some Princeton and Institute news, and any gossip, which can be transmitted. I am rather fed up with global ("im Grossen") news and yearning for local ("in kleinem") news. (I know, this *is* the lowest form of using human intellect ...).

My mother is well, although the excitements of the past 3 weeks disturbed her very badly.

Please remember me to Elizabeth. Best greetings from

I am yours truly,

John

P.S. Peculiarly enough, I did some mathematics. I can make a unitary spectral theory for non-Hermitean matrices, which seems to be new even in 2 dimensions, and looks quite amusing. It is obtained by analytical methods, and connects with the Fejér-Carathéodory-Schur theory of bounded analytical functions.

It runs like this: A (closed) set (of complex numbers) S is said to be a *spectral set* of the matrix A, if this is true: For every (complex) rational function $p(z)$ (z a complex variable) which is regular and $|p(z)| \leq 1$ in all of S, the matrix $p(A)$ can be formed, and $|||p(A)||| \leq 1$. (For a matrix B we define $|||B|||$ as the smallest c, such that for all vectors η $\|B\eta\| \leq c\|\eta\|$.)

Now I can prove:

1) A has the circle $|z| \leq r$ as a spectral set, if and only if [189] $\|A\| \leq r$.
2) A has the half-plane $\Im z \geq 0$ as a spectral set, if and only if $\frac{1}{2}(A + A^*)$ is definite.
3) A has the real axis [unit circle] as a spectral set, if and only if A is Hermitean [unitary].
4) Given a closed set S, call a function $g(z)$ on S quasi-analytic, if it is the uniform limit on S of rational functions. (If S is an open set + its closure, then quasi-analytic means: Analytic in each connected part of S, but not necessarily the same analytic function in all of them. If S is a Jordan arc or curve, then quasi-analytic means: Continuous.)

   If A has the spectral set S, then $g(A)$ can be defined in a natural way

---

[189] In view of the introduced notation and especially of the definition of $|||A|||$ von Neumann should write here $|||A||| \leq r$. The Editor.

for all $g(z)$, which are quasi-analytic in S. And it has those algebraical properties, which one would expect.

5) Call a (closed) set S "minor", if $\bar{z}$ is quasi-analytic on S. (Then every continuous function is, too.) (Every Jordan curve is "minor".) If A has a "minor" spectral set S, then A is normal. (That is: $A^*A = AA^*$, and the ordinary spectral theory applies to A.)
6) The intersection of all spectral sets of A is precisely the set of all proper values of A.
7) Let A be fixed. Then the intersection of two spectral sets of A is always again a spectral set of A, if and only if the set of all proper values of A (cf. 6) above) is a spectral set of A – i.e., if and only if a minimum spectral set of A exists.
8) This is equivalent to normalcy of A (cf. 5) above.)

1)-7) hold in any (finite or infinite) number of dimensions, with literally the same proof. I could only prove 8) for a finite number of dimensions, although I expect it to hold generally, too. The really decisive theorem is 1).

What happened at the sesquicentennial meeting of the Amer. Math. Soc.?

Original in O. Veblen papers, Library of Congress, language of original: English

---

The Penn-Harris
Harrisburg, Pennsylvania

November 27, [????]

Dear Oswald,

Excuse me please for the hurry in which I left, but I hope that you will understand, that this was the only way to make our daily mileage.

As you see, we did not drive very fast or very far, but we hope to be in Cleveland before 5 p.m. tomorrow.

I am writing now, to give a precise account of the construction I mentioned to you two days ago.

I) Consider an $n-1$-dimensional projective space P_{n-1}. When considering its linear subspaces $a, b, c, \ldots$, the usual way of counting dimensions is this: A point has 0 dimensions, a line 1, a plane 2, ..., the full space P_{n-1} has $n-1$. Consequently the empty set has -1 dimensions. Denoting this dimensionality with $D(a)$, the well-known law

(i) $\quad D(a \cup b) + D(a \cap b) = D(a) + D(b)$

holds, and we have the "normalisation"

(ii) $\quad D(0) = -1, \qquad D(P_{n-1}) = n-1.$

Now (i) remains true, if we replace $D(a)$ by any

(iii) $\quad d(a) = AD(a) + B$

where A, B are real numbers and constants, and A, B might be determined by requiring some other normalisation (ii)' $\quad d(0) = \alpha, \qquad d(P_{n-1}) = \beta.$

instead of (ii). (α, β are real numbers). I choose the normalisation

(ii)^0 $\quad d^0(0) = 0, \qquad d^0(P_{n-1}) = 1$

(That is $\alpha = 0, \beta = 1$, (ii) was $\alpha = -1, \beta = n-1$.) Thus clearly

(iii)^0 $\quad d^0(0) = \frac{D(a)+1}{n}.$

II) Compare now a P_{n-1} and a P_{2n-1}. The general point of P_{n-1} is characterised by n coordinates $x_1, \dots, x_n$, or more precisely: by their ratio $x_1 : \dots : x_n$. Similarly the general point of P_{2n-1} has $2n$ coordinates $x_1, \dots, x_{2n}$, or more generally: a ratio $x_1 : \dots : x_{2n}$.

A linear subset of P_{n-1} is a set of ratios $x_1 : \dots : x_n$. A linear subset of P_{2n-1} is a set of ratios $x_1 : \dots : x_{2n}$.

Now let to each *linear* subset a of P_{n-1} correspond the following subset a^* of P_{2n-1}: a^* is the set of all those ratios $x_1 : \dots : x_{2n}$, for which both ratios $x_1 : \dots : x_n$ and $x_1 : \dots : x_{2n}$ belong to a. a^* is obviously a *linear* subset of P_{2n-1}. It is easy to see:

α) If a has K ($=-1, 0, 1, 2, \dots, n-1$) dimensions in the normal sense, then a^* has $2K+1$ such dimensions. That is:

(iv) $\quad D(a^*) = 2D(a) + 1$

Therefore by (iii)^0

(iv)^0 $\quad d^0(a^*) = d^0(a).$

β) If a is empty, so is a^*.
If a is a point, a^* is a line.
If a is a line, a^* is a 3-space
If a is a plane, a^* is a 5-space

$\vdots$

If a is all P_{n-1}, a^* is all P_{2n-1}.

III) Consider now a sequence of spaces

$$P_1, P_3, P_7, P_{15}, \dots$$

Between each pair of successive ones set up the correspondence described in II):

$$P_1 \to P_3 \to P_7 \to P_{15} \to \dots$$

Identify each a in a P_{2^m-1} with the a^* which corresponds to it in $P_{2^{m+1}-1}$. So each a which occurs "first" in P_{2^l-1} (that is: which occurs in P_{2^l-1}, but is not the b^* corresponding to a b in $P_{2^{l-1}-1}$), is identified with its a^* in $P_{2^{l+1}-1}$, its a^{**} in $P_{2^{l+2}-1}, \dots$, and with nothing else.

With these identifications $D(a)$ looses its meaning (as $D(a) \neq D(a^*)$ by (iv)), but $d^0(a)$ conserves its meaning (as $d^0(a) = d^0(a^*)$ by (iv)^0).

Similarly 0, $a \cup b, a \cap b$ are unaffected by these identifications.

IV) In any P_{n-1} (irrespectively of the discussion in **III)**) a "dimensional distance" may be defined for any two linear subsets a, b:

(v)

$$\begin{aligned} \text{Dimensional Distance } (a,b) &= \\ DD(a,b) = d^0(a \cup b) - d^0(a \cap b) &= \frac{D(a \cup b) - D(a \cap b)}{n} \end{aligned}$$

One verifies easily, that $DD(a,b)$ is a distance, that is:

a) $DD(a,a) = 0$
b) $DD(a,b) > 0$ if $a \neq b$
c) $DD(a,a) = DD(b,a)$
d) $DD(a,c) \leq DD(a,b) + DD(b,c)$.

Besides $a \cup b, a \cap b$ have these continuity-like properties:

e) $DD(a \cup b, c \cup d) \leq DD(a,c) + DD(b,d)$.
f) $DD(a \cap b, c \cap d) \leq DD(a,c) + DD(b,d)$.

Of course P_{n-1} is "discrete" with this distance:

g) If $DD(a,b) \neq 0$, then $DD(a,b) \geq \frac{1}{n}$.

V) Now consider the sequence

$$P_1 \to P_3 \to P_7 \to P_{15} \to \dots$$

in **III)**. If a is in P_{2^k-1} and b in P_{2^l-1}, then both belong to every P_{2^m-1} with $m \geq Max(k,l)$. In every such P_m, $DD(a,b)$ may be formed, and as $(\text{iv})^0$ gives $DD(a,b) = DD(a^*,b^*)$, so $DD(a,b)$ is the same in $P_{2^{m+1}-1}$ as in P_{2^m-1}. Thus $DD(a,b)$ does not depend on m at all. Therefore the rules a)-f) of **IV)** hold again. But g) becomes void, as n may be any number $1, 3, 7, 15, \dots$.

Now proceed "á la Cantor":

A) Consider all sequences $a_1, a_2, \dots$, where a_k is a linear subset of P_{2^k-1}.
B) Call such a sequence *fundamental*, if

$$\lim_{k,l \to \infty} DD(a_k, a_l) = 0$$

C) Call two fundamental sequences $a_1, a_2, \dots$ and $b_1, b_2, \dots$ *equal*, if

$$\lim_{k \to \infty} DD(a_k, b_k) = 0$$

After these preliminaries it is obvious, how $\cup, \cap, d^0$ and DD have to be defined for fundamental sequences by "continuity".

The fundamental sequences give the desired example of a "continuous dimensional geometry". They represent, of course, the linear subspaces, while points do not exist.

Please pardon the somewhat lengthy expectorations.

Hoping to see you again on Monday, and with many greetings

Yours,

John

Original in O. Veblen Papers, Library of Congress, language of original: English

The Institute for Advanced Study
School of Mathematics
Princeton, New Jersey

November 30, [????]

Dear Oswald,

I talked yesterday to Aydelotte about Gödel. I assured him that

(1) Gödel's work on Leibniz was important in the history of mathematics. He might still do better work in mathematics proper, but this was solid work too, and a man of his caliber and record ought to be the sole judge of what he does.
(2) Gödel's whole intellectual behavior at present is such, that he may easily do more work in mathematics proper. In fact I judged, that his probability of doing some is no worse than that of most mathematicians past 35.
(3) Gödel's two past papers ("Widerspruchfreiheit" and Continuumhypothesis), are in any case worth more than the total literary output, past, present and future, of most mathematicians, including many who would be good enough to be our colleagues.
(4) Gödel did some of of his best work (Cont. hyp.) at the Institute – actually at a time when he was *less* normal than now. The Institute is clearly committed to support him, and it is ungracious and undignified to continue a man of Gödel's merit in the present arrangement forever. Hence it is just as well to correct the situation as quickly as feasible.
(5) (Upon Aydelotte's question.) Gödel's *technical* performance is *vastly* greater than B. Russel's.

After some discussion Aydelotte seemed to agree with me on all counts. He stated:

a) Gödel's situation is now very much on his mind and he proposes to correct it.
b) He recognizes that giving him Mayer's position is the necessary minimum.
c) He feels that he has to bring this matter, however, before the Faculty.

I agreed with c), and took the line, that Gödel's cause will only gain if the existence of the problem is publicly recognized and his objective scientific importance is publicly discussed – that is before the Faculty.

I think that it will be very good if you repeat these thoughts to Aydelotte. He definitely committed himself to taking the action indicated.

With best regards

yours as ever,

John

Original in O. Veblen Papers, Library of Congress, Washington, language of original: English

Letters to N. Wiener

November 20, 1945

Professor Norbert Wiener
Mathematics Department
Massachusetts Institute of Technology
Cambridge 39, Massachusetts

Dear Norbert:

The negotiations at Princeton have come to a conclusion, with the result that the Institute of Advanced Study and the Radio Corporation of America (whose research laboratory is, as you know, in Princeton), with the cooperation of Princeton University, have decided to undertake a joint high-speed, automatic, electronic computer development. While this is to be a community effort, it is agreed between all parties that the completed computer is to be located at the Institute, and used exclusively as a research tool. I have been offered the over-all direction of this project.

As far as I can see, the project is adequately financed and has the necessary guarantees of manpower, availability of the RCA's experience in various pertinent fields etc.

Since I have always urged such an enterprize on the Institute for Advanced Study authorities, and insisted that it is of great importance in several vital respects, I feel that it is now my prime responsibility to see it through here.

I need not tell you how much I regret that this means that I cannot join you at Tech[190], especially since I am sure that without the decisive encouragement that I received from you and from the Tech authorities I would have hardly had the perseverance and the strength of conviction which are the minimum requirements in this project. On the other hand, I hope that we shall work together in the field just the same. I think that we should discuss the modalities as quickly as possible.

I am staying in Princeton continuously until November 30. Then I am going for $2\frac{1}{2}$-3 weeks to Los Alamos. This trip, plus the Christmas vacation eliminate the month of December. Could you find the time to visit here before the end of this month, say between November 26 (Monday) and 29 (Thursday), both inclusive? I hope that you can do this and will accept our invitation to come as a guest of the Institute.

Hoping to see you soon, and very much looking forward to it,

---

[190]MIT. The Editor.

Yours as ever

John von Neumann

Original in von Neumann Papers, Library of Congress, language of original: English

---

November 29, 1946

Dear Norbert:

This letter represents an effort to do better than I estimated in my letter of November 25, in which I proposed that we might get together for the afternoon or evening of December 4 in Cambridge, and indicated only somewhat vaguely what the subject was that I would like to discuss with you. I am now trying to give you a more detailed advance notice, hoping that this will make our discussion on December 4 more specific.

Our thoughts – I mean yours and Pitts' and mine – were so far mainly focused on the subject of neurology, and more specifically on the human nervous system, and there primarily on the central nervous system. Thus, in trying to understand the function of automata and the general principles governing them, we selected for prompt action the most complicated object under the sun – literally. In spite of its formidable complexity this subject has yielded very interesting information under the pressure of the efforts of Pitts and McCulloch, Pitts, Wiener and Rosenblueth. Our thinking – or at any rate mine – on the entire subject of automata would be much more muddled than it is, if these extremely bold efforts – with which I would like to put on one par the very un-neurological thesis of R. Turing – had not been made. Yet, I think that these successes should not blind us to the difficulties of the subject, difficulties, which, I think, stand out now just as - if not more-forbiddingly as ever.

The difficulties are almost too obvious to mention: They reside in the exceptional complexity of the human nervous system, and indeed of any nervous system. What seems worth emphasizing to me is, however, that after the great positive contribution of Turing - cum - Pitts - and-McCulloch is assimilated, the situation is rather worse than better than before. Indeed, these authors have demonstrated in absolute and hopeless generality, that anything and everything Brouwerian can be done by an appropriate mechanism, and specifically by a neural mechanism – and that even one, definite mechanism can be "universal". Inverting the argument: Nothing that we may know or learn about the functioning of the organism can give, without "microscopic", cytological work any clues regarding the further details of the neural mechanism. I know that this was well known to Pitts, that the "nothing" is not wholly fair, and that it should be taken with an appropriate dose of salt, but I think that you will feel with me the type of frustration that I am trying to express. (H. N. Russell used to say, or to quote, that if the astrophysicist found a general theory uniformly corroborated, his exclamation should be "Foiled again" since no experimenta crucis would emerge.) After these devastatingly general and positive results one is therefore thrown back on microwork and cytology – where one might have remained in the first place. (This "remaining there" is, of course, highly figurative in my case, who have never been there.) Yet, when we are in that field, the complexity of the subject is overawing. To understand the brain with neurological methods seems to me about as hopeful as to want to understand the ENIAC with no instrument at one's disposal that is smaller than about 2 feet across its critical organs, with no methods of intervention more delicate than playing with a fire hose (although one might fill it with kerosene or nitroglycerine instead of water) or dropping cobblestones into the circuit. Besides the system is not even purely digital (i.e. neural): It is intimately connected to a very complex analogy (i.e. humoral or hormonal) system, and, almost every feedback loop goes through both sectors, if not through the "outside" world (i.e. the world outside the epidermis or within the digestive system) as well. And it contains, even in its digital part, a million times more units than the ENIAC. And our intellectual possibilities relatively to it are about as good as some bodies vis-a-vis the ENIAC, if he has never heard of any part of arithmetic. It is true that we know a little about the syndromes of a few selected breakdowns – but that is not much.

My description is intentionally exaggerated and belittling, but don't you think that there is an element of truth in it?

Next: If we go to lower organisms from man with 10^{10} neurons to ants with 10^6 or to some sub-animal with, say, 10^2 neurons – we lose nearly as much as we gain. As the digital (neural) part simplifies, the analogy (humoral) part gets less accessible, the typical malfunctions less known, the subject less articulate, and our possibilities of communicating with it poorer and poorer in content.

Further: I doubt that the "Gestalt" theory, or anybodies verbal theory will help any. The central nervous system is complicated, and therefore its attributes and characteristics have every right to be complicated. Let not our facile familiarity with it, through the medium of the subjective consciousness, fool us into illusions in this respect.

What are we then to do? I would not have indulged in such a negative tirade if I did not believe that I see an alternative. In fact, I have felt all these doubts for the better part of a year now, and I did not talk about them because I had no idea as to what one might say in a positive direction.

I think now that there is something positive to be said, and I would like to indicate in which direction I see it.

I feel that we have to turn to simpler systems. It is a fallacy, if one argues, that because the neuron is a cell (indeed part of its individual insulating wrapping is multicellular), we must consider multicellular organisms only. The cell is clearly an excellent "standard component", highly flexible and suited to differentiation in form and in function, and the higher organisms use it freely. But its self-reproductivity indicates that it has in itself some of the decisive attributes of the integrated organisms – and some cells (e.g. the leukocytes) are self-contained, complete beings. This in itself should make one suspicious in selecting the cells as the basic "undefined" concepts of an axiomatism. To be more par terre: Consider, in any field of technology, the state of affairs which is characterized by the development of highly complex "standard components", which are at the same time individualized, well suited to mass production, and (in spite of their "standard" character) well suited to purposive differentiation. This is clearly a late, highly developed style, and not the ideal one for a first approach of an outsider to the subject, for an effort towards understanding. For the purpose of understanding the subject, it is much better to study an earlier phase of its evolution, preceding the development of this high standardization – with differentiation. I.e. to study a phase in which these "elegant" components do not yet appear. This is especially true, if there is reason to suspect already in that archaic stage mechanisms (or organisms) which exhibit the most specific traits of the simplest representatives of the above mentioned "late" stage.

Now the less-than-cellular organisms of the virus or bacteriophage type do possess the decisive traits of any living organism: They are self-reproductive and they are able to orient themselves in an unorganized milieu, to move towards food, to appropriate it and to use it. Consequently a "true" understanding of these organisms may be the first relevant step forward and possibly the greatest step that may at all be required.

I would, however, put on "true" understanding the most stringent interpretation possible: That is, understanding the organism in the exacting sense in which one may want to understand a detailed drawing of a machine – i.e. finding out where every individual nut and bolt is located, etc.

It seems to me that this is not at all hopeless. A typical bacteriophage, which can be multiplied at will (and hence "counted" – by the colonies it forms on a suitable substrate – as reliably as elementary particles can be "counted" by a Geiger-counter) is a phage that is parasitic, I think, on the Bacillus Coli. It has been extensively worked with, e.g. by Delbrueck at Vanderbilt. It is definitely an animal, with something like a head and a tail. Its dimensions are I think, ca. $60m\mu \times 25m\mu \times 25m\mu$, i.e. its volume is $60 \times 25 \times 25 \times 10^{-21}cm^3 = 3.7 \times 10^{-17}cm^3$. The density may be taken to be 1, hence its mass is about $3.7 \times 10^{-17}gr$. i.e. the same as about 2.5×10^7 H atoms. Since the average chemical composition of these things is usually about one C or N or O per one or two H, the average atomic weight of its constituents is about 6. Hence the number of atoms in it is about 4×10^6. Furthermore, it is known from the behavior of physiological membranes, that they are monomolecular – or oligomolecular – Langmuir-layers, which exercise their function in a highly mechanical way. E.g. the so-called "active permeability": The peculiar ability to "permit" ions to pass through the membrane against an electrical field - an activity which clearly must, and demonstrably does, require an energy supply from the metabolism - and which is therefore better described as "pushing the ions across" than as "permitting them to pass". I understand that here an ion simply gets seized by the opposite-polarity end of one of the rare (charged) radicals in the membrane, which then turns around and deposits the ion on the other side of the membrane. Very similar things can be said about the functioning of the "phosphate bond", which seems to be the main physiological device for localized, short time energy storage – i.e. the equivalent of a spring. Thus one can really talk of "mechanical elements", each of which may comprise 10 atoms or more. Thus the organism in question consists of six million atoms, but probably only of a few hundred thousand "mechanical elements". I suppose (without having done it) that if one counted rigorously the number of "elements" in a locomotive, one might also wind up in the high ten thousands. Consequently this is a degree of complexity which is not necessarily beyond human endurance.

The question remains: Even if the complexity of the organisms of molecular weight $10^7 - 10^8$ is not too much for us, do we have the observational means to ascertain all the facts? Or to be more lenient: If we do not possess such means now, can we at least conceive them, and could they be acquired by developments of which we can already foresee the character, the caliber, and the duration? And are the latter two not excessive and impractical?

I feel that the answer to these questions can already be given, and that it need not be unfavorable. Specifically:

I am talking of molecular weights $10^7 - 10^8$. The major proteins have molecular weights $10^4 - 10^5$. (the lowest one known actually appears to be only about 7,000) and the determination of their exact structure may not be hopeless. This is controversial, but some good authorities like Langmuir and Dorothy Wrinch consider it promising. Langmuir asserts that a 2-4 year efforts with strong financial backing should break the back of the problem. His idea of an attack is: Very high precision X-ray analysis, Fourier transformation with very massive fast computing, in combination with various chemical substitution techniques to vary the X-ray pattern. I realize that this is in itself a big order, and that it is still by a factor 10^3 off our goal but it would probably be more than half the difficulty.

In addition there is no telling what really advanced electron-microscopic techniques will do. In fact, I suspect that the main possibilities may well lie in that direction. The best (magnetic) electron-microscope resolutions at present are a little better than $10\mathring{A} = 1m\mu$. With 4×10^6 atoms in a volume of $3.7 \times 10^{-17} cm^3 = 4 \times 10^4 m\mu^3$, the average atomic volume is $10^{-2} m\mu^3$, and hence the average atomic distance about $1/5m\mu$. Hence the 1 $m\mu$ resolution is inadequate – but not very far from what might be adequate. A resolution that is improved by a factor of $10 - 20$ might do. It is dubious whether electron lenses can be improved to this extent. On the other hand, the proton microscope need not be more than 2-4 years in the future, and it would certainly overcome these difficulties.

Besides, all these developments might be pushed and accelerated. Of course, everybody knows what a $1 - \frac{1}{2}\mathring{A}$ resolution would mean: One could "look" at an H atom, and with a little more, say $1/5\mathring{A}$, one could "see" the Schroedinger-charge-cloud of the orbital electrons! But the physiological implications are even more extraordinary, and they should receive a great deal of emphasis in the immediate future.

At any rate, I think that we could do these things:

Study the main types of evidence: Physiology of viruses and bacteriophages, and all that is known about the gene-enzyme relationship. (Genes are probably much like viruses and phages, except that all the evidence concerning them is indirect, and that we can neither isolate them nor multiply them at will.)

Try to learn a reasonable amount about the present state of knowledge and opinions concerning protein structure.

Study the methods of organic-chemical structure determination by X-ray analysis and Fourier-analysis with their necessary complement of chemical manipulations.

Study the principles and methods of electron-microscopy, both in the (direction of electron optics and in the direction of object-manipulation). Try to get oriented as to the possibilities of proton-microscopy.

Finally: Compile for our common use two lists: (1) relevant publications with the main emphasis for the immediate future, considering our lack of education, on books and survey articles. (2) Persons from whom we might learn most about the state of affairs and the outlook in these fields.

I did think a good deal about self-reproductive mechanisms. I can formulate the problem rigourously, in about the style in which Turing did it for his mechanisms.

I can show that they exist in this system of concepts. I think that I understand some of the main principles that are involved. I want to fill in the details and to write up these considerations in the course of the next two months. I hope to learn various things in the course of this literary exercise, in particular, the number of components required for self-reproduction. My (rather uninformed) guess is in the high ten thousands or in the hundred thousands, but this is most unsafe. Besides, I am thinking in terms of components based on several rather arbitrary choices. At any rate, it will be necessary to produce a complete write-up before much discussing is possible.

Certain traits of the gene-enzyme relationship, of the behavior of some mutants, as well as some other phenomena, seem to emphasize some variants of self-reproductivity, which one would be led to investigate on purely combinatorial grounds as well. E.g.: Self-reproductivity may be symbolized by the schema $A \to A$. What about schemata like $A \to B \to C \to A$, or $A \to B \to C \to C$, or $A \to B \to C \to D \to E \to C$, etc.?

This is as far as my ideas go at this moment. I hope you will not misinterpret the anti-neurological tirade at the beginning of this letter. Of course I am greatly interested in that approach and I have the greatest respect for the important results that have been obtained in that field, and in our border area with it. I certainly hope that these efforts will continue. I wanted to point out, however, that I felt that the decisive "break" was more likely to come in another theater. I was trying to formulate and to systematize my motives for believing this, and the simplest literary mode to do this is the controversial one. I hope therefore that I have not given you a false impression of the spirit in which I am starting a "controversy".

I am most anxious to have your reaction to these suggestions. I feel an intense need that we discuss the subject extensively with each other.

Hoping that this letter has not been unbearable just by its sheer length, and hoping to hear from you and to see you again soon, I am, with the best regards,

Yours, as ever

John von Neumann

Original in Von Neumann Papers, Library of Congress, language of original: English

Letter to H. Wold

October 28, 1946

Professor Herman Wold
Statistiska Institutionen vid
Uppsala Universitet
Uppsala, Sweden

Dear Professor Wold:

I was very glad to receive your letter of October 14, and to see the the misunderstanding which had existed between us is now clarified.

Your remarks concerning numerical and non-numerical utilities interest me greatly. In this connection I would like to add the following comments:[191]

(1) I think that it will be difficult to get away from some rigidly determined (i.e., "up to a linear transformation" uniquely determined) connection between utilities and numbers in any *complete* system of concepts: Probability is hard to eliminate from the world with which we are dealing, and probability seems to be an essentially numerical concept.

(2) The general comparability of utilities, i.e. the completeness of their ordering by (one person's) subjective preferences, is, of course, highly dubious in many important situations. Moreover, since it is nevertheless very desirable to work with a system into which events affected with probabilities can be incorporated, the incompleteness of the ordering does not remove a certain numerical character from the utilities. In this order of ideas one will probably be led to *vectorial utilities*. I.e., I would expect that utilities $U, V \ldots$ can then be made to correspond to aggregates of numbers: $U = (u_1, u_2, \ldots)$, $V = (v_1, v_2, \ldots)$, $\ldots$ $(u_i, v_i, \ldots$ real numbers), so that then

$$\alpha U + (1-\alpha)V = (\alpha u_1 + (1-\alpha)v_1, \alpha u_2 + (1-\alpha)v_2, \ldots) \tag{1}$$

$(0 < \alpha < 1$, α is a probability, $\alpha U + (1-\alpha)V$ corresponds to the expectation of U, V with probabilities $\alpha, 1-\alpha)$, and U is preferred to V if, and only if, $u_i > v_i$ for all $i = 1, 2, \ldots$ (or something similar).

(3) Regarding the "dynamic theory of utility" my feeling is this: Something more elaborate than the above concepts is certainly necessary to understand the behaviour of people in lotteries, etc. I doubt, however, that the distinctions of an *ex ante* and *ex post* utilities is fruitful in this respect. In fact, I doubt that the ideas which may bring the solution have as yet

[191] See Section 9 of the Introductory Comments for notes on the utility function. The Editor.

appeared anywhere in the literature. I think that this problem is really rather deep.

You may have noticed that Morgenstern and I repeatedly pointed out that we were fully aware of such limitations (Cf., e.g., pp. 19 and pp. 28-29).[192]

The Cramers are now in Princeton, but, to my great regret, I have not met them yet, mainly because I have been away a good deal. I expect, however, to meet them shortly, and I will not fail to transmit your regards.

I am,

Very sincerely yours,

John von Neumann

Original in Von Neumann Papers, Library of Congress, language of original: English

[192] O. Morgenstern and J. von Neumann: *Theory of Games and Economic Behavior*, Princeton University Press, 1944. The Editor.

Notes on addressees of von Neumann's letters

(1) Aronszajn, N. (1907-1980)
Mathematician. While at Kansas University, he published a joint paper with K.T. Smith containing a proof of existence of invariant subspaces for compact operators on Banach spaces. Aronszajn corresponded with von Neumann about the invariant subspace problem, see footnote 9.
(July 15, 1948)

(2) Aydelotte, F. (1880-1956)
Professor of English, educator, President of Swarthmore College between 1921 and 1940, where he introduced the honors program. Von Neumann had contact to Aydelotte while he was the director of Institute for Advanced Study during the years 1939-1947.
(March 23, 1940), (July 23, 1941)

(3) Beckenbach, E.F. (1906-1982)
Mathematician, worked on the theory of minimal surfaces, subharmonic functions, convex functions and inequalities. First Managing Editor of Pacific Journal of Mathematics.
(March 24, 1949)

(4) Bethe, H.A. (1906-)
German-born physicist, emigrated to the U.S.A. in 1934. Bethe was awarded the Nobel Prize in 1967 for his work on energy production in stars. He was head of the theoretical division of the Manhattan Project in Los Alamos. Von Neumann met Bethe during his visits to Los Alamos as consultant to the Manhattan Project.
(May 16, 1941)

(5) Birkhoff, G. (1911-1966)
Mathematician, especially known for his work in algebra and lattice theory. Birkhoff was a friend and coauthor of von Neumann: their joint paper in 1936 created quantum logic (see Section 6 of the Introductory Comments). Birkhoff was involved in war work at the Ballistic Research Laboratory at the Aberdeen Proving Ground, where von Neumann also was a consultant.
(January 15, 1935), (January 19, 1935 ?), (November 6, 1935 ?), (November 13, 1935 ?), (November 21, 1935), (November 27, 1935 ?), (December 10, 1941), (April 7, 1947)

(6) Blaschke, W.J.E. (1885-1962)
Austrian-born mathematician, worked mainly on differential and integral geometry. After appointments at various European universities he became professor in Hamburg, Germany, and had built a strong department. Blaschke was the president of the German Mathematical Society (DMV)

for a short period of time during the thirties. Von Neumann announced to Blaschke his resignation from the Society (see Section 1 of the Introductory Comments).
(January 28, 1935)

(7) Burington, R.S. (1882-1981)
Burington was a professor of mathematics at Case School of Applied Sciences (now Case Western Reserve University) and Ohio State University between 1926-1941. He became a civilian employee of the U.S. Navy Bureau of Ordnance in 1941, and in 1946 was named its chief mathematician. In 1959, Burington became chief mathematician of the Bureau of Naval Weapons and in 1966 was named the chief mathematician of the Naval Air Systems Command.
(January 19, 1951)

(8) Bush, V. (1890-1974)
Engineer and science administrator, adviser to the U.S. government. Bush served as president of the Carnegie Institution, as chair (appointed by president Roosevelt) of the National Defense Research Committee (NDRC) and as director of the Office of Scientific Research and Development (OSRD), organizations which Bush himself helped to create. The OSRD and NDRC were coordinating military research towards the end of World War II. Von Neumann's involvement in military related research brought him in touch with Bush.
(November 15, 1949), (December 6, 1949), (January 25, 1950)

(9) Carnap, R. (1891-1970)
German-born philosopher of science, key member of the group of philosophers known as "Vienna Circle", which – among other prominent scientists and philosophers – also included K. Gödel. Carnap emigrated to the U.S.A. in 1935 and spent two years at the Institute for Advanced Study (1952-1954). Von Neumann met Carnap on several occasions, among them during the conference on epistemology in Königsberg in 1930, see Section 2 of the Introductory Comments.
(June 7, 1931)

(10) Cattell, W.
No data available.
(April 8, 1940)

(11) Cherry, T.M. (1898-1966)
Australian mathematician. Cherry was chairman of the Applied Mathematics Department of the University of Melbourne in Australia (1952-1963) and became Fellow of the Royal Society in 1954. Cherry worked on differential equations and on hydrodynamics; his exchange of letters with von Neumann concerns these issues.
(November 27, 1959)

(12) Cirker, H. (1917-2000)
Cirker was founder and president of Dover Publications. Dover published a reprint of von Neumann's book *Mathematishe Grundlagen der Quantenmechanik* (Springer 1932) in 1943, and in 1949 von Neumann was apparently in negotiations with Dover about publishing a revised, English

translation of the book. (The first English translation was published finally by Princeton University Press in 1955), see Section 5.1 of the Introductory Comments.
(October 3, 1949)

(13) Crocker, Mrs., F.W.
No data available.
(September 1, 1955)

(14) Davie, M.R. (1914-1975)
Sociologist, specialist on emigration issues. Davie served as chair of the Department of Sociology at Yale University and he is the (co)author of *Refugees in America* (New York, London, Harper and brothers, 1947). It was in connection with his research on refugees in the U.S. that he contacted von Neumann in 1946, seeking information about von Neumann's immigration to the U.S.A.
(May 3, 1946)

(15) Deming, W.E. (1900-1993)
Mathematician-statistician. Deming worked as a mathematical physicist for the U.S. Department of Agriculture, he was a statistical advisor for the U.S. Census Bureau, and then from 1946-1993 was a statistics professor at New York University and Columbia University. He is best known for his work on statistical methods, especially in connection with quality control.
(January 23, 1945)

(16) Destouches, J.L. (1909-1980)
French physicist and philosopher. Destouches was professor of mathematical physics at the faculty of science, Paris (and taught logic and methodology of science at the faculty of letters, as well). His main works include "Corpuscules et systémes de corpuscule" (Paris, Gauthier-Villars, 1941) and "Principes fondamentaux de physique théorique" (Paris, Hermann, 1942).
(December 18, 1951)

(17) Dirac, P.A.M. (1902-1984)
English-born theoretical physicist, mainly worked on quantum mechanics and on relativistic quantum field theory, the relativistic generalization of Schrödinger equation is named after him. Dirac predicted the existence of positron on theoretical grounds in 1930, the particle was observed later (1932). Dirac received (with E. Scrödinger) the 1933 Nobel prize in physics. Dirac spent a year in Princeton (1934). Von Neumann corresponded with Dirac about quantizing spacetime to avoid some divergencies in quantum electrodynamics (see Section 5.4 of the Introductory Comments).
(January 27, 1934)

(18) Dixmier, J. (1924-)
French mathematician. Dixmier wrote the first monograph on von Neumann algebras (see footnote 65).
(November 24, 1953), (June 18, 1954)

(19) Dodd, P.A. (1902-1992)
Professor of economics at University of California at Los Angeles. Dodd

was Dean of the College of Letters and Science of UCLA between 1946-1961; this was the time when von Neumann was negotiating with Dodd for a position of professorship at UCLA shortly before his death (see Section 1 of the Introductory Comments).
(February 24, 1956)

(20) DuMond, W.M. (1892-1976)
French-born experimental physicist, emigrated to the U.S.A. with his parents in 1899. DuMond worked on spectroscopy; during the thirties he refined measurements of fundamental physical constants (charge and mass of electron, Planck's constant). He did war work at California Institute of Technology on rocket technology. His war work brought him in touch with von Neumann.
(April 17, 1944)

(21) Duncan, R.E.
No data available.
(December 18, 1947)

(22) Farquharson, R.
No data available.
(November 28, 1955)

(23) Flexner, A. (1866-1959)
Educational reformer, famous for his critical 1910 report of U.S. medical schools that recommended closing 120 of the 155 medical schools in the U.S.A. Founding director (1931-1939) of the Institute for Advanced Study (IAS); Flexner negotiated the appointment of the first professors of the IAS, including the appointment of von Neumann.
(September 27, 1939)

(24) Fornaguera, R.O. (1915-1974)
Spanish physicist, he had a PhD in both physics and mathematics. From 1954 Fornaguerra was the head of the "Theoretical Physics Division" of the "Nuclear Energy Council" of Spain. Fornaguerra was the translator into Spanish of von Neumann's book "Mathematishe Grundlagen der Quantenmechanik" (Springer 1932). In 1946 von Neumann and Fornaguerra corresponded about issues related to the translation of the book.
(December 10, 1947)

(25) Goldsmith, N.H. (1907-1949)
Physicist, participated in building the first chain reacting nuclear pile in Chicago. Goldsmith was the founder in 1945 and first co-editor of the Bulletin of the Atomic Scientists. Goldsmith invited von Neumann to join the Board of the Bulletin (which von Neumann refused).
(October 5, 1948)

(26) Gottschalk, W.H. (1918-????)
Mathematician. Gottschalk was member of the Institute for Advanced Study in 1947-1948. Between 1955-1958 he was chairman of the mathematics department at the University of Pennsylvania. During his chairmanship he sought von Neumann's advice concerning whether a computer science department should be established as part of the mathematics department or as an independent unit (see Section 1 of the Introductory

Comments).
(November 30, 1955)

(27) Gödel, K. (1906-1978)
Austrian born mathematician and logician. Gödel made fundamental discoveries in logic, set theory and also in general relativity. Most significant of these discoveries were the incompleteness theorems in logic and the independence result in connection with the continuum hypothesis. Von Neumann met Gödel in Königsberg during the 1930 Conference on Epistemology of the Exact Sciences (see Section 2 of the Introductory Comments). In 1934 Gödel lectured in Princeton on undecidable propositions of formal mathematical systems. From 1940 through 1953 Gödel was von Neumann's colleague at the Institute for Advanced Study. Von Neumann had the highest respect for Gödel and his work.
(November 20, 1930), (November 29, 1930), (January 12, 1931), (February 14, 1933)

(28) Haberler, G. (1901-1955)
Austrian-born economist. Haberler was a member of the circle of influential economists around the Austrian economist Ludwig von Mises. He emigrated to the U.S.A. in 1936 to become professor of economics at Harvard University. Haberler worked on the theory of international trade and business cycles. Haberler asked von Neumann to review the book by Samuelson, see Section 9 of the Introductory Comments.
(October 31, 1947)

(29) Halperin, I. (1911-)
Canadian born mathematician, a student and coauthor of von Neumann. (While von Neumann did not have a teaching position, hence he did not have students officially, he was the supervisor (from 1934 on) of Halperin's PhD dissertation (Princeton University, 1936).) When Halperin was charged in Canada with spying for the Soviet Union, von Neumann tried to help – an action which was unfavorably mentioned during von Neumann's confirmation hearing for the Atomic Energy Commissioner position. It was Halperin to whom von Neumann confided in 1939 his discovery of type III "rings of operators" and other discoveries in the field of operator algebras, see Section 3 of the Introductory Comments.
(June 1, 1939 ?), (December 17, 1939), (February 22, 1940)

(30) Harrison, G.B. (1898-1979)
Experimental physicist. After his appointment in Stanford, Harrison joined the MIT faculty in 1930 and became Dean of the School of Science of MIT in 1942. Von Neumann received an offer from MIT in 1945 when Harrison was the Dean, see Section 1 of the Introductory Comments.
(November 20, 1945)

(31) de Horvath, M.
No data are available.
(September 20, 1955)

(32) Householder, A.S. (1904-1993)
Mathematician, worked on mathematical biology and numerical analysis. Householder was senior mathematician at the Oak Ridge National Laboratory (1946-1969). In 1948 von Neumann corresponded with Householder

on generating random numbers by computers.
(February 3, 1948)

(33) Hurd, C.C. (1946-1992)
Mathematician, computer company executive. Hurd had been head of technical research of the Union Carbide and Carbon Corporation in Oak Ridge (1945-1947) before he joined the IBM corporation in 1949, where he formed (and became director of) IBM's Applied Science Department and later director of the IBM Electronic Data Processing Machines Division. Hurd played a major role in hiring von Neumann as consultant to IBM. Von Neumann worked with Hurd on specific technical problems related to IBM.
(December 3, 1948), (October 13, 1950)

(34) Husimi, K. (1909 -)
Japanese physicist. Husimi joined the faculty of Osaka Imperial University in 1939, where he became professor in 1940. He served as Director of the Plasma Institute at Nagoya University from 1961 and was the President of the Science Council of Japan. Husimi was very mathematically minded as a physicist and his interest in the foundations of quantum mechanics brought him in touch with von Neumann.
(September 29, 1937)

(35) Jordan, P. (1902-1980)
German physicist, coauthor of von Neumann. Jordan studied with M. Born and, after a four year stay in Göttingen that overlapped with von Neumann's stay there in 1926-1928, he became professor in Rostock and in Hamburg. Jordan worked on the foundations of quantum mechanics. His joint paper with Wigner and von Neumann on the algebraic generalization of quantum mechanics established the theory of Jordan algebras.
(December 11, 1949), (January 12, 1950)

(36) Kaplansky, I. (1917-)
Canadian-born mathematician, with work mainly in algebra, game theory and continuous geometry. Kaplansky's density theorem is of central importance in the theory of von Neumann algebras. Von Neumann corresponded with Kaplansky on operator algebra theory and game theory (see Sections 3 and 9 of the Introductory Comments).
(February 1, 1945), (March 1, 1950)

(37) Kemble, C.E. (1889-1984)
Physicist, received his PhD from Harvard in 1917 in theoretical physics, the first such degree in the U.S.A. Kemble was the first distinguished U.S. physicist specializing in quantum theory and he did important work in teaching quantum mechanics. During the war (1940-1945) he was chair of the physics department at Harvard, after the war he was deputy director of the program that evaluated the German atomic bomb project ("ALSOS Mission").
(December 6, ????)

(38) Killian, J.R. (1904-1988)
College administrator, military and government adviser. Killian was president of MIT during 1948-1959, and it was during his tenure that MIT offered a position to von Neumann, which von Neumann finally declined

(see Section 1 of the Introductory Comments).
(February 24, 1956)

(39) Kloosterman, H.D. (1900-1968)
Dutch mathematician. After appointments in German Universities (Göttingen, Hamburg and Münster) Kloosterman became a faculty member of the University of Leiden, where he became full professor in 1947. Kloosterman worked mainly on representation theory of groups. He contacted von Neumann on behalf of the organizing committee of the International Congress of Mathematicians (Amsterdam, 1954), see the end of Section 6 of the Introductory Comments.
(March 25, 1953), (April 10, 1953)

(40) Kuhn, H. (1925-)
Mathematician, Professor Emeritus of Mathematical Economics at Princeton University. Kuhn got his PhD from Princeton University in 1950. He worked on game theory and non-linear programming. In 1953-1954 he served as president of SIAM (Society of Industrial and Applied Mathematics). He was winner (shared) of the John von Neumann Prize in 1980.
(April 14, 1953)

(41) Lederberg, J. (1925-)
Geneticists, computer scientist, also worked in the field of artificial intelligence. Lederberg won/shared the Nobel Prize in 1958 in Physiology/Medicine for his discoveries concerning genetic recombination and the organization of the genetic material of bacteria. He conceived (with E. Feigenbaum) and created DENDRAL, the first computer based expert system in chemistry in 1969. It was von Neumann's work on mathematical modelling of self-reproduction that prompted Lederberg to get in touch with von Neumann; they exhanged a few letters in the summer of 1955.
(August 15, 1955)

(42) Lingelbach, W.E.
No data available.
(April 23, 1938)

(43) MacLane, S. (1909-)
American mathematician, creator (with S. Eilenberg) of the major concepts of category theory. Von Neumann tried to convince MacLane to accept membership on advisory committees for the military (see Section 1 of the Introductory Comments).
(May 17, 1948)

(44) McKinsey, J.C.C. (1908-1953)
Mathematical logician. McKinsey worked at the Rand Corporation in Santa Monica, California before joining the philosophy department at Stanford University in 1951. He was the author of *Introduction to Game Theory*; his interest in game theory brought him in touch with von Neumann.
(February 18, 1949)

(45) Mitchell, M.M.
No data available.
(December 14, 1934)

(46) Moore, T.V.
No data available.
(April 9, 1953)

(47) Morgenstern, O. (1902-1977)
German-born economist, educated in Vienna, Austria. He met von Neumann in Princeton after he emigrated to the U.S.A. in 1938. Morgenstern and von Neumann were discussing game theory in Princeton during 1939-1943, those discussions led to the treatise *Theory of Games and Economic Behavior, Princeton University Press, 1944* coauthored by von Neumann and Morgenstern, see Section 9 of the Introductory Comments.
(January 2, 1953)

(48) Morse, M. (1892-1977)
Mathematician, best known for his work on the calculus of variations ("Morse Theory"); for his contribution he was awarded the Bôcher prize in 1933 (shared with N. Wiener). After appointments in several universities in the U.S.A., Morse became a member of the Institute for Advanced Study in 1935, he retired from the Institute in 1962.
(April 3, 1952), (March 2, 1955)

(49) Nagel, E. (1901-1985)
Philosopher of science, born in the Austro-Hungarian Monarchy (Bohemia). Nagel emigrated to the United States in 1911 and became professor of philosophy at Columbia University (1931-1970).
(December 9, 1953)

(50) Oppenheimer, R. (1904-1967)
Physicist, director of Manhattan Project, chairman of the General Advisory Panel to the Atomic Energy Commission and director of the Institute for Advanced Study (1947-1952). Von Neumann and Oppenheimer disagreed strongly on a number of issues, including the need to develop thermonuclear bombs; yet von Neumann testified in favor of Oppenheimer in the notorious hearings that resulted in revoking Oppenheimer's security clearance in 1954.
(February 19, 1948)

(51) Ortvay, R. (1885-1945)
Hungarian physicist, friend of the von Neumann family, professor of theoretical physics of Pazmany Peter University in Budapest, Hungary, during the thirties. The famous physics colloquia organized and led by Ortvay were a major means through which the results of modern physics were disseminated in the Hungarian science community. Ortvay was a friend and supporter of von Neumann in Hungary; they were corresponding during the thirties, their correspondence touched upon very diverse topics from sciences to politics. The current volume contains just a selection of letters by von Neumann to Ortvay.
(April 17, 1934), (March 17, [1938 ?]), (January 26, 1939), (February 2, 1939), (February 26, 1939), (March 29, [1939 ?]), (July 18, [1939 ?])

(52) Overbeck, W. (????-????)
Von Neumann addresses Overbeck as "William", and W. Aspray identifies William Overbeck as an MIT engineer working on electronic storage tubes during the war [**3**, p. 57]; however, only *Wilcox* Overbeck can be found

in the MIT archives (G. Hon, personal communication to the Editor.) So it remains ambiguous who precisely W. Overbeck was.
(November 28, 1945)

(53) Rankin, H.H.
No data available.
(August 26, 1954)

(54) Robertson, H.P. (1903-1961)
Mathematician and physicist, worked mainly on quantum mechanics, differential geometry and relativity theory (cosmology). Robertson was professor at Princeton University between 1929-1947 (from 1938 as full professor of mathematical physics) and he met von Neumann and von Neumann's colleagues at the Institute for Advanced Study during this time. From 1939 on Robertson was involved in war work, for which he received the Medal for Merit in 1946. He had key scientific, military and government advisory positions after the war. Von Neumann told Robertson about the story of discovering the mean ergodic theorem, see Section 7 of the Introductory Comments.
(January 16, 1932)

(55) Schrödinger, E. (1887-1961)
Austrian-born theoretical physicist, recipient of the 1933 Nobel Prize (jointly with P.M. Dirac) for his pioneering work in quantum mechanics. In 1925 he discovered the wave equation (named after him) governing the dynamical evolution of quantum systems. Schrödinger had many positions across Europe and had also visited Princeton University in 1934. In the mid thirties he published a number of papers on the conceptual foundations of physics; the letter by von Neumann to Schrödinger published in this volume was von Neumann's reaction to some of the conceptual points Schrödinger raised at that time (see Sections 5.3 and 5.4 of the Introductory Comments).
(April 11, 1936)

(56) Segre, E. (1905-1989)
Italian-born nuclear physicist, co-discoverer (in 1955) of the antiproton, for which he received (jointly with O. Chamberlain) the Nobel Prize in physics in 1959. Segre emigrated to the U.S.A. in 1938 and also participated in the Manhattan Project.
(October 24, 1955)

(57) Silsbee, F.B. (????-????)
Silsbee worked in the National Bureau of Standards and served as President of the Washington Philosophical Society (WPS) in 1936. It was in his capacity as the representative of WPS that he contacted von Neumann, inviting him to deliver the Joseph Henry Lecture in 1945 on quantum logic (see the end of Section 6 of the Introductory Comments).
(November 3, 1944), (February 14, 1945), (June 11, 1945), (July 2, 1945), (October 22, 1945), (April 20, 1946), (December 23, 1946)

(58) Spitzer, L. (1914-1997)
Physicist-astronomer. Spitzer became professor and chairman of the astronomy department at Princeton University and director of the observatory in Princeton in 1947. Spitzer was the chairman of the Scientist's Committee on Loyalty Problems (1948-1951). Von Neumann corresponded with Spitzer as chair of this loyalty committee about loyalty clearance from scientists.
(June 9, 1949)

(59) Stone, M. (1903-1989)
Mathematician, friend and co-author of von Neumann. Stone's results were in classical and functional analysis, (especially spectral theory of Hilbert space operators), algebra and topology. The large intersection of the professional interests of von Neumann and Stone is reflected in the substantial letters von Neumann wrote to him (see Sections 3 and 5.2 of the Introductory Comments). Stone served as President of the American Mathematical Society (1943-1944), and he is credited with establishing a superb mathematics department in the University of Chicago after taking up the chairmanship there (on the recommendation of von Neumann) in 1946.
(October 8, 1930), (January 22, 1932), (March 31, 1935), (April 19, 1935), (October 4, 1938)

(60) Strauss, L.L. (1896-1974)
Banker, financier and U.S. government official. An admirer, supporter and confidant of von Neumann. Strauss helped to raise funds for the Institute for Advanced Study (IAS) computer project directed by von Neumann at IAS in Princeton (see Section 8 of the Introductory Comments). Strauss served as Chairman of the Atomic Energy Commission (1953-1958), and it was he who initiated the appointment of von Neumann as an Atomic Energy Commissioner.
(October 20, 1945), (October 24, 1945), (December 9, 1946)

(61) Stroux, J. (1886-1954)
German philologist, first rector after World War II of Humboldt University in East Berlin. Stroux was the founding president of the East German Academy of Sciences.
(September 19, 1950)

(62) Tannaka, T. (1908-1987)
Japanese mathematician. Tannaka studied mathematics in Tohoku Imperial University, where he became professor in 1945 and remained there until his retirement in 1972. The most famous work of Tannaka is the generalization of the Pontrjagin duality to non-abelian compact groups; von Neumann's letter to Tannaka concerns this result.
(August 8, 1939)

(63) Teller, P. (1908-2004)
Hungarian-born physicist, he emigrated to the U.S.A. in 1935. Teller was a major figure in the Manhattan Project, and later in the development of the thermonuclear bomb. Teller and von Neumann met on many occasions during von Neumann's war work and consultancy.
(April 8, 1947)

(64) Tuckerman, L.B. (1915-2002)
Mathematician. After teaching at Cornell and Oberlin College, Tuckerman worked for the Mathematics Department at the IBM T.J. Watson Research Center. For 5 years he also worked with von Neumann at IAS on numerical computations.
(August 25, 1950), (September 25, 1950)

(65) Ulam, S. (1909-1984)
Polish-born mathematician, a close friend of von Neumann. Von Neumann arranged inviting Ulam to the Institute for Advanced Study for the academic year 1934-1935, after which Ulam took up appointments at Harvard. In 1943 Ulam became a U.S. citizen. It was on von Neumann's recommendation that Ulam was invited to work for the Manhattan Project, to which he made major contributions. Ulam also contributed substantially to the development of the thermonuclear bomb. He is credited with discovering the Monte Carlo method.
(October 3, 1936), (February 3, 1937), (October 4, [1937 ?]), (March 15, [1939 ?]), (April 2, 1942), (November 9, 1943)

(66) Van Kampen, E.R. (1908-1942)
Dutch-born mathematician. Van Kampen left Europe in 1933 to take up a position at Johns Hopkins University. He spent the year 1933 in Princeton, where his interest in almost periodic functions brought him in touch with von Neumann.
(November 21, 1934)

(67) Veblen, O. (1880-1960)
Mathematician, geometer and topologist, president of the American Mathematical Society (1923-1924). Veblen was the first professor to be appointed at the Institute for Advanced Study (IAS) in 1932 and he had strongly influenced the decision on who the other five appointees were (among them von Neumann). Veblen was close to von Neumann, they corresponded from the early thirties on both political and scientific matters (see Sections 1 and 6 of the Introductory Comments). Veblen remained von Neumann's colleague at IAS until his retirement in 1950.
(September 23, 1930), (January 11, 1931), (April 3, 1933), (July 6, 1935), (October 3, 1938), (November 27, [????]), (November 30, [????])

(68) Wiener, N. (1894-1964)
Mathematician, with a wide range of intellectual interests and activities. The most significant of his mathematical work is his contribution to harmonic analysis, for which he received the Bôcher Prize in 1933. Von Neumann corresponded with Wiener from about 1933 on different scientific matters and they met on several occasions. Von Neumann regarded highly Wiener's work in designs of early (analog) computers and in information theory. Von Neumann kept Wiener informed about his decision to decline MIT's offer of professorship in mathematics and to stay in Princeton to direct the computer project at the Institute for Advanced Study (see Section 1 of the Introductory Comments.
(November 20, 1945), (November 29, 1946)

(69) Wold, H. (1908-1992)
Swedish econometrician and statistician, worked on time series. Von Neumann criticized Wold's attempt to introduce non-numerically valued utility functions, see Section 9 of the Introductory Comments.
(October 28, 1946)

Bibliography

[1] W. Aspray and A. Burks (eds.): *Papers of John von Neumann on Computing and Computer Theory* (MIT Press, Cambridge, and Tomash Publishers, Los Angeles, 1987)

[2] W. Aspray: The origins of John von Neumann's theory of automata in [**14**] 289-309

[3] W. Aspray: *John von Neumann and the origins of modern computing* (MIT Press, Cambridge, Mass., 1990)

[4] G. Birkhoff: Lattices in Applied Mathematics, in *Lattice Theory* (Proceedings of the Second Symposium in Pure Mathematics of the American Mathematical Society, April, 1959, ed. by R.P. Dilworth) (Providence, 1961)

[5] F. Brody and T. Vámos (eds.): *The Neumann Compendium. World Scientific Series of 20th Century Mathematics Vol. I.*, (World Scientific, Singapore, 1995)

[6] S. Brush: *The Kind of Motion we Call Heat.* vol. 2. (North Holland, Amsterdam, New York, Oxford, 1976)

[7] E. Carson, R. Huber (eds.): *Intuition and the Axiomatic Method*, (Western Ontario Series in Philosophy of Science) (forthcoming)

[8] M.L. Dalla-Chiara, R. Giuntini: The logics of orthoalgebras *Studia Logica* **55** (1995) 3-22

[9] J. Czelakowski: Logics based on partial Boolean σ-algebras *Studia Logica* **33** (1974) 371-396

[10] J. Earman, M. Rédei: Why ergodic theory does not explain the success of equilibrium statistical mechanics *The British Journal for the Philosophy of Science* **47** (1996) 63-78

[11] A. Einstein, B. Podolsky and N. Rosen: Can quantum mechanical description of physical reality be considered complete? *Physical Review* **47** (1935) 777-780

[12] M. Frechet: Emile Borel, initiator of the theory of psychological games and its application *Econometrica* **21** (1953) 95-96

[13] M. Frechet: Commentary on the three notes of Emile Borel *Econometrica* **21** (1953) 118-124

[14] J. Glimm (ed.): *The Legacy of John von Neumann* (American Mathematical Society, Providence, R.I., 1990) Proceedings of Symposia in Pure Mathematics; 50.

[15] K. Gödel: On formally undecidable propositions of *Principia Mathematica* and related systems I in [**16**] 144-195

[16] K. Gödel: *Collected Works*, Vol. I. Publications 1929-1936, S. Feferman, J. W. Dawson, Jr., S.C. Kleene, G. H. Moore, R. Solovay, J. van Heijenort (eds.) (Oxford University Press, New York, 1986)

[17] K. Gödel: *Collected Works*, Vol. V. Correspondence H-Z, S. Feferman, J. W. Dawson, Jr., W. Goldfarb, C. Parsons, W. Sieg (eds.) (Oxford University Press, New York, 2003)

[18] R. Haag: *Local Quantum Physics. Fields, Particles, Algebras* (Springer-Verlag, Berlin Heidleberg, 1992)

[19] P. Halmos: *Lectures on Ergodic Theory* (Mathematical Society of Japan, Tokyo, 1956)

[20] P.R. Halmos: *Algebraic Logic* (Chelsea Publishing Company, New York, 1962)

[21] P. Halmos: The legend of John von Neumann *American Mathematical Monthly* **50** (1973) 382-394

[22] M. Heidelberger, F. Stadler (eds.): *History of Philosophy of Science. New Trends and Perspectives* (Institute Vienna Circle Yearbook 2001) (Kluwer Academic Publishers, Dordrecht, Boston, London, 2002)

[23] M. Jammer: *Philosophy of Quantum Mechanics* (Wiley, New York, 1974)

[24] M. Jammer: The EPR problem in its historical development
in [**29**] 129-149

[25] R.V. Kadison: Operator algebras – An overview
in [**14**] 61-89

[26] T. Kato: Fundamental properties of hamiltonian operators of Schrödinger type
Transactions of the American Mathematical Society **70** (1951) 195-211

[27] H.W. Kuhn, A.W. Tucker: John von Neumann's work in the theory of games and mathematical economics
in [**37**] 100-122

[28] R.J. Leonard: Form parlor games to social science: von Neumann, Morgenstern and the creation of game theory 1928-1944
Journal of Economic Literature **XXXIII** 730-761

[29] P. Lahti and P. Mittelstaedt (eds), *Symposium on the Foundations of Modern Physics. 50 years of the Einstein-Podolsky-Rosen Gedankenexperiment* (World Scientific, Singapore, 1985)

[30] P.D. Lax: *Functional Analysis* Pure and Applied Mathematics. Wiley-Interscience (John Wiley & Sons, New York, 2002)

[31] G.W. Mackey: Von Neumann and the early days of ergodic theory
in [**14**] 25-38

[32] R. Mane: *Ergodic Theory and Differentiable Dynamics* (Springer-Verlag, Berlin, Heidelberg, New York, 1987)

[33] R. von Mises: Grundlagen der Wahrscheinlichkeitsrechnung
Mathematische Zeitschrift **5** (1919) 52-99

[34] Richard von Mises, *Probability, Statistics and Truth* (second English edition of *Wahrscheinlichkeit, Statistik und Wahrheit*, Springer, 1928) (Dover Publications, New York, 1981)

[35] F.J. Murray: The rings of operator papers
in [**14**] 57-60

[36] F. Nagy (ed.): *Neumann János és a "Magyar Titok" a dokumentumok tükrében* (Országos Műszaki Információs Központ és Könyvtár, Budapest, 1987)

[37] J.C. Oxtoby, B.J. Pettis and G.B. Price (eds.): John von Neumann 1903-1957
Bulletin of the American Mathematical Society **64** (1958) 1-129

[38] D. Petz: *An invitation to the algebra of canonical commutation relation* (Leuven University Press, Leuven, 1990)

[39] D. Petz and M. Rédei: John von Neumann and the theory of operator algebras
in [**5**] 163-181

[40] J. von Plato: Boltzmann's ergodic hypothesis
Archive for History of Exact Sciences **42** (1991) 71-89

[41] M. Rédei: Why John von Neumann did not like the Hilbert space formalism of quantum mechanics (and what he liked instead)
Studies in the History and Philosophy of Modern Physics **27** (1996) 493-510

[42] M. Rédei: *Quantum Logic in Algebraic Approach* (Kluwer Academic Publishers, Dordrecht, 1998)

[43] M. Rédei: "Unsolved problems in mathematics" J. von Neumann's address to the International Congress of Mathematicians, Amsterdam, September 2-9, 1954
The Mathematical Intelligencer **21** (1999) 7-12

[44] M. Rédei: John von Neumann's concept of quantum logic and quantum probability
in [**45**] 153-172

[45] M. Rédei, M. Stöltzner (eds.): *John von Neumann and the Foundations of Quantum Physics* (Vienna Circle Institute Yearbook 8), Kluwer, Dordrecht, 2001.

[46] M. Rédei: Mathematical physics and philosophy of physics (with special consideration of J. von Neumann's work)
in [**22**] 239-243

[47] M. Rédei: John von Neumann on mathematical and axiomatic physics in *The Role of Mathematics in Physical Sciences: Interdisciplinary and Philosophical Aspects* G. Boniolo, P. Budinich and M. Trobok (eds.) (Springer, Dordrecht, 2005) 43-54

[48] M. Rédei: On the history of quantum logic (submitted for publication)

[49] M. Rédei, M. Stöltzner: Soft axiomatism: von Neumann on method and von Neumann's method in the physical sciences in [**7**] (forthcoming)

[50] P.A. Samuelson: *Foundations of Economic Analysis*, (Cambridge, Mass. Harvard University Press, 1947)

[51] E. Schrödinger: Discussion of probability relations between separated systems *Proceedings of the Cambridge Philosophical Society* **31** (1935) 555-563

[52] E. Schrödinger: Probability relations between separated systems *Proceedings of the Cambridge Philosophical Society* **32** (1936) 446-452

[53] M. Stone: Linear transformations in Hilbert space, III.: Operational methods and group theory. *Proceedings of the National Academy of Sciences U.S.A.* **16** (1930) 172-175

[54] M. Stöltzner: Opportunistic axiomatics: John von Neumann on the methodology of mathematical physics in [**45**] 35-62

[55] D. Szasz: Boltzmann's Ergodic Hypothesis: A Conjecture for Centuries? Studia Scientia Mathematica Hungaria **31** (1996) 299-322

[56] S.J. Summers: On the Stone-von Neumann uniqueness theorem and its ramifications in [**45**] 135-152

[57] M. Takesaki: *Theory of Operator Algebras, I.* (Springer-Verlag, New York, 1979)

[58] P. Walters: *An Introduction to Ergodic Theory* (Springer-Verlag, New York, Heidelberg, Berlin, 1982)

[59] H. Weyl: David Hilbert and his mathematical work *Bulletin of The American Mathematical Society* **50** (1944) 612-654

[60] A. S. Wightman: Hilbert's sixth problem in *Proceedings of Symposia in Pure Mathematics* vol. 28, AMS, 1976, pp. 147-240.

[61] J.D. Zund: Georg David Birkhoff and John von Neumann: A question of priority and the ergodic theorems, 1931-1932 *Historia Mathematica* **29** (2002) 138-156

[62] Transcript of J. von Neumann's confirmation hearings in the congress related to his appointment as Atomic Energy Commissioner, John von Neumann Papers, Library of Congress, Washington, D.C., U.S.A.

Publications by J. von Neumann referred to in the *Introductory Comments*

[Neumann 1922] M. Fekete, J. von Neumann: Über die Lage der Nullstellen gewisser Minimumpolynome *Jahresbericht der Deutschen Mathematiker Vereinigung* bf 31 (1922) 125-138 in [**Neumann1962I**] No. 2

[Neumann1927a] D. Hilbert, L. Nordheim, J. von Neumann: Über die Grundlagen der Quantenmechanik (1926) *Mathematische Annalen* **98** (1927) 1-30 in [**Neumann1962I**] No. 7

[Neumann1927b] J. von Neumann: Mathematische Begründung der Quantenmechanik *Göttinger Nachrichten* (1927) 1-57 in [**Neumann1962I**] No. 9

[Neumann1927c] J. von Neumann: Wahrscheinlichkeitstheoretischer Aufbau der Quantenmechanik *Göttinger Nachrichten* (1927) 245-272 in [**Neumann1962I**] No. 10

[Neumann1927d] J. von Neumann: Thermodynamik quantenmechanischer Gesamtheiten *Göttinger Nachrichten* (1927) 245-272 in [**Neumann1962I**] No. 11

[Neumann1927e] J. von Neumann: Zur Hilbertschen Beweistheorie
Mathematische Zeitschrift **26** (1927) 1-46
in [**Neumann1962I**] No. 12

[Neumann1928a] J. von Neumann: Zur Theorie der Gesellschaftspiele
Mathematische Annalen **100** (1928) 295-320
in [**Neumann1962VI**] No. 1

[Neumann1928b] J. von Neumann: Die Axiomatisierung der Mengenlehre
Mathematische Zeitschrift **27** (1928) 669-752
in [**Neumann1962I**] No. 16

[Neumann1929a] J. von Neumann: Allgemeine Eigenwertstheorie Hermitescher Funktionaloperatoren
Mathematische Annalen **102** (1929) 49-131
in [**Neumann1962II**] No. 1

[Neumann1929b] J. von Neumann: Zur Algebra der Funktionaloperatoren und Theorie der normalen Operatoren
Mathematische Annalen **102** (1929) 370-427
in [**Neumann1962II**] No. 2

[Neumann1931] J. von Neumann: Die Eindeutigkeit des Schrödingerschen Operatoren
Mathematische Annalen **104** (1931) 570-578
in [**Neumann1962II**] No. 7

[Neumann1932a] J. von Neumann: Über adjungierte Funktionaloperatoren
Annals of Mathematics **33** (1932) 294-310
in [**Neumann1962II**] No. 11

[Neumann1932b] J. von Neumann: Proof of the Quasi-ergodic Hypothesis
Proceedings of the National Academy of Sciences **18** (1932) 70-82
in [**Neumann1962II**] No. 12

[Neumann1932c] J. von Neumann: Physical applications of the ergodic hypothesis
Proceedings of the National Academy of Sciences **18** (1932) 263-266
in [**Neumann1962II**] No. 13

[Neumann1932d] J. von Neumann: *Mathematische Grundlagen der Quantenmechanik*
(Dover Publications, New York, 1943) (first American Edition; first edition: Springer-Verlag, Heidelberg, 1932; first English translation: Princeton University Press, Princeton, 1955.)

[Neumann1936a] F.J. Murray, J. von Neumann: On rings of operators
Annals of Mathematics **37** (1936) 116-229
in [**Neumann1962III**] No. 2

[Neumann1936b] J. von Neumann: On an algebraic generalization of the quantum mechanical formalism (Part I)
Matematicseskij Sbornik **1** (1936) 721-734
in [**Neumann1962III**] No. 9

[Neumann1936c] G. Birkhoff and J. von Neumann: The logic of quantum mechanics
Annals of Mathematics **37** (1936) 823-843
in [**Neumann1962IV**] No. 7

[Neumann1936d] J. von Neumann: Continuous geometry
Proceedings of the National Academy of Sciences **22** (1936) 101-108
in [**Neumann1962IV**] No. 8

[Neumann1937a] F.J. Murray, J. von Neumann: On rings of operators, II
American Mathematical Society Transactions **41** (1937) 208-248
in [**Neumann1962III**] No. 3

[Neumann1937b] J. von Neumann: Quantum logics (strict- and probability logics), Unfinished manuscript, John von Neumann Archive, Libarary of Congress, Washington, D.C. reviewed by A. H. Taub in [**Neumann1962IV**] 195-197

[Neumann1940] J. von Neumann: On rings of operators, III
Annals of Mathematics **41** (1940) 94-161
in [**Neumann1962III**] No. 4

[Neumann1943] F.J. Murray, J. von Neumann: On rings of operators, IV
Annals of Mathematics **44** (1943) 716-808
in [**Neumann1962III**] No. 5

[Neumann1944a] J. von Neumann: The mathematician, in *The Works of Mind*, ed. by R.B. Heywood (University of Chicago Press, 1946) 180-196
in [**Neumann1962I**] No. 1

[Neumann1944b] J. von Neumann, O. Morgenstern: *Theory of Games and Economic Behavior* (Princeton University Press, Princeton, 1944)

[63] [Neumann1945] J. von Neumann: First Draft of a Report on EDVAC. Report prepared for the U.S. Army Ordnance Department under Contract W-670-ORD-4926.
in [**1**] 17-82

[Neumann1953] J. von Neumann: Communication on the Borel notes
Econometrica **21** (1953) 124-125
in [**Neumann1962VI**] No. 2

[64] J. von Neumann: *The Computer and the Brain* (Yale University Press, New Haven, 1958)

[Neumann2001a] J. von Neumann: Quantum mechanics of infinite systems
in [**45**] 249-268

[Neumann2001b] J. von Neumann: Unsolved problems in mathematics
in [**45**] 231-245

Von Neumann's Collected Works

[Neumann1962I] J. von Neumann: *Collected Works Vol. I. Logic, Theory of Sets and Quantum Mechanics*, A.H. Taub (ed.) (Pergamon Press, 1962)

[Neumann1962II] J. von Neumann: *Collected Works Vol. II. Operators, Ergodic Theory and Almost Periodic Functions in a Group*, A.H. Taub (ed.) (Pergamon Press, 1962)

[Neumann1962III] J. von Neumann: *Collected Works Vol. III. Rings of Operators*, A.H. Taub (ed.) (Pergamon Press, 1962)

[Neumann1962IV] J. von Neumann: *Collected Works Vol. IV. Continuous Geometry and Other Topics*, A.H. Taub (ed.) (Pergamon Press, 1962)

[Neumann1962V] J. von Neumann: *Collected Works Vol. V. Design of Computers, Theory of Automata and Numerical Analysis*, A.H. Taub (ed.) (Pergamon Press, 1962)

[Neumann1962VI] J. von Neumann: *Collected Works Vol. VI. Theory of Games, Astrophysics, Hydrodynamics and Meteorology*, A.H. Taub (ed.) (Pergamon Press, 1961)